Architectural Acoustics

Architectural Acoustics

Principles and Practice

Edited by

WILLIAM J. CAVANAUGH AND JOSEPH A. WILKES

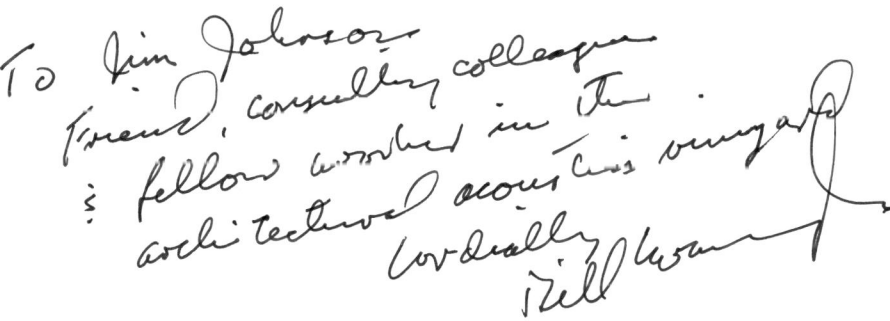

JOHN WILEY & SONS, INC.
New York • Chichester • Weinheim • Brisbane • Singapore • Toronto

This book is printed on acid-free paper. ∞

Copyright © 1999 by John Wiley & Sons, Inc. All rights reserved.

Published simultaneously in Canada.

No part of this publication may be reproduced, stored in a retrieval system or transmitted in any form or by any means, electronic, mechanical, photocopying, recording, scanning or otherwise, except as permitted under Sections 107 or 108 of the 1976 United States Copyright Act, without either the prior written permission of the Publisher, or authorization through payment of the appropriate per-copy fee to the Copyright Clearance Center, 222 Rosewood Drive, Danvers, MA 01923, (978) 750-8400, fax (978) 750-4744. Requests to the Publisher for permission should be addressed to the Permissions Department, John Wiley & Sons, Inc., 605 Third Avenue, New York, NY 10158-0012, (212) 850-6011, fax (212) 850-6008, E-Mail: PERMREQ @ WILEY.COM.

This publication is designed to provide accurate and authoritative information in regard to the subject matter covered. It is sold with the understanding that the publisher is not engaged in rendering professional services. If professional advice or other expert asistance is required, the services of a competent professional person should be sought.

Library of Congress Cataloging-in-Publication Data:

Architectural acoustics : principles and practice / William J.
   Cavanaugh and Joseph A. Wilkes, editors.
      p. cm.
   Includes bibliographical references and index.
   ISBN 0-471-30682-7 (cloth : alk. paper)
   1. Architectural acoustics.   I. Cavanaugh, William J.
II. Wilkes, Joseph A.
NA2800.A685   1998
690′.2—dc21   98-6418

Printed in the United States of America.

10 9 8 7 6 5 4 3 2

This book is dedicated to Richard H. Bolt, Leo L. Beranek, and the late Robert B. Newman* who taught generations of students and clients the fundamentals of acoustics and its practical applications through their pioneering research and consulting firm, Bolt, Beranek, and Newman Incorporated.†

---

*Robert Bradford Newman died in 1983 at the peak of his teaching and consulting career.

†In May, 1984, the American Institute of Architects awarded BBN its prestigious Institute Honor. The AIA Citation read:

> *Pioneering acoustical consultants, who in thirty-five years have almost single-handedly invented an entire profession by creating an awareness of acoustical considerations in design and by integrating technical solutions based on scientific principles with architectural and artistic concepts.*

Contents

Preface		xi
About the Authors		xv
Acknowledgments		xix

CHAPTER 1: Introduction to Architectural Acoustics and Basic Principles   **1**

William J. Cavanaugh

1.1	Introduction	1
1.2	Basic Concepts	2
1.3	Design Criteria	35
1.4	Selected Standards in Building Acoustics	44
	References and Further Reading	46
	Case Study: Fogg Art Museum Lecture Hall, Harvard University (1895–1973): The Beginnings of Modern Architectural Acoustics	47

CHAPTER 2: Acoustical Materials and Methods   **55**

Rein Pirn

2.1	Introduction	55
2.2	Sound Attenuation	56
2.3	Sound Absorption	59
2.4	Common Building Materials	63
2.5	Acoustical Materials	65
2.6	Special Devices	71

vii

2.7	Performance Tables	76
References and Further Readings		94
Case Study: Duke University Chapel: A Lesson on Acoustical Materials		95

CHAPTER 3: Building Noise Control Applications — 100
Gregory C. Tocci

3.1	Introduction	100
3.2	Acoustical Analysis	100
3.3	Building Noise Criteria	110
3.4	Noise Control Methods	120
References and Further Reading		147
Case Study: Mechanics Hall, Worcester, Mass.: Cooling Tower Sound Isolation		149

CHAPTER 4: Acoustical Design: Places for Listening — 151
L. Gerald Marshall and David L. Klepper

4.1	Introduction	151
4.2	Sound Propagation in Listening Places—Outdoors and Indoors	152
4.3	Concert Halls and Recital Halls	155
4.4	Opera Houses, Theaters, General-Purpose Auditoria, and Worship Spaces	168
4.5	Other Places for Speech and Music Activities	178
References and Further Reading		181
Case Study: Hitchcock Presbyterian Church, Scarsdale, New York		182
Case Study: Lenna Hall at the Chatauqua Institution, Chatauqua, New York		183
Case Study: Ben Franklin Hall, American Philosophical Society, Philadelphia, Pennsylvania		185

CHAPTER 5: Sound Reinforcement Systems — 187
J. Jacek Figwer

5.1	Introduction	187
5.2	Loudspeaker Systems	189

5.3	Equipment	191
5.4	Examples of Sound Reinforcement and Reproduction Systems	197
5.5	Special Sound System Installations	203
References and Further Readings		204
Case Study: Concord-Carlisle Regional High School, Auditorium Sound System, Concord, Massachusetts		205
Case Study: Cargill 97 Lecture Auditorium Sound System, Northeastern University School of Law, Boston, Massachusetts		206
Case Study: The Louisiana Superdome Sound System, New Orleans, Louisiana		208
Case Study: The Community Church of Vero Beach, Florida, Sound System		212
Case Study: Queen Sirikit National Convention Center Sound System, Bangkok, Thailand		215
Case Study: Hong Kong Cultural Center, Sound Systems, Tsim Sha Tsui, Kowloon, Hong Kong		223

CHAPTER 6: Recent Innovations in Acoustical Design and Research — 233

Gary W. Siebein and Bertram Y. Kinzey, Jr.

6.1	Introduction	233
6.2	Understanding and Measuring Room Acoustic Qualities	235
6.3	Acoustical Modeling and Aural Simulation	264
6.4	Other Directions in Architectural Acoustics Research	269
6.5	Conclusions	273
References and Further Reading		273
Case Study: Segerstrom Hall, Orange County Performing Arts Center		277
Case Study: McDermott Concert Hall, Morton H. Meyerson Symphony Center, Dallas, Texas		278

Case Study: Evangeline Atwood Concert Hall,
Alaska Center for the Performing Arts,
Anchorage, Alaska   293
Case Study: Acoustical Model Tests   297

APPENDIXES   305

Appendix A: Conversion Factors, Abbreviations, and Unit Symbols   305

Appendix B: Acoustical Societies Throughout the World   309

Appendix C: Selection of an Acoustical Consultant   316

GLOSSARY   318

INDEX   327

Preface

In this era of ever-increasing technological development, the architectural design team includes the services of many specialists. Depending on the nature of the project, these may include soils and foundation engineers, structural engineers, mechanical systems designers, civil and landscape consultants, interior and lighting designers, and, increasingly on practically all building types, acoustical consultants.

Buildings where the services of acoustical consultants are most often used include airports and other transportation structures, places of assembly including places of worship, theaters, opera houses, concert halls, educational structures, including classrooms, lecture halls, libraries, music practice rooms, athletic buildings, and sports stadia, residential structures of all types including apartment buildings, hotels, multifamily units, and single family residences, and commercial and industrial buildings of all types.

Acoustical consultants play a major role in the materials selection and detailing of construction components and interior surface materials, design and specification of sound and communications systems, and detailing of components for noise and vibration control in mechanical systems.

The art and science of acoustics has also made great strides in the technical development and understanding of the ways sound is produced, measured, enhanced, and controlled in many of the man-made environments in and around buildings.

The purpose of this book is to make the fruits of the science of acoustics available to the designers and users of many types of structures, where the generation of desired or undesired sounds, whether as voice, instrumental music, or mechanical systems, is an important factor in the utility and enjoyment of that structure. To that end, the book is divided into six chapters authored by a group of experienced consultants in the still-developing field of architectural acoustics. Each chapter includes case studies that demonstrate the application of the acoustical principles in the design of buildings. These chapters are based on material that was originally published in Wiley's five volume *Encyclopedia of Architecture: Design, Engineering, and Construction*.

Chapter 1 introduces the field of architectural acoustics and some fundamentals of sound and its control, as well as the development of acoustics as a science and technology and its role in contemporary design of structures. The basic measurement of sound, wave lengths, the audible frequencies range, units of sound intensity (decibels), and frequency measurement of musical instruments are explained. Increasing attention by building codes to acoustical aspects of building design, and the need for the early establishment of acoustical design criteria in building spaces, are shown to be essential for adequate sound insulation for privacy between spaces in residential and office spaces, and sound enhancement and reverberation control in spaces for listening such as auditoria, theaters, and, places of worship.

Chapter 2 provides a description of the vast array of construction materials and their ability to absorb, reflect, or transmit sound in the structures created by their use. All materials, it is observed, are basically "acoustical" in that they all affect the acoustical environment of the building to some degree. Materials used for sound attenuation, including impact noises in multistory residential structures, vibration of motors, and air movement in ducts, are discussed.

Performance ratings are provided for the sound attenuation of common building materials such as brick, concrete, gypsum, and glass, as well as for materials and systems for walls and floor/ceiling systems. A section of this chapter describes the special devices such as air-springs, duct silencers, flexible pipe connectors, resilient clips, channels, and hangers used for sound attenuation of the noise-producing components of the mechanical systems such as ducts, pipes, and motors.

Chapter 3 analyzes noise from its multiple sources through their paths to the ultimate receivers and the controls necessary to assure acoustical comfort and privacy. In all buildings control of background noise is needed to some extent, while in some critical listening spaces it is essential. Noise control methods in the form of special construction systems, sound isolating details, and privacy measurement have become industry standards for some building types and are often required by building codes, even international codes. Systems described include building partitions, floor/ceiling constructions, building exterior envelope, and HVAC mechanical systems and their distribution systems. Included here is a description of motor mounts and bases, fan blade and housing design and pipe, and duct noise reduction devices.

Chapter 4 deals with the man-made indoor and outdoor structures where the sounds produced and heard are largely the reason for their existence. These are here called places for listening. They include structures for a variety of uses such as concert halls, opera houses, legitimate and movie theaters, houses of worship, sports arenas, and, in educational facilities, lecture halls, classrooms, and practice rooms. The desire to make these structures better able to serve the performers and their audiences has led to extensive research into better understanding the acoustics of these spaces, and the use of improved materials and design techniques has raised the enjoyment levels of these spaces for the users and listeners alike.

To improve the quality of sound in most listening places discussed in the previous chapter, sound reinforcement systems are integrated with the basic architectural design. Chapter 5 discusses the basics of sound reinforcement system design and the components of these systems, including microphones, preamplifiers, control consoles, signal processing equipment, power amplifiers, loudspeakers, and sound control devices, along with applications in the many types of listening places. The goal of these systems is to make every desired sound created by the performer audible in every part of the listening space by the audience, including even those with impaired hearing. In very large listening spaces an adequate sound amplification system is essential to hear speech and music. The technology employed in the design of these sound reinforcement systems is under constant development toward greater efficiency and cost effectiveness.

Chapter 6 seeks to record the most recent progress in the field of acoustics in developing methods of evaluating, modeling, and predicting the acoustical qualities of buildings. It is anticipated that new methods for understanding and quantifying the sonic qualities of spaces will prove of great benefit in the design process of all buildings. Also reviewed are new developments in modeling and aural simulation, research in new materials for sound diffusion, noise control devices for mechanical systems, impact sounds, the noises in air distribution systems, and advances in other building systems.

To improve the design of building spaces with the most desirable characteristics of sound, many studies have been conducted to establish the reverberation time most acceptable for audiences of different kinds of musical performances. Other tests have measured speech intelligibility in order to establish the characteristics of the space most expected to increase or decrease this phenomenon.

In summary, co-editor Bill Cavanaugh and I hope that *Architectural Acoustics: Principles and Practice* fulfills our vision of producing an up-to-date, useful, and practical treatment of this subject for architects, engineers, consultants, teachers and students, and building owners and managers, that they may produce buildings with good acoustics.

Joseph A. Wilkes, FAIA
Annapolis, MD
January, 1998

About The Authors

Each of the six chapters in *Architectural Acoustics: Principles and Practice* is authored or co-authored by an experienced acoustical consultant still in practice as this book is published. Between them are represented almost three centuries of experience and, perhaps more importantly, they represent literally thousands of projects throughout the world, ranging from simple "one-conference" consultations to several years of close contact with a project design team, from the "schematic-programming" phase through "final construction–initial uses," including acoustical conformance testing. Since acoustics, and more particularly architectural acoustics, is a discipline that crosses many of the traditional building design and engineering disciplines, every consultation becomes a thread in the overall fabric of experience in the field.

The late Professor Robert B. Newman of MIT's School of Architecture used to tell his students that the real world and its buildings, where people work, study, play, and live, is our laboratory. Yes, independent psychoacoustical experiments are needed, too, to fully understand the many interdependent variables involved, but the knowledge of how people respond to their acoustical environments demands both research in the real world as well as in the laboratory.

Each of the chapter contributors has been heavily involved in acoustical research simultaneously with his consulting. Additionally, the authors have been deeply involved in teaching, lecturing, the presentation of technical papers, as well as the publication of articles for architectural journals, and otherwise diffusing knowledge about the field of acoustics, as the objective of their principal professional society, the Acoustical Society of America, mandates. Their individual professional resumes would fill a book, of course. However, at the risk of significant omission, the following short biographies are presented.

William J. Cavanaugh has been continuously in practice as an acoustical consultant since early 1954, when he had, as he says, "the good fortune to land a paying job" with two of his former teachers at the MIT School of Architecture, Professor Richard Bolt and his then-graduate assistant, Robert Newman. The consulting research firm of Bolt, Beranek, and Newman (BBN) had just been established in 1948 to undertake its first big assignment, the acoustics of the prestigious new United Nations headquarters in New York City. Cavanaugh, delighted to be working on projects with the world's leading architectural firms, rose to become Director of BBN's Architectural Technologies Division and a Divisional Vice President. He established an independent acoustical consulting practice in 1970. In 1975, he formed with Gregory Tocci, Cavanaugh Tocci

Associates, Inc. of Sudbury, Massachusetts, to deal with a diverse range of architectural, environmental, mechanical, and structural acoustical problems in and around buildings. Cavanaugh has taught acoustics courses and lectured at many schools of architecture. From 1961 to 1993, he was an adjunct faculty member at the Division of Architectural Studies, Rhode Island School of Design, in Providence. Cavanaugh is a Fellow of the Acoustical Society of America (FASA), a board-certified member of the Institute of Noise Control Engineering (INCE, Bd. Cert.), a member of the National Council of Acoustical Consultants (NCAC), the Audio Engineering Society (AES), and is associated with several other professional groups in the building technologies and design fields.

Rein Pirn, a native of Estonia, was educated at the University of Natal in South Africa where he received the B. Arch. degree, and at Harvard University's Graduate School of Design from where he graduated with the Master's degree in Architecture. In 1963, following graduation from Harvard, Pirn joined the consulting staff of Bolt, Beranek and Newman (BBN). Presently, he is a Consultant with Acentech Incorporated of Cambridge, Massachusetts, the successor firm to BBN's Architectural and Environmental Acoustics Division. A Fellow of the Acoustical Society of America, Pirn has lectured and published widely and served as a Visitng Assistant Professor in the Department of Architecture at the University of Notre Dame. He is a noted authority on open-plan acoustics, having pioneered the principles that to this day apply in the design of open offices and schools. In his 35 years as a practitioner, Pirn has consulted on the design of virtually every type of building. His principal expertise and love lies, however, with spaces made for music, where, next to his rich background in architecture, he can draw upon years of music making as an accomplished amateur violinist.

Gregory C. Tocci earned his BS in Mechanical Engineering from Tufts University in 1970 and his MS in Mechanical Engineering from the Massachusetts Institute of Technology in 1973. Tocci studied acoustics with Dr. Richard H. Lyon and other distinguished faculty at MIT in acoustics and vibration. In 1975, with William Cavanaugh, he co-founded Cavanaugh Tocci Associates, Inc. of Sudbury, Massachusetts, and serves as its President and Chief Executive Officer. He is a registered professional engineer in Massachusetts and Rhode Island. Tocci was President of the National Council of Acoustical Consultants, later receiving NCAC's C. Paul Boner Medal for contributions to acoustical consulting. He is a Fellow of the Acoustical Society of America (FASA), a board-certified member of the Institute of Noise Control Engineering (INCE, Bd. Cert.), and serves as an Associate Editor of INCE's professional journal, the *Noise Control Engineering Journal (NCEJ)*, as well as INCE Vice President for Board Certification. He has published numerous technical papers and is the author of a widely used design manual, Monsanto's *Acoustical Glazing Design Guide*. Tocci consults on a diverse range of projects, including many types of noise and vibration studies, speech privacy and intelligibility studies, and environmental noise impact assessments for residential, commercial, and industrial developments and for transportation systems.

L. Gerald Marshall has been involved in acoustical consulting since 1965 when he joined the consulting staff of Bolt, Beranek, and Newman, after graduate studies at the Massachusetts Institute of Technology with the late Professor Robert B. Newman. Marshall holds a B. Arch. Eng. from the University of Colorado, as well as a master's degree in Music from the University of Oklahoma. He was a professional musician, playing trumpet with the Denver Symphony, the Buffalo Philharmonic, the Oklahoma Symphony, and the United States Military Academy Band at West Point. He is a Fellow of the Acoustical Society of America (FASA) and has lectured, taught, and published widely on architectural acoustics and musical acoustics. In 1971, Marshall, with co-author David Klepper and Larry King, formed the architectural acoustics consulting firm of KMK Associates. Marshall currently serves as President and Principal Consultant of KMK Associates, now located in Chappaqua, New York. He is involved in a broad range of acoustical consulting projects and research studies, with a special emphasis on performing arts facilities and worship spaces.

David L. Klepper is a co-founder of the architectural acoustics consulting firm of KMK Associates. He received his BS and MS degrees in Electrical Engineering from the Massachusetts Institute of Technology. While a graduate student at MIT, Klepper served as a research assistant at the MIT Acoustics Laboratory, then directed by Professor Richard Bolt. He joined the Bolt, Beranek, and Newman consulting staff in 1957, where he was principal consultant on hundreds of consulting and research assignments, including major music performance facilities, theatres, music education and rehearsal facilities, churches and synagogues, and multiuse buildings of all types. Klepper is a Fellow of the Acoustical Society of America (FASA) and the Audio Engineering Society (FAES). He is also a member of the Institute of Noise Control Engineering (INCE), the Institute of Electronic and Electrical Engineering (IEEE), the American Guild of Organists (AGO), and an affiliate member of the Interfaith Forum on Religion, Art, and Architecture (INFRAA). In 1987, the Audio Engineering Society awarded Klepper its prestigious Silver Medal for contributions to building acoustics. He presently resides and consults independently in Israel. He also serves as associate cantor at the Hebrew University and, of course, plays the organ whenever and wherever the opportunity arises.

J. Jacek Figwer has been consulting in acoustics since 1955. He received a master's degree in Electrical Engineering from the Silesian Polytechnic Institute in Gilwice, Poland, and a Ph.D. in Applied Acoustics from the NIKFI in Moscow. Figwer has provided consulting services for the design, construction, and commissioning of hundreds of projects in North and South America, the Far and Middle East, Africa and Australia, and in Europe. He is a Fellow of the Acoustical Society of America (FASA) and the Audio Engineering Society (FAES). He has presented numerous technical papers and lectures widely on acoustics. He has long been associated with motion picture and broadcast engineering professional groups and served on many audio standards development committees. From 1951 to 1960, Figwer worked in research and development and headed the Acoustical Laboratory of the Motion Picture Industry Laboratories in Warsaw, Poland. From 1960 to 1962 he worked in research and development, studio and control room design, for Deutsche Grammophon Gesellschaft m.b.H. in Hannover, Germany. He then became a Supervisory Consultant with Bolt, Beranek, and Newman. In 1978, he formed his own consulting firm, Jacek Figwer Associates, Inc. in Concord, Massachusetts, where he consults in acoustics and sound system and audio-visual system design. He speaks five languages.

Gary W. Siebein received his BS and B. Arch. at the Rensselaer Polytechnic Institute. A master's thesis on acoustical modeling under co-author Professor Bertram Y. Kinzey led to his M. Arch. degree from the University of Florida, and his immediately joining the university faculty. In 1985, he founded and assumed directorship of the Architecture Technology Research Center at the University of Florida. He continues to wear three "hats" as a teacher, researcher, and acoustical consultant. Since 1981, he has headed his own independent consulting firm, Siebein Associates, in Gainesville, Florida. He is a Registered Architect in the state of Florida. He is a Fellow of the Acoustical Society of America (FASA), a member of the National Council of Acoustical Consultants (NCAC), and the Institute of Noise Control Engineers (INCE). Siebein has directed major research and consulting efforts, which has led to his preeminence as one of the leading researchers in acoustics, highly regarded by his professional colleagues.

Bertram Y. Kinzey is a Professor of Architecture Emeritus at the University of Florida and has been a consultant in architectural acoustics since 1961. He has an independent acoustical consulting practice in Gainesville, Florida, and is a Registered

Architect in the states of Florida and Virginia. He is a Fellow emeritus of the Acoustical Society of America (FASA emeritus), and a longtime member of the National Council of Acoustical Consultants (NCAC), the Institute of Noise Control Engineering (INCE), the American Society of Heating, Refrigerating, and Air-conditioning Engineers (ASHRAE), and the American Guild of Organists (AGO). Professor Kinzey has inspired generations of architectural students to better understand acoustical environments in buildings, and for some, like co-author Gary Siebein, was instrumental in their choosing the field of architectural acoustics as their major professional career path. Kinzey has lectured and published extensively in architecture and in architectural acoustics, making significant contributions to their expanding bodies of literature.

Joseph A. Wilkes has been an architect in private practice in the Washington D.C. area for thirty years he also taught in schools of architecture at the University of Florida and the University of Maryland. Wilkes was a major contributor to the sixth edition of *Architectural Graphic Standards,* published by John Wiley & Sons, in 1970 and served as head of the Editorial Review Board for the seventh edition of *Graphic Standards* in 1981. In the late 1980's Wilkes with associate Robert T. Packard edited the five volume *Encyclopedia of Architecture, Design, Engineering And Construction,* published by John Wiley & Sons. Wilkes is a Fellow of the American Institute of Architects and has been active in building research organizations including the Building Research Advisory Board (BRAB).

Acknowledgments

The editors of *Architectural Acoustics: Principles and Practice* wish to acknowledge with appreciation the countless former teachers, colleagues, and others who have contributed directly or indirectly to this work. We wish to thank the chapter authors in particular, knowing that no single individual possesses all the technical knowledge on the diverse field of architectural acoustics or has the ultimate overall grasp of total experience in the even more diverse fields of the building technologies, yet they have generously contributed their current views on the state of the art in the field as we approach the twenty-first century. You might say that this book is merely a snapshot, with as wide an angle viewer as technology permits us today in 1998. Gary Siebein, in Chapter 6, does, however, give us a pretty good idea of where current and future research methodologies may lead us.

We also want to express our deep appreciation for the ever patient and gentle prodding of our editors at John Wiley & Sons, including Dan Sayre, Amanda Miller, Janet Feeney, Matt Van Hattem and Bob Hilbert. Their task of dealing with so many chapter authors, whose ever-increasing consulting workloads always had precedence, was no small one.

Finally, we acknowledge with gratitude the enormous help we have been given by architects, consulting engineers, building materials providers, and many others who have taught all of us the way that practical, cost-effective, aesthetically pleasing buildings are produced. If this book helps toward better hearing conditions and optimum acoustical environments in and around buildings, our objectives will have been met.

William J. Cavanaugh and Joseph A. Wilkes

Introduction to Architectural Acoustics and Basic Principles

William J. Cavanaugh

1.1 INTRODUCTION

The acoustical environment in and around buildings is influenced by numerous interrelated and interdependent factors associated with the building planning–design–construction process. From the very outset of any building development, the selection of the site, the location of buildings on the site, and even the arrangement of spaces within the building can, and often does, influence the extent of the acoustical problems involved. The materials and construction elements that shape the finished spaces will also determine how sounds will be perceived in that space as well as how they will be transmitted to adjacent spaces. The architect, the engineer, the building technologist, and the constructor all play a part in the control of the acoustical environment. With some fundamental understanding of basic acoustical principles, how materials and structures control sound, many problems can be avoided altogether or, at least, solved in the early stages of the project at greatly reduced cost. "Corrective" measures are inevitably most costly after the building is finished and occupied, if indeed, a solution is possible at all.

Increasingly, federal, state, and local building codes and standards require attention to the acoustical aspects of building design. The need for special attention to acoustics is obvious in a concert auditorium or radio studio building. However, most of the problems involve the ordinary spaces where people work and live. In response to the Environmental Protection Act of 1970, most major federal agencies in the United States have developed criteria and standards toward promoting safe and comfortable working and living environments. Almost all of these have implications for the building design professionals. For example, the U.S. Department of Labor is concerned with the protection of workers' hearing in the industrial environment and has established standards of maximum worker noise exposure levels. Industrial buildings can have significant effects on an industrial worker's individual environment. The U.S. Department of Housing and Urban Development is similarly responsible for ensuring that federally subsidized housing developments are not located in excessively noisy environments or, if they must be, that suitable sound attenuation features are incorporated in the building design. Some state agencies require special sound attenuation features on all public buildings constructed near major airports or major highways. Local municipal building codes are, with increasing frequency, adopting provisions that require adequate attention to acoustical privacy between dwelling units and adequate control of noise transmission from building systems equipment. The U.S. General Services Administration, the largest builder of office space for the federal government, has adopted as a matter of policy, the "open-plan" concept for future office constructions. Acoustics is perhaps the major concern in the ultimate acceptability of such office working environments.

Finally, as people became more aware of their own wants and needs concerning their living and working environments and realize that something can be done to improve conditions, these demands initially are reflected in engineering design criteria and ultimately in building codes and standards.

1.2 BASIC CONCEPTS

Every building acoustics consideration can be thought of as a system of sources, paths, and receivers of sound. Even the most complex problem can be broken down into one or more sources to be studied along with the paths over which the sound will be transmitted to the eventual receptors of the sound. Whether a source is one we want to hear or an undesired source (i.e., noise), control can be exercised at each element of the system. Figure 1.1 illustrates that even in a simple lecture auditorium both desired (speech from the lecturer as well as from and between the listeners) and undesired sounds (air-conditioning system sounds, etc.) may be present and must be controlled. Naturally the building design and technology has most influence on the transmission paths. However, understanding the source and receiver aspects of a given situation may be essential to realize an effective overall resolution of the problem. For example, the selection and specification of the quietest available types of mechanical/electrical equipment may obviate the need for later design of special noise and vibration control building elements. Or the location of a particularly noisy operation or activity within a building so that it is remote from critical occupancies can save later concern, as well as the considerable cost of extraordinary sound attenuation features in the enclosing construction.

For the most part, effective control of the acoustical environment in buildings involves at least a conceptual understanding of the basic properties of sound, how it is propagated throughout typical building spaces, and how it is influenced by various building materials and construction systems. Such understanding is essential for those concerned with the complete building design/construction process who will influence the fundamental decisions concerning the building to be constructed. And just as with the numerous other disciplines involved in the overall building environment, thermal comfort, lighting, energy conservation, and so forth, the solutions to the acoustics problems require no small amount of experienced judgment and just plain common sense. After all, people do not respond to just one aspect of their environment. Acoustics, therefore, is rarely the most important aspect, but is a significant part of that environment and its effective control will help produce good buildings.

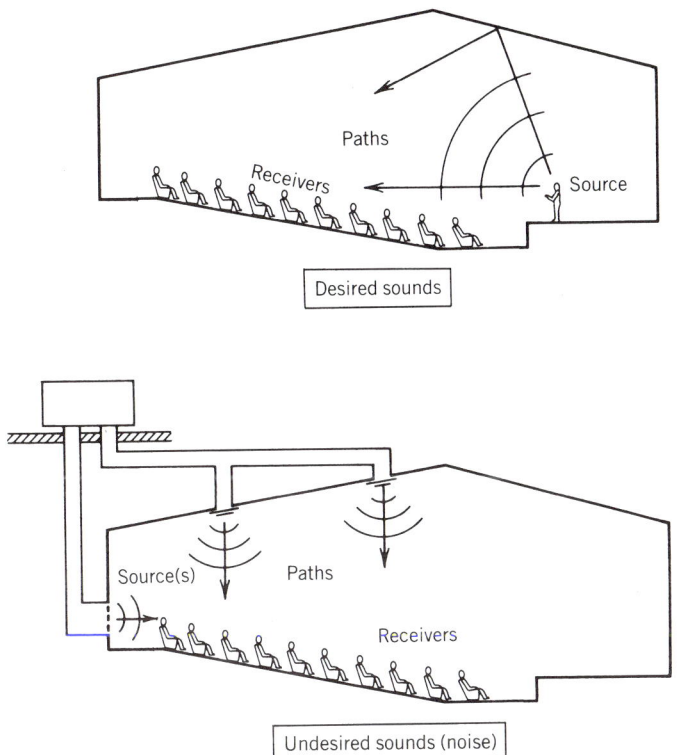

Figure 1.1 Every building acoustics problem, whether the enhancement of desired sounds or the control of undesired sounds (noise), can be considered in terms of a system of sound sources, paths, and receivers. (From William J. Cavanaugh, "Acoustics—General Principles," in Encyclopedia of Architecture: Design, Engineering & Construction, *Joseph A. Wilkes, Ed. Copyright © 1988 John Wiley & Sons. Reprinted by permission of John Wiley & Sons.)*

Fundamentals of Sound and Its Control

Sound has certain measurable physical attributes that must be understood, at least in a conceptual way, in order to understand the basic procedures for controlling sound in buildings. Sound is generated whenever there is a disturbance of an elastic medium. Once this disturbance occurs, whether it is in air by the vibrating string of a musical instrument or in a solid floor surface by the impact of a dropped object, the sound wave will propagate away from the source at some rate depending on the elastic properties of the medium.

Sound, in perhaps its simplest form, can be generated by striking a tuning fork as illustrated in Figure 1.2. The arms of the tuning fork are set into vibration and the air molecules immediately adjacent to the vibrating surface are alternately compressed and rarefied as the surface goes through each complete to-and-fro movement. This cyclical disturbance (compression and rarefaction of the air molecules) is passed on to the adjacent molecules and thus travels outward from the source. The outwardly progressing sound may be thought of as a "chain reaction" of vibrations constantly being transferred to adjacent molecules much like the disturbance created in a crowded subway train when a few more people try to squeeze on. The originally disturbed air molecules do not continue to move away from the source. Instead, they move back and forth within a limited zone and simply transfer their energy to the adjacent

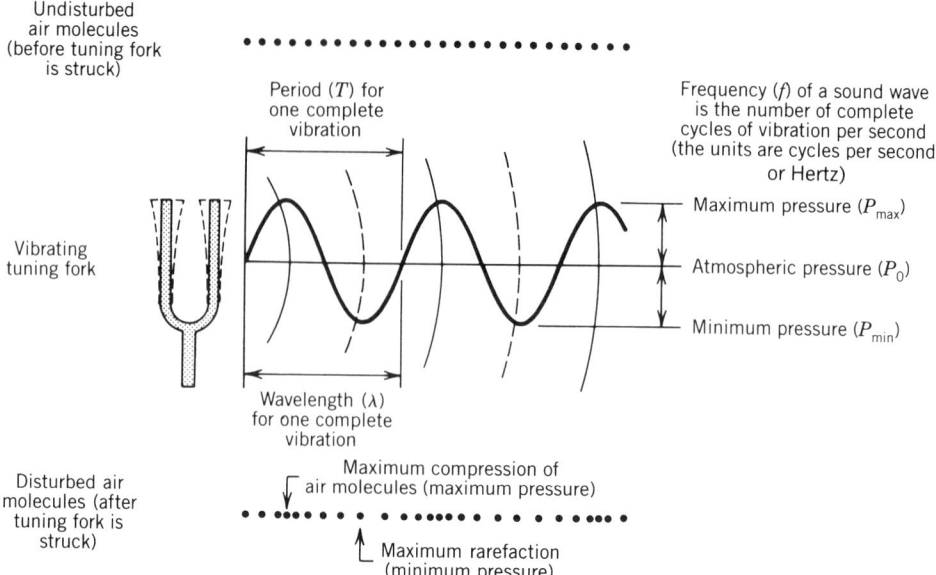

Figure 1.2 Tuning fork illustrates how a simple pure tone develops. (From William J. Cavanaugh, "Acoustics—General Principles," in Encyclopedia of Architecture: Design, Engineering & Construction, *Joseph A. Wilkes, Ed. Copyright © 1988 John Wiley & Sons. Reprinted by permission of John Wiley & Sons.*)

molecules. Although the last person squeezing on the train cannot move very far, the disturbance created can be felt by people at some distance.

The pressure disturbance created by the vibrating tuning fork cannot be seen by the naked eye, but ultimately the sound wave may reach a human ear, causing the eardrum to vibrate and, through a marvelously complex mechanism, finally produce the sensation of hearing in that person's brain. While our own ears are perhaps the most sophisticated sound measuring device available, humans have developed some useful measuring instruments that approximate closely the sensitivity of the ear and give us numerical quantities necessary for scientific experimentation and engineering applications. With a simple sound wave generated in air by a vibrating tuning fork (as with all other more complex sound waves), there are basically two measurable quantities of interests: the *frequency* of the sound wave and its *magnitude*.

Frequency

The frequency of a sound wave is simply the number of complete vibrations occurring per unit of time. Musicians refer to this as pitch, and this basic frequency or rate of repetition of the vibration defines its character. Low-frequency sounds such as a deep bass voice are classified as "boomy." High-frequency sounds such as a steam jet may have a "hissing" character.

The unit of measure is the hertz and is abbreviated Hz (older acoustical textbooks and publications may use cycles per second or cps). The tuning fork described above generates sound at just a single frequency. A simple musical tone would have a fundamental tone along with one or more harmonically related tones. All other common sounds, music, speech, and noise are more complex because they contain sound energy (i.e., vibrations) over considerably wider ranges of the human audible spectrum (about 20 to 20,000 Hz for young persons with normal healthy ears). Figure 1.3 illustrates how these simple and more complex common sounds compare.

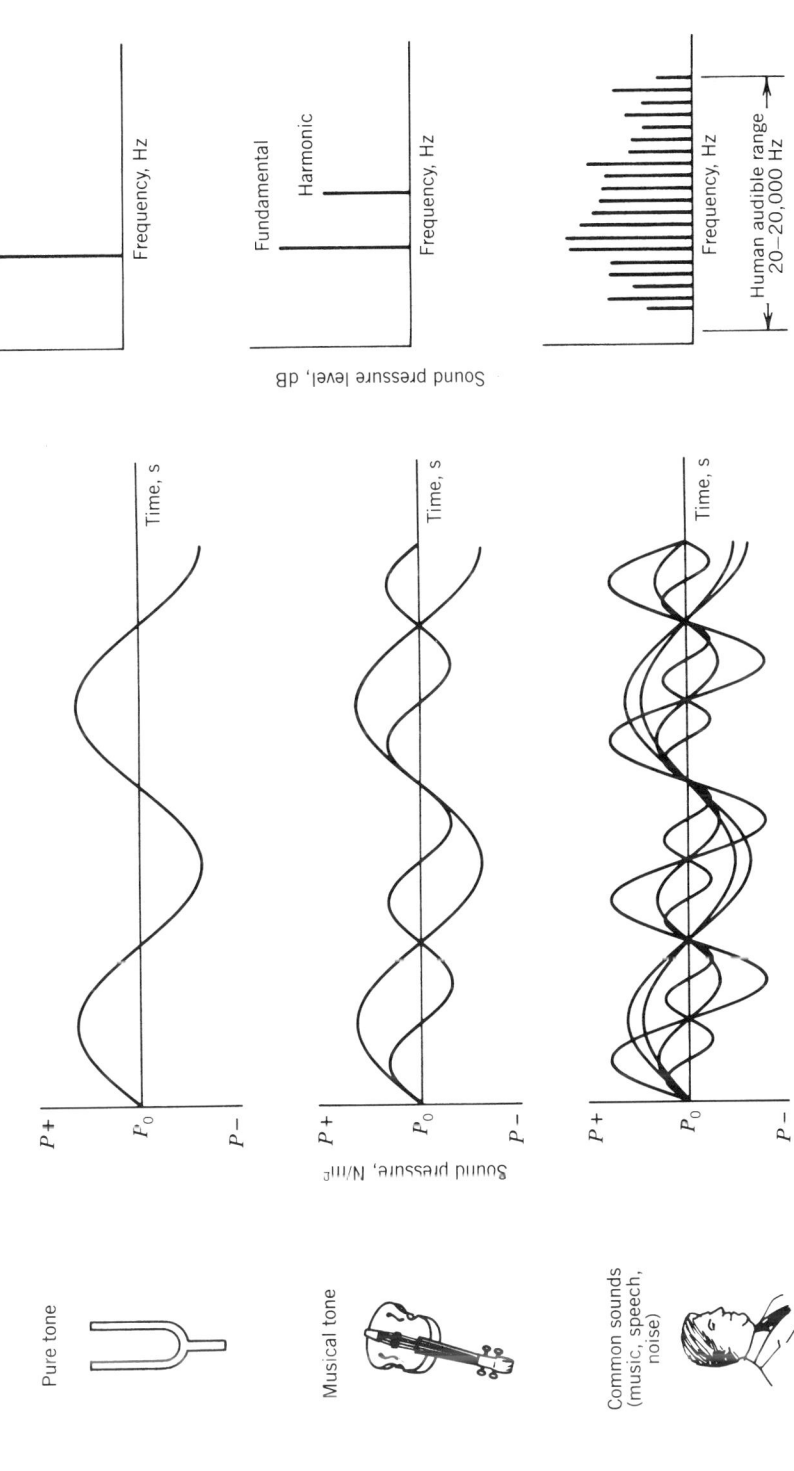

Figure 1.3 *Comparison of simple and more complex everyday sounds. Simple, pure, and musical tones contain sound energy at a fundamental frequency or fundamental plus harmonically related frequencies only. Common everyday sounds contain sound energy over a wide range of the human audible spectrum. (From William J. Cavanaugh, "Acoustics—General Principles," in Encyclopedia of Architecture: Design, Engineering & Construction, Joseph A. Wilkes, Ed. Copyright © 1988 John Wiley & Sons. Reprinted by permission of John Wiley & Sons.)*

Figure 1.4 illustrates the frequency ranges for some typical sounds including the frequencies where peak or predominant intensities are likely to occur. For comparison, the piano keyboard and its frequency range is shown. Thus, most of the sounds around us generally contain energy to some degree over rather wide ranges of the audible frequency range.

Frequency Bands

For measurement purposes, the audible frequency range may be divided into convenient subdivisions such as shown in Figure 1.5. Measurements may be made over the entire range or, utilizing electronic filters in the measurement system, the frequency range may be divided into segments such as octave bands or $\frac{1}{2}$-, $\frac{1}{3}$-, $\frac{1}{10}$-, octave bands. Octave bands generally yield sufficient frequency information about a sound source. In some laboratory measurements, such as in measuring the sound transmission loss characteristics of walls, however, $\frac{1}{3}$-octave-band measurements are made. The sound sources commonly encountered in buildings, as well as the acoustical performance of products and materials for sound control, are frequency dependent (i.e., vary with frequency). It is important to keep in mind that a wide range of frequencies is involved, even if simple averages or single number values are ultimately used to describe sound levels or to specify products.

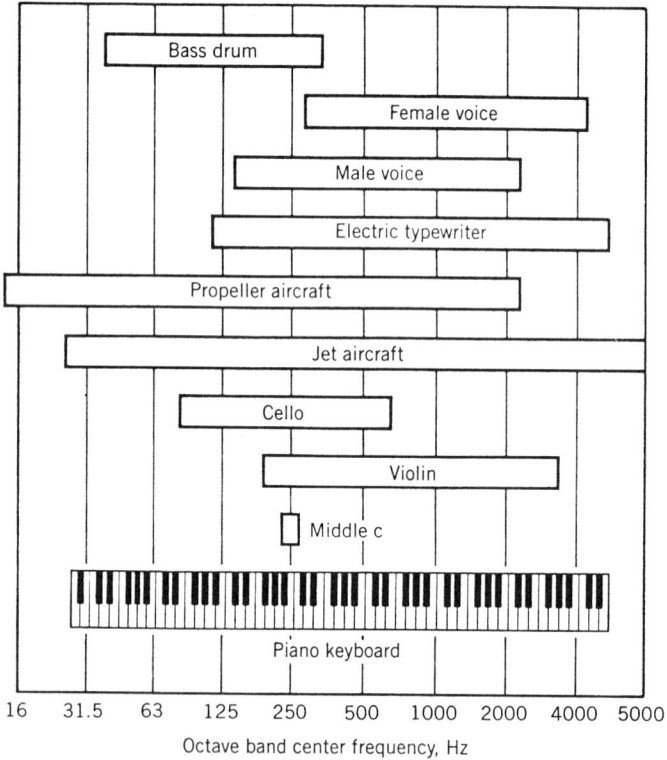

Figure 1.4 Comparison of frequency ranges for some common sounds with that of a piano keyboard. (From William J. Cavanaugh, "Acoustics—General Principles," in Encyclopedia of Architecture: Design, Engineering & Construction, *Joseph A. Wilkes, Ed. Copyright © 1988 John Wiley & Sons. Reprinted by permission of John Wiley & Sons.)*

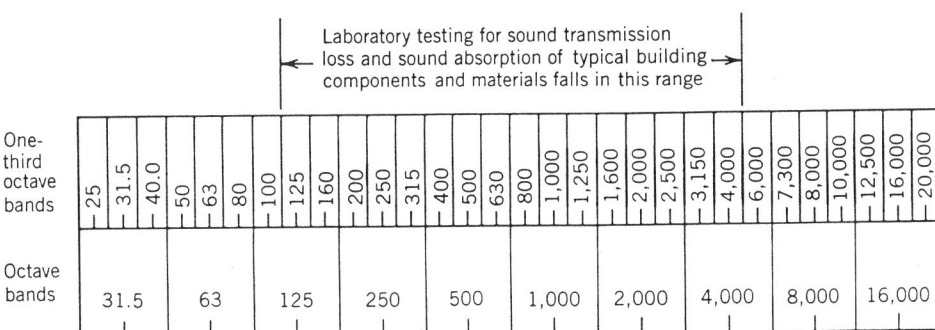

Figure 1.5 Audible frequency range divided into standard octave and ⅓ octave frequency bands, which are convenient segments for measurement and analysis. Laboratory test standards for the acoustical performance of many building components extend from bands centered at 100 Hz to those at 4000 Hz. (From William J. Cavanaugh, "Acoustics—General Principles," in Encyclopedia of Architecture: Design, Engineering & Construction, *Joseph A. Wilkes, Ed. Copyright © 1988 John Wiley & Sons. Reprinted by permission of John Wiley & Sons.*)

Wavelength of Sound

Another fundamental property of a sound wave that is related to its frequency is its wavelength. This is the distance within which the complete cycle of disturbance takes place. There is a basic relationship between the velocity of sound in a medium (e.g., air or concrete) and its frequency and wavelength given by the expression:

$$c = f\lambda$$

where

c = velocity of sound
f = frequency
λ = wavelength

For example, middle C on the piano has a frequency of 256 Hz. In air, where sound travels about 1100 ft/sec, the wavelength would be 4.3 ft. If a 256-Hz sound wave were excited in water where the speed of sound increases to 4500 ft/sec, the wavelength would be 12 ft. Correspondingly, in a solid concrete structure where sound travels even faster (10,200 ft/sec), the wavelength would be 24 ft.

It is useful to keep in mind the range of wavelengths encountered in the audible frequency range for various building acoustics problems. For example, in the laboratory, sound transmission loss and other measurements on building components are usually made starting at the ⅓-octave band centered at 125 Hz up through the ⅓-octave band centered at 4000 Hz. The wavelengths corresponding to these frequency limits are approximately 8.8 and 0.28 ft, respectively. Generally speaking, it takes rather massive, large elements to control low-frequency sound where the wavelengths are large. On the other hand, thinner, smaller building elements can provide effective sound control by absorption, for example, at high frequencies where the wavelengths are smaller.

Magnitude of Sound

In addition to the character (i.e., frequency) of a sound, also of concern is the intensity or magnitude of acoustical energy contained in the sound wave. Sound intensity is proportional to the amplitude of the pressure disturbance above and below the undisturbed atmospheric pressure (refer to Figure 1.2). The pressure fluctuations may be minute, yet a healthy ear has the ability to detect very faint sound pressure differences down to as little as 0.000000003 psi. At the same time, the human ear can tolerate for short periods the painful roar of a jet engine at close range that may be a million times as intense, say 3×10^{-2} psi. While sustained exposure to such intense sounds can cause hearing damage, the range of intensities or pressures that define the magnitude of sound energy is, like the wide range of frequencies, very large nevertheless. Because of the wide range, as well as the fact that the human ear responds roughly in a logarithmic way to sound intensities, a logarithm-based measurement unit called the decibel has been adopted for sound level measurements. The decibel unit is abbreviated dB.

Decibel Scale

The decibel scale starts at 0 for some chosen reference value and compares other intensities or pressures to that reference value. For sound pressure level measurements, a reference value of 0.00002 newtons/square meter (2×10^{-5} N/m²) is chosen. This is the threshold of hearing for a typical healthy young person. The sound pressure level in decibels for any sound for which the pressure is known is given by the following expression:

$$L_p = 20 \log \frac{p}{p_0}$$

where

L_p = sound pressure level in decibels (dB)
p = measured sound pressure of concern
p_0 = preference sound pressure usually taken to be 2×10^{-5} N/m² (older texts and publications may show the equivalent reference values of 0.0002 microbar or 2×10^{-4} dyne/cm²

Fortunately, acoustical instruments give the measured decibel values directly. However, it is important to realize that since this is basically a logarithmic scale, there are a few precautions to be observed when combining decibel units as will be discussed later in this chapter.

Figure 1.6 shows an "acoustical thermometer" of common sounds compared in terms of a measure of pressure [pounds per square inch (psi)], as well as in terms of sound pressure level (in dB). The convenience of the "compressed" decibel scale is obvious in dealing with this enormous range of sound magnitudes that can be accommodated rather well by a healthy human ear. Also shown on Figure 1.6 is the relative subjective description a typical listener might assign to the various levels of sound pressure from "very faint" (below 20 dBA) to "painful" (above 120 dBA).

Figure 1.7 shows frequency spectra for three common types of sound in octave bands of frequency compared to upper and lower threshold limits. For example, the air-conditioning spectrum contains a great deal of low-frequency sound compared to the mid- and high-frequency range, which results in its sounding "boomy" to an observer. An air jet, on the other hand, is generally just the reverse and contains predominantly high-frequency sound energy. Human speech not only covers a relatively wide range of frequencies but at the same time has fluctuating levels with time in the process of continuous speech. The dynamic range of speech is some 30 dB between the lowest and highest speech sound levels produced.

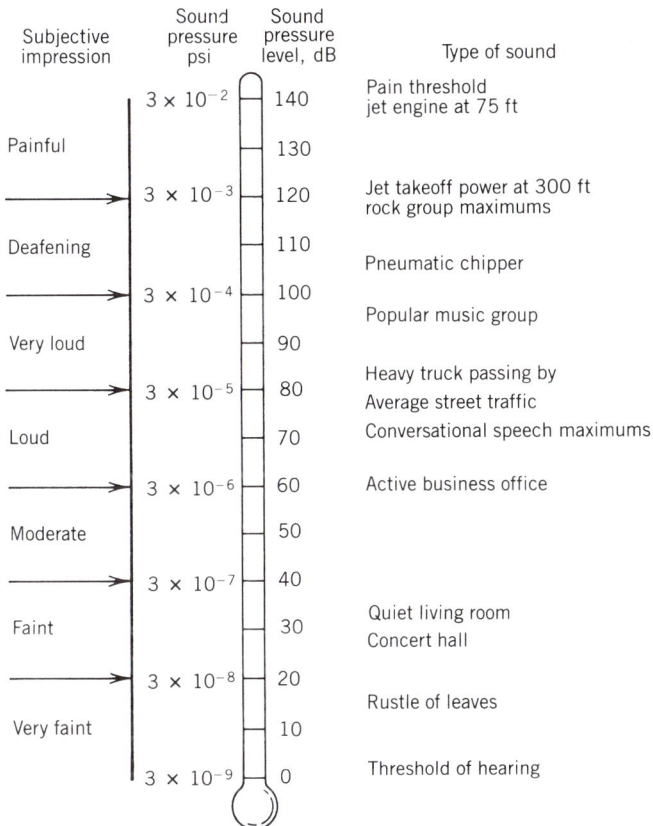

Figure 1.6 Acoustical thermometer compares the magnitude of sound pressures of sounds, in pounds per square inch, with the equivalent logarithmic quantities, decibels, used in acoustical standards. (From William J. Cavanaugh, "Acoustics—General Principles," in Encyclopedia of Architecture: Design, Engineering & Construction, Joseph A. Wilkes, Ed. Copyright © 1988 John Wiley & Sons. Reprinted by permission of John Wiley & Sons.)

Fortunately, it is not always necessary to deal with the full frequency range of various sounds of concern in many building acoustics problems. When the frequency characteristics are known for a type of sound source and are generally repeatable and/or are constant, simple single number sound level values may be adequate. Figure 1.8 shows typical octave band spectra for various transportation noise sources along with their simple sound level equivalent values. Over the past several decades an enormous amount of measured data on aircraft, rail, and highway transportation sources as well as on other environmental sounds has been accumulated by international and national agencies. The automobile, aircraft, and truck sound level spectra illustrated in Figure 1.8, for example, are from the U.S. Environmental Protection Agency. Chapter 3 describes an application of the use of such data in the acoustical design of the outside enclosing walls, windows, and roofs of buildings.

"Simple" Frequency-Weighted Sound Levels

The human ear does not simply add up all the energy for a sound over the entire audible range and interpret this value as the *loudness* of the sound. The human ear

10 • Introduction to Architectural Acoustics and Basic Principles

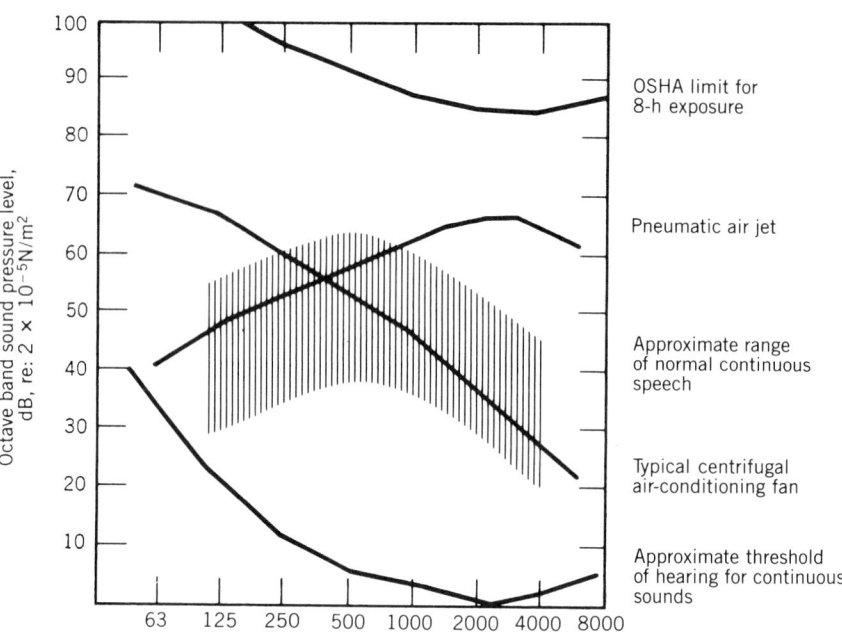

Figure 1.7 Typical octave band spectra of common sounds compared to the threshold of audibility for healthy, young ears and hearing damage risk criteria. (From William J. Cavanaugh, "Acoustics—General Principles," in Encyclopedia of Architecture: Design, Engineering & Construction, *Joseph A. Wilkes, Ed. Copyright © 1988 John Wiley & Sons. Reprinted by permission of John Wiley & Sons.)*

discriminates against low-frequency sounds (i.e., it "weights" or ignores some of the low-frequency sound energy). A given sound level will appear to be louder in the mid- and high-frequency range than that same level at lower frequencies. Electronic filters or "weighting networks" can be incorporated in a sound level meter to permit the instrument to approximate this characteristic and to read out sound level values that correspond well with the way the human ear judges the relative loudness of sounds.

Figure 1.9 illustrates the conversion of a sound source spectrum measured over the full frequency range to single-number values. Two frequency weightings are commonly used on standard simple sound level meters: C scale and A scale. The C scale is a "flat" frequency weighting; essentially, all the sound energy is summed up and converted to an overall value. The unit is usually identified as dBC to denote the frequency-weighting network used. The A scale network corresponds to the way a human ear responds to the loudness of sounds; the low-frequency sounds are filtered out or "ignored," just as the ear does, and a weighted sound level value is read on the meter. Such simple A scale sound levels are actually the most common and useful descriptors for many of the sounds encountered in buildings and are expressed in dBA units. They are adequate for the simplified analysis of many problems and for the specification of simple sound tests as long as the frequency content of the noise sources of concern are known or implied beforehand. Figure 1.10 shows the range of many common interior and exterior sound sources as would be measured with A frequency weighting using a standard sound level. For example, in a quiet residence one might expect sound levels in the 30- to 50-dBA range. In a typical factory environment, however, a worker could be exposed to levels from as low as 60 dBA to over 100 dBA.

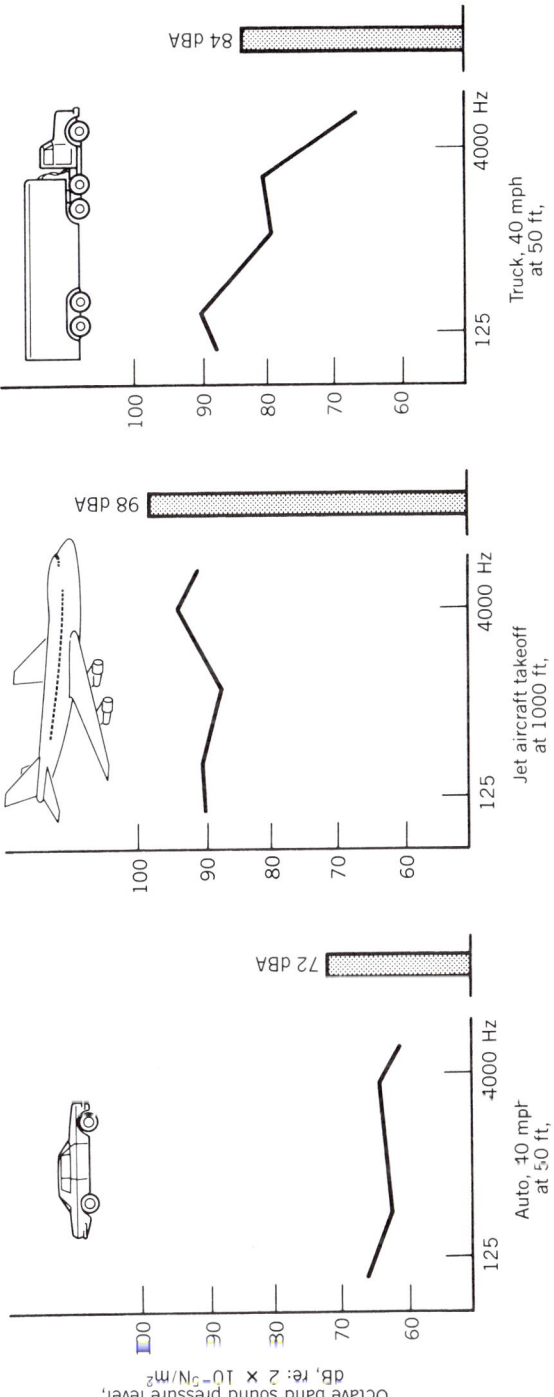

Figure 1.8 Examples of common exterior noise levels from transportation sources. (Data from U.S. Environmental Protection Agency report, EPA 560/9-79-100, Nov 1978.) (From William J. Cavanaugh, "Acoustics—General Principles," in Encyclopedia of Architecture: Design, Engineering & Construction, Joseph A. Wilkes, Ed. Copyright © 1988 John Wiley & Sons. Reprinted by permission of John Wiley & Sons.)

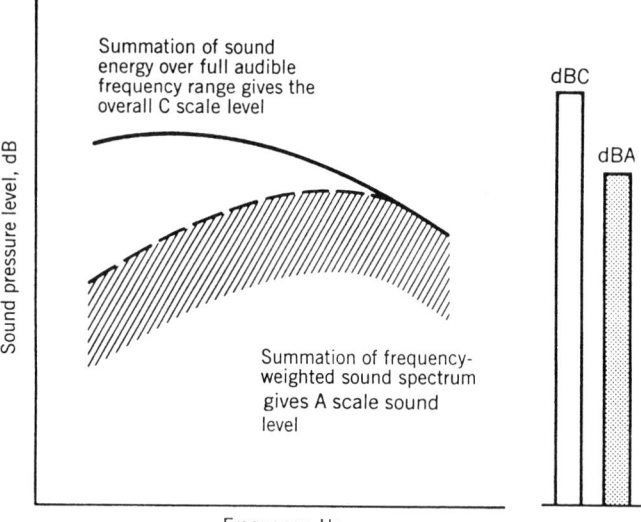

Figure 1.9 Frequency-weighting characteristic of standard sound level meters, which yield simple, commonly used overall sound levels (decibels with A scale weighting in dBA and decibels with C scale, or essentially unweighted flat frequency weighting, in dBC). (From William J. Cavanaugh, "Acoustics—General Principles," in Encyclopedia of Architecture: Design, Engineering & Construction, Joseph A. Wilkes, Ed. Copyright © 1988 John Wiley & Sons. Reprinted by permission of John Wiley & Sons.)

Time-Varying Sound Levels

Both indoor and outdoor environmental sound levels usually vary markedly with time whether in a relatively quiet setting such as remote rural areas or in highly developed downtown communities. With such time-varying sounds, as with the weather, there is no single, simple convenient metric to completely describe the quality and quantity of sound energy present.

Figure 1.11 from a U.S. Environmental Protection Agency report shows a 10-min time history of typical outdoor sound as would be measured on a quiet suburban street on a typical, otherwise uneventful, afternoon. The maximum sound level of 73 dBA occurs instantaneously when a sports car passes on a nearby street. The generally lowest sound levels of this 10-min sample, that is, those exceeded 90% of the sample time, are about 44 dBA. This is referred to as the 90 percentile level of L_{90}. The one percentile level (L_1) is the level exceeded only 1% of the sample observation period and is generally taken to be representative of the maximum sound levels expected during an observation period (1% of this 10-min sample is 6 sec).

Clearly, most outdoor sounds like those shown in Figure 1.11 must be described in statistical terms, such as the above, to properly describe the sound environment. Indeed, many community noise standards written with simple unqualified limiting values, not properly defined, are not only difficult to evaluate but encourage situations where the noise code is unenforceable and largely ignored. Unrealistically low ordinance limits often cannot be enforced as a practical matter since many normal activity sounds would be in violation. In other words, an arbitrary low limiting value would not be reasonable and would end up being disregarded.

In recent decades, largely as the result of the passage of the U.S. Environmental Protection Act of 1970 mandating that all federal agencies develop environmental

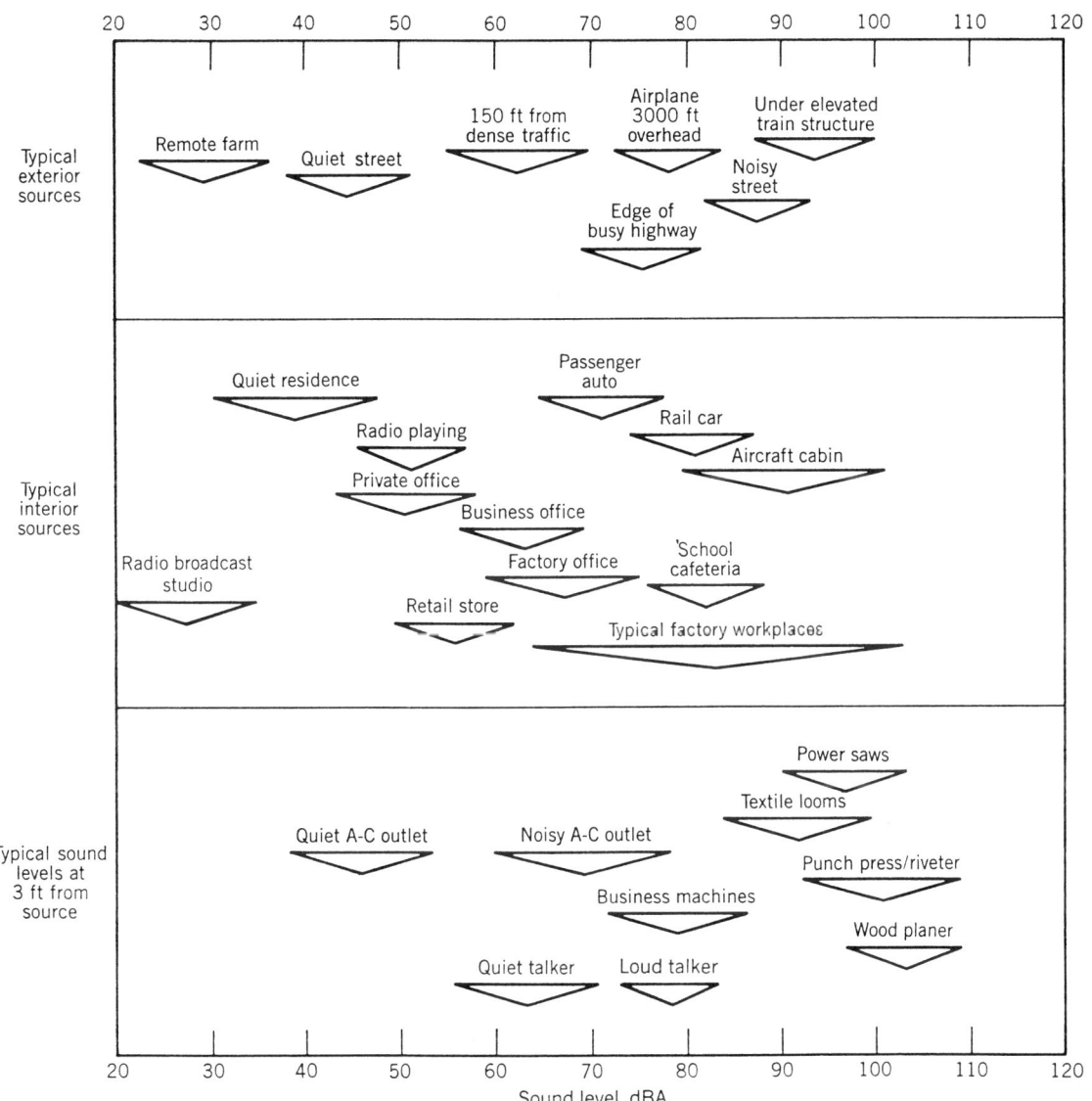

Figure 1.10 Ranges of sound levels in decibels with A scale frequency weighting, in dBA, for common interior and exterior sound sources. (From William J. Cavanaugh, "Acoustics—General Principles," in Encyclopedia of Architecture: Design, Engineering & Construction, *Joseph A. Wilkes, Ed. Copyright © 1988 John Wiley & Sons. Reprinted by permission of John Wiley & Sons.*)

standards and, with recent availability of sophisticated sound measurement instrumentation, a more meaningful and straightforward metric for measuring and evaluating time-varying sounds has been in use—*the energy equivalent sound level* (L_{eq}). The L_{eq} is the hypothetical equivalent steady sound level containing all of the acoustical energy in an actual time-varying sound sample over a given time period. For the time-varying sound of Figure 1.11, the corresponding L_{eq} value is 58 dBA. Thus, the L_{eq} more accurately represents the actual acoustical energy present in a fluctuating sound over the observation period. The duration of the observation period must be stated. The

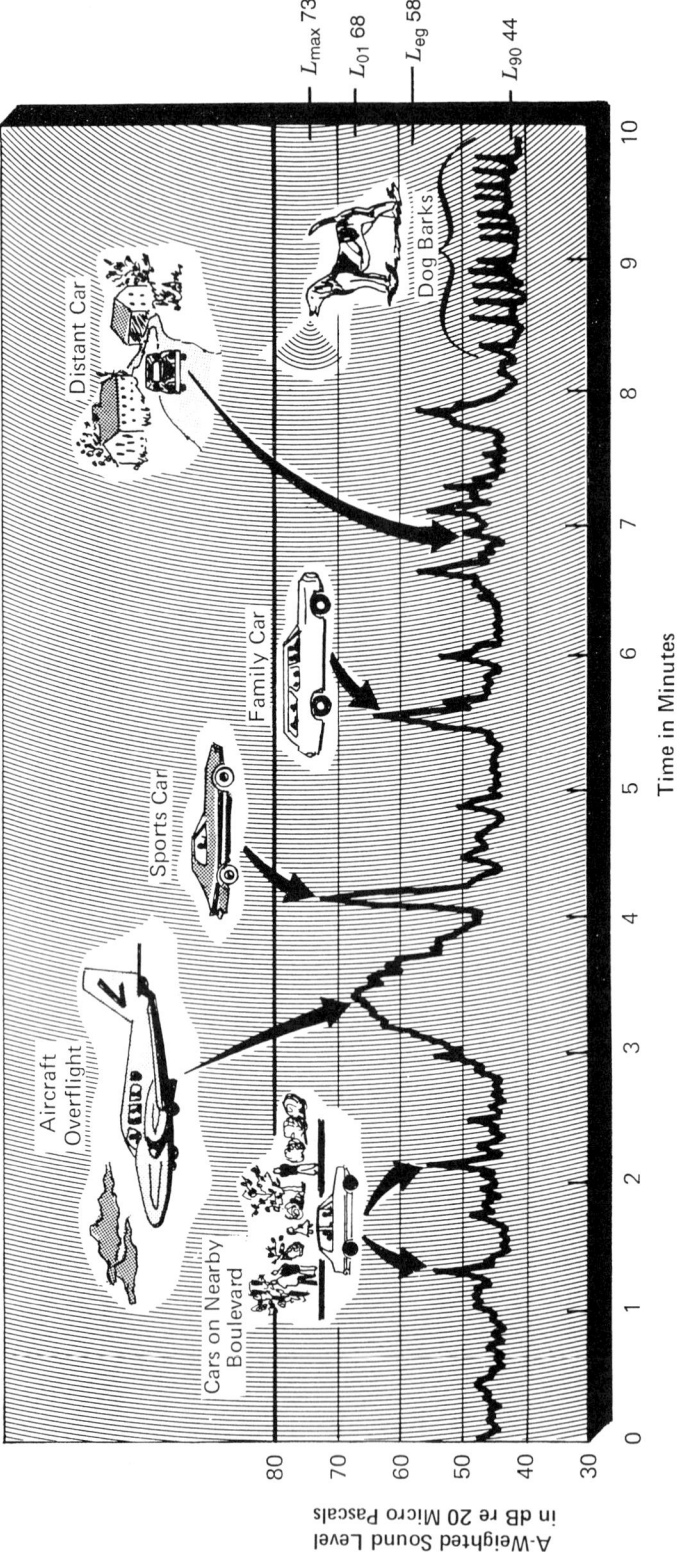

Figure 1.11 Typical outdoor sound measured on a quiet suburban street (from *U.S. Environmental Protection Agency Report, EPA 560/9–79–100, November 1978*).

TABLE 1.1 Land-Use Categories and Metrics for Transit Noise Impact Criteria

LAND-USE CATEGORY	NOISE METRIC (dBA)	DESCRIPTION OF LAND-USE CATEGORY
1	Outdoor $L_{eq}(h)^a$	Tracts of land where quiet is an essential element in their intended purpose. This category includes lands set aside for serenity and quiet, and such land uses as outdoor amphitheaters and concert pavilions, as well as national historic landmarks with significant outdoor use.
2	Outdoor L_{dn}	Residences and buildings where people normally sleep. This category includes homes, hospitals, and hotels where a nighttime sensitivity to noise is assumed to be of utmost importance.
3	Outdoor $L_{eq}(h)^a$	Institutional land uses with primarily daytime and evening use. This category includes schools, libraries, and churches where it is important to avoid interference with such activities as speech, meditation, and concentration on reading material. Buildings with interior spaces where quiet is important, such as medical offices, conference rooms, recording studios, and concert halls fall into this category. Places for meditation or study associated with cemeteries, monuments, and museums. Certain historical sites, parks, and recreational facilities are also included.

[a] L_{eq} for the noisiest hour of transit-related activity during hours of noise sensitivity.
From U.S. Federal Transit Administration Report DOT-T-95-16 Transit Noise and Vibration Impact Assessment, April 1995.

use of the descriptor L_{eq} alone is insufficient. One must always indicate the time period for which the L_{eq} applies (e.g., a "worst" hour, $L_{eq(1h)}$).

Practically all federal standards [the Department of Housing and Urban Development (HUD), the Federal Highway Administration (FHWA), the Federal Aviation Administration (FAA), and other agencies] now rely upon L_{eq} values in their standards for environmental sound. In addition, many local and state codes have adopted L_{eq} values in their environmental sound ordinances. The L_{eq} metric as well as other statistical measures are normally used in studies in addressing environmental sound issues.

A further refinement of the L_{eq} methodology for analysis of time-varying sounds in communities is the *day–night equivalent sound level* (L_{dn}). The L_{eq} values would be summed up over a 24-hr period and a 10-dB penalty would be added for the more sensitive sleeping hours. In other words, noise events occurring during the nighttime hours (usually, 10 PM to 7 AM) would be considered to be 10 dB higher in level than they actually measure. This methodology is extensively used in dealing with airport, transit system, and other outdoor noise events, and a significant body of research shows that L_{dn} values correlate quite well with a community's response to noise impact. Table 1.1 shows an example of categorization of various land uses where $L_{eq(hourly)}$ and L_{dn} noise metrics may be used in assessing transit system noise impact.

Combining Decibels

Sound energy levels in decibel units from independent sound sources may not be added directly. The sound pressure levels must be converted back to arithmetic units and added and then reconverted to decibel units. For example, if two sound sources each measured 50 dB when operated independently, they would measure 53 dB when operated together. Figure 1.12 is a nomogram for easily "adding" (i.e., combining) two sound energy levels. From Figure 1.12 it can be seen that two identical sources (difference between the two sound levels is 0 dB) will result in an increase in sound level of 3 dB with both sources operating. Similarly, if there is a 10-dB or greater difference between two sources, there would be negligible contribution from the "qui-

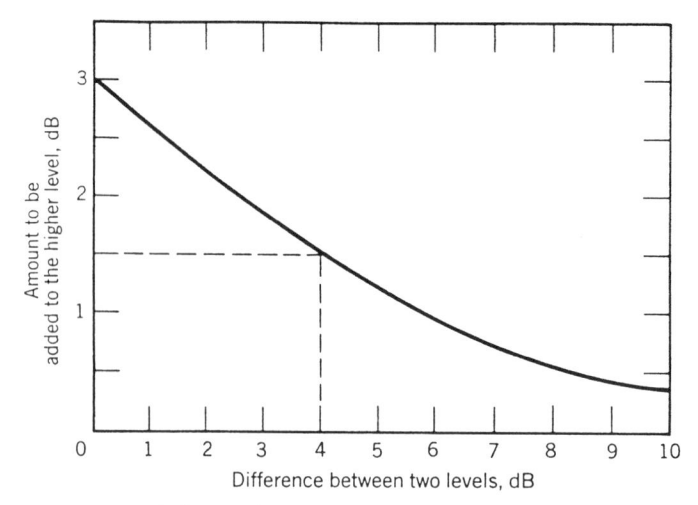

Figure 1.12 Nomograph for combining two sound sources in decibels. In the example shown, two sound sources produce sound levels of 50 and 54 dB, respectively. What level would be produced with both sources operating together? Difference, 54 − 50 = 4 dB; amount to be added to the higher level, 1.5 dB; sound level with both sources operating, 54 + 1.5 = 55.5 or 56 dB. (From William J. Cavanaugh, "Acoustics—General Principles," in Encyclopedia of Architecture: Design, Engineering & Construction, *Joseph A. Wilkes, Ed. Copyright © 1988 John Wiley & Sons. Reprinted by permission of John Wiley & Sons.*)

eter" source. Figure 1.12 also illustrates the addition of a source measuring 54 dB and one measuring 50 dB. The "combined" sound level of 55.5 dB is always higher than the higher value. In other words the louder source dominates. Thus, whenever multiple sound sources are involved, the total sound output may be estimated using the nomogram of Figure 1.12 simply by combining two sources at a time.

Relative Change in Sound Levels

The relative subjective change between two sound source levels or conditions is often of interest in evaluating the effectiveness of various sound control measures. Figure 1.13 shows that a 1-dB change in sound level is just detectable in a controlled laboratory environment. A 3-dB change (which is actually a doubling of the sound energy level) would be just perceptible in a typical room environment. On the other hand, a 10-dB change is required to cause a subjective sensation of doubling of loudness (or halving). These rather unique characteristics of human hearing response must be borne in mind in dealing with practical sound control problems in buildings. In other words, a 1- or 2-dB improvement alone may not represent a significant result and may not be worth the cost of the control measure.

Sound Outdoors versus Sound Indoors

In order to fully appreciate how sound behaves inside rooms and is transmitted from space to space within buildings, it is helpful to consider first how sound behaves outdoors (see Figure 1.14). With a simple nondirective source, the sound intensity will fall off as the distance from the source is increased. The sound wave moving outward from the source spreads its energy over an ever-increasing spherical area.

```
Difference                  Subjective
between two                 change in
sound levels, dB             loudness


    20                    Much louder (or quieter)




    10                    Twice (or half) as loud

                          Clearly noticeable
     5                    in most spaces

                          Just perceptible:
     3                      in typical rooms
     1                      in laboratory environment
```

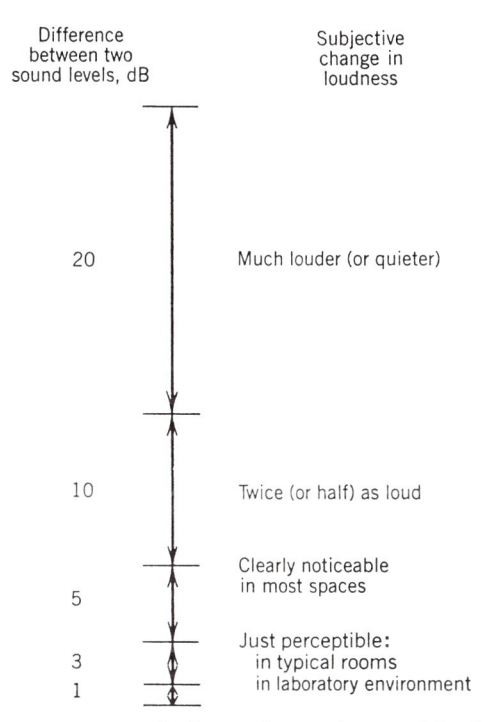

Figure 1.13 Subjective meaning of relative changes in sound levels measured in decibels. (From William J. Cavanaugh, "Acoustics—General Principles," in Encyclopedia of Architecture: Design, Engineering & Construction, *Joseph A. Wilkes, Ed.* Copyright © 1988 John Wiley & Sons. Reprinted by permission of John Wiley & Sons.)

This commonly observed decay of sound level with distance in a "free-field" acoustical environment follows the so-called inverse square law. For simple point sources, the falloff rate is 6 dB per doubling of distance from the source. In effect, as the radius of the sphere over which the sound has spread has doubled, this results in a spherical area four times greater and a sound level reduced by 10 log 4, or 6 dB. If the source is a long narrow cylindrical radiator of sound (as might be the case with a steady stream of road traffic), the rate of falloff would be reduced to 3 dB per doubling. In any case, typical sources outdoors generally fall within the 6 or 3 dB per doubling of distance falloff rate. In addition, some further losses (or gains) may be present in real life situations, due to atmospheric effects, wind, temperature, ground foliage, and so forth. However, these effects can usually be neglected for first-order approximation of expected sound losses outdoors where distances are not very large.

Indoors, on the other hand, sound intensity will fall off with distance only very near the source (in most building situations, within several feet). As one continues to move away from the source, the reflected sound from the floor, walls, and ceiling of the room begins to overwhelm the direct sound component that continues to be emitted from the source. Within the reflected or so-called reverberant sound field, the sound level remains generally constant throughout the room no matter how far away from the source a listener is located. If the room surfaces are basically hard and sound reflective (plaster, concrete, glass, etc.), there will be very little loss of sound at each impact of the sound wave with the room surfaces, and the built-up reflected sound level will be relatively high. If soft, porous materials (rugs, draperies, acoustical tiles, etc.) are placed on the room surfaces, there will be appreciable losses each time the

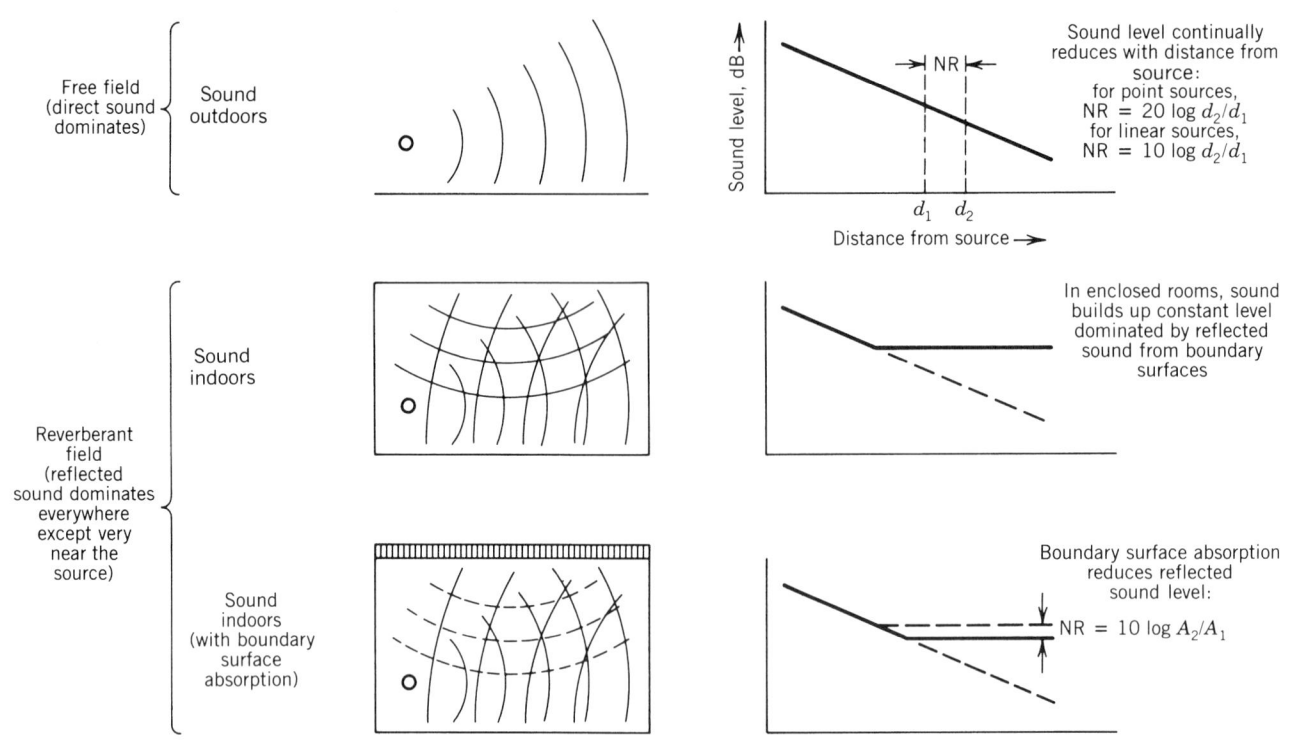

Figure 1.14 Diagrams showing the relative differences in sound behavior outdoors (free field) vs. indoors (reverberant field). (From William J. Cavanaugh, "Acoustics—General Principles," in Encyclopedia of Architecture: Design, Engineering & Construction, *Joseph A. Wilkes, Ed. Copyright © 1988 John Wiley & Sons. Reprinted by permission of John Wiley & Sons.*)

reflected sound waves encounter the room surfaces. Accordingly, the built-up reflected sound levels will be lower. This is the principal effect of placing sound-absorbing materials on the surfaces of rooms (i.e., to lower the sound level in the reverberant acoustic field dominated by reflected sound). Ultimately, if completely efficient sound-absorbing materials are placed on all boundary surfaces of a room, outdoor conditions would be approximated where only the direct sound remains.

Note: The application of absorbing materials on the room surfaces does not affect in any way the direct sound that continues to decay with distance from the source.

Sound-Absorbing Materials

Sound-absorbing materials, carpeting, acoustical tiles, and other specially fabricated absorbing products can absorb appreciable amounts of sound energy. The sound-absorbing efficiency of a material is given by its sound absorption coefficient (α). The sound absorption coefficient is a ratio of the incident sound to the reflected sound and may vary from 0 (no absorption, or perfect reflection) to 1 (complete absorption, or no reflection). Sound absorption coefficients are determined from laboratory measurements. For typical building applications, the most meaningful sound absorption data is obtained from relatively large samples of a material measured in a large reverberant chamber in accordance with standardized test procedures [American Society for Testing and Materials (ASTM) Method of Test C 423].

Figure 1.15 illustrates the typical sound-absorbing characteristics of various generic types of sound-absorbing materials. Porous materials (fibrous or interconnecting cellular plastic forms, etc.) account for most of the prefabricated factory-finished products available. The overall thickness including any spacing of the material from a backup surface influences the absorption in the low-frequency range. The thicker the porous material and/or deeper the air space behind the absorbing layer, the higher will be the low-frequency sound absorption coefficients. The surface facing applied to or on the porous material for architectural finish reasons (durability, light reflectance, appearance, etc.) influences the high-frequency absorption of the assembly. The more open and acoustically transparent the assembly, the less will be the effect on the mid- and high-frequency sound absorption coefficients. Sound reflection from the solid areas between the openings, perforations, or fissures of a surface facing material tend to reduce absorption efficiency of the material at high frequencies.

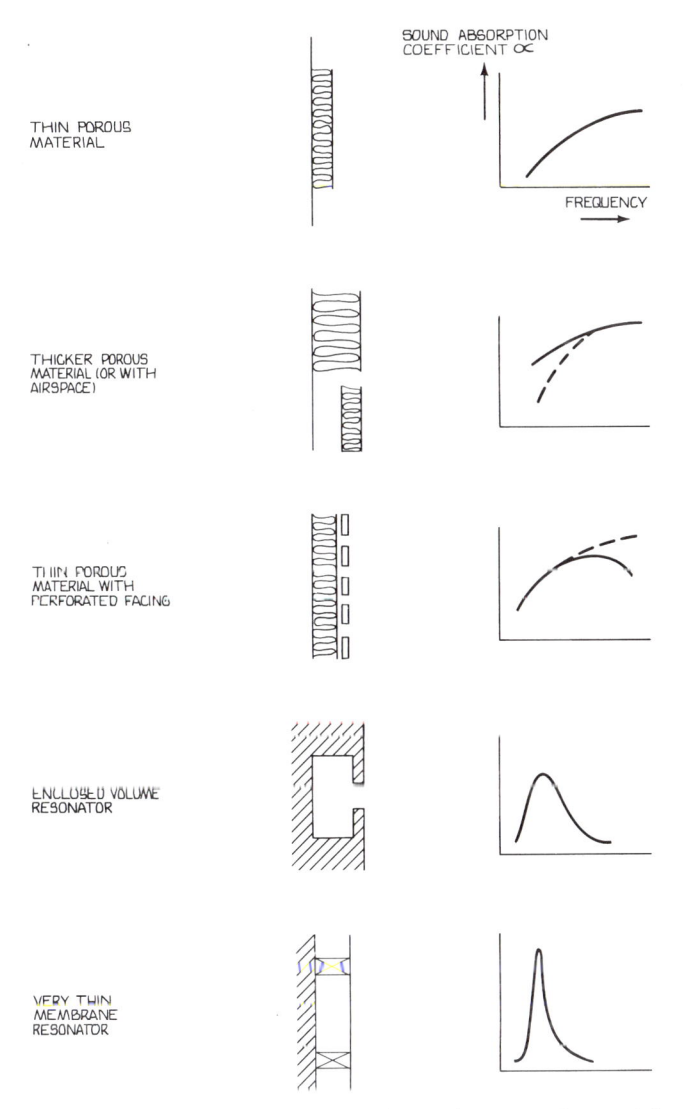

Figure 1.15 Sound-absorbing characteristics of typical acoustical materials. (From W. J. Cavanaugh, Building Construction: Materials and Types of Construction, 5th ed. by Whitney Huntington and Robert Mickadeit. Copyright © 1981 John Wiley & Sons. Reprinted by permission of John Wiley & Sons.)

Volume, or "cavity"-type absorbers and thin panel membrane absorbers, also indicated on Figure 1.15, are effective primarily in the low-frequency range. In all sound-absorbing materials and assemblies, however, the basic mechanism is friction. Sound energy is dissipated as the incident sound moves through the porous material or neck of the cavity or as it sets a thin membrane into vibration. Chapter 2 describes the absorption performance characteristics of a wide range of common building materials and for materials designed specifically for high sound absorption.

Noise Reduction Coefficient

An industrywide accepted method of describing the "average" sound absorption characteristics of an acoustical material is the noise reduction coefficient (NRC). The NRC is the arithmetic average of the measured sound absorption coefficients at 250-, 500-, 1000-, and 2000-Hz test frequencies, rounded off to the nearest 0.05.

Further discussion on sound-absorbing materials may be found in Chapter 2 including tables of acoustical performance values for common building materials and for products and systems specifically designed to provide efficient sound-absorbing surfaces. In general, effective sound absorption is achieved when the sound absorption coefficients exceed about 0.4 (i.e., 40% of the incident sound is absorbed and 60% is reflected back into the room). On the other hand, materials having coefficients of 0.8 or greater (80% absorbed and 20% reflected) are considered very effective absorbers. The average NRC values may be considered in the same manner as absorption coefficients at specific frequencies. However, when using NRC values remember that the average value is obtained using coefficients from 250 through 2000 Hz. If sound absorption is needed above or below this range, particularly at 125 and 63 Hz, NRC values may not be adequate. For example, if low-frequency echoes from an auditorium rear wall present a problem, the NRC values will not provide an indication of sound absorptivity below 250 Hz. Sound absorption coefficients on the low-frequency performance of the rear wall material being considered would be needed.

Reduction of Room Sound Levels

The reduction of reverberant sound levels in rooms may be determined from the following expression (see also Figure 1.14):

$$\text{NR} = 10 \log \frac{A_2}{A_1}$$

where

NR = reduction in reverberant sound level in decibels between two different conditions of room absorption
A_1 = total absorption in square feet or square meters initially present in room (sum of room surface areas times their absorption coefficients)
A_2 = total absorption in square feet or square meters after new absorbing material is added

Typically, room reverberant sound levels can be reduced by up to about 10 dB over an initial "hard" room condition by application of efficient sound-absorbing ceiling treatment and floor carpeting. The simple nomogram of Figure 1.15 may be used for estimating this reduction in typical rooms. For example, a classroom finished

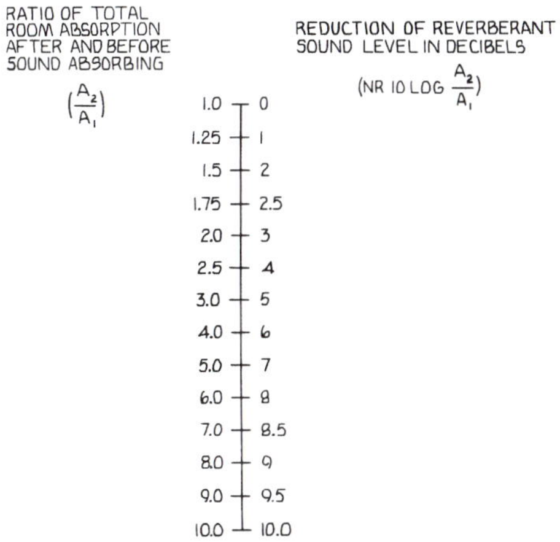

```
RATIO OF TOTAL
ROOM ABSORPTION        REDUCTION OF REVERBERANT
AFTER AND BEFORE       SOUND LEVEL IN DECIBELS
SOUND ABSORBING
                                        A₂
   (A₂/A₁)                    (NR 10 LOG ──)
                                        A₁
         1.0  ┬  0
        1.25  ┼  1
         1.5  ┼  2
        1.75  ┼  2.5
         2.0  ┼  3
         2.5  ┼  4
         3.0  ┼  5
         4.0  ┼  6
         5.0  ┼  7
         6.0  ┼  8
         7.0  ┼  8.5
         8.0  ┼  9
         9.0  ┼  9.5
        10.0  ┴  10.0
```

Figure 1.16 Reduction of room reverberant sound level by added sound-absorbing material. Example: given a room with 100 ft² (sabins) of total sound absorption, estimate NR with 800 ft² of new absorption added. $A_1 = 100$ ft²; $A_2 = 100 + 800 = 900$ ft²; NR = $10 \log \frac{900}{100} = 10 \log 9 = 9.5$ dB. (From W. J. Cavanaugh, Building Construction: Materials and Types of Construction, *5th ed. by Whitney Huntington and Robert Mickadeit. Copyright © 1981 John Wiley & Sons. Reprinted by permission of John Wiley & Sons.*)

basically in hard materials may have a total initial absorption (A_1) of 100 ft² (determined by adding all the surface areas times their absorption coefficients). If a new acoustical tile ceiling added 800 ft² of new absorption, A_2 would correspondingly increase ($A_2 = 800 + 100 = 900$ ft²). From Figure 1.16, a ratio of $A_2/A_1 = 900/100 = 9$ would indicate a reduction of 9.5 dB. In other words, the classroom would be almost 10 dB quieter with the new sound-absorbing ceiling no matter how loud the source is. Reduction of room activity noise levels of this order of magnitude can be significant (refer to Figure 1.13.)

However, remember that the surface treatment does not affect or reduce the direct sound in any way. In other words, the best we can do in any room is to approximate "outdoors" where the direct sound coming from the source will always remain. For example, an "outdoor picnic" may still be a noisy affair even though just about all of the "outdoor" room may be totally absorptive.

Reverberation in Rooms

In addition to providing control of continuous room sound levels, surface-applied sound-absorbing materials in a room affect the persistence of "lingering" of sound after a source is stopped. The reverberation period (time in seconds for the sound level to decay 60 dB after the source is turned off) is directly proportional to the cubic volume of the space and inversely proportional to the total sound absorption present:

$$T = 0.05 \frac{V}{A} \quad \text{(English units), or}$$

$$T = 0.16 \frac{V}{A} \quad \text{(Metric units)}$$

where

> T = reverberation time in seconds
> V = volume in cubic feet (or cubic meters)
> A = total absorption in square feet (or square meters)
> (sum of room surfaces times their sound absorption coefficients plus the sound absorption provided by furnishings or audience, etc.)

Using the sound absorption coefficients of the performance tables in Chapter 2 the reverberation period may be computed for most building spaces where the room dimensions are within about a 1:5 aspect ratio. The sound field in very wide rooms with low ceilings, for example, does not decay in a manner that permits direct use of the above expression. Similarly in highly absorbent "outdoorlike" spaces, the expression does not apply since, by definition, the concept of reverberation becomes meaningless where a sound field is not dominated by repeated reflections from the bounding surfaces. However, for most typical rooms in buildings, the expression can yield a good estimate of the reverberation period. Note too that since the sound absorption coefficients of most building materials vary with frequency, the reverberation calculations must be carried out at representative low-, mid-, and high-frequency ranges (e.g., in octave bands from 125 through 4000 Hz). For less critical rooms a single computation at a representative mid-frequency range (e.g., 1000 Hz) may be adequate. Needless to say new computer technology makes extensive and rapid calculation of reverberation-based metrics very convenient especially for critical music performance halls and spaces for organ and liturgical music (see Chapters 4 and 6).

Sound Transmission between Rooms

When greater reduction of sound than is possible by room sound-absorbing treatment alone is required, full enclosure of the receiver by means of separate rooms may be necessary. Figure 1.17 illustrates schematically the simple case of sound transmission between adjacent enclosed rooms. In essence, a sound source will develop a reverberant sound field in one room (the source room) and its sound pressure level will depend on the total absorption provided by the source room boundary surfaces. In this simple case, assuming that the sound can travel to the adjacent room (the receiving room) only via the common separating wall, the transmitted sound level to the receiving room will depend on three factors: (1) the sound-isolating properties of the wall (i.e., sound transmission loss), (2) the total surface area of the common wall that radiates

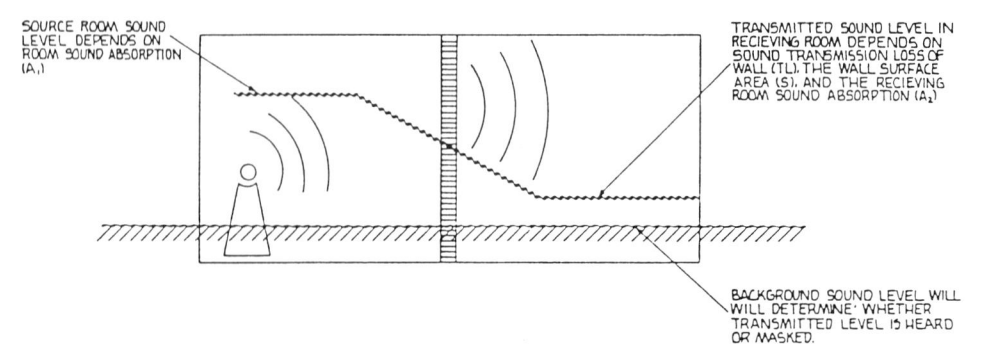

Figure 1.17 Essential elements in sound transmission between rooms. (From W. J. Cavanaugh, Building Construction: Materials and Types of Construction, 5th ed. by Whitney Huntington and Robert Mickadeit. Copyright © 1981 John Wiley & Sons. Reprinted by permission of John Wiley & Sons.)

sound into the adjacent receiving room, and (3) the total sound absorption present in the receiving room. The reduction of sound between rooms is given by the expression:

$$L_1 - L_2 = \text{TL} + 10 \log \frac{A_2}{S}$$

where

L_1 = sound pressure level in the source room in decibels
L_2 = sound pressure level in the receiving room in decibels
TL = sound transmission loss of the common wall in decibels
A_2 = total sound absorption in the receiving room in square feet (or square meters)
S = common wall surface area in square feet (or square meters)

The transmitted sound level L_2 in any given situation will be audible and possibly disturbing to the receiving room occupants if it exceeds the ambient or background sound level in the room. Thus the background sound level is an extremely important part of any sound isolation problem. The background sound level may be thought of as the residual sound level present whether or not the offending noise is present in the source room. The common wall may be thought of as a large diaphragm radiating sound into the receiving room—the larger it is, the more sound is radiated. On the other hand, absorbing material in the receiving room tends to reduce the built-up reflected sound radiated into the receiving room. Thus, the A_2/S correction term in the room-to-room sound reduction expression accounts for the particular environment in which a wall construction is used. This correction is rarely more than ±5 dB but can be significant especially at low frequencies.

The major loss in sound energy from room to room is, however, provided by the common wall (or floor/ceiling) construction itself. Typical lightweight partition or floor systems may have sound transmission losses of the order of 20 dB. Massive and/or double constructions can achieve sound transmission loss values of 40 to 60 dB or greater. (See also the performance tables in Chapter 2).

Sound Transmission Loss

A basic acoustical property of a sound-isolating wall or floor/ceiling system is then its ability to resist being set into vibration by impinging sound waves and thus to dissipate significant amounts of sound energy. The heavier and more complex the construction, the greater its ability to reduce sound transmission from one side to the other. The sound-reducing capability of a construction is measured by its sound transmission loss (TL). The sound transmission loss is a logarithmic ratio of the transmitted sound power to the sound power incident on the source room side of the construction. A construction that transmits or lets through only small amounts of the incident sound energy will have a high sound transmission loss. For example, a 4-in.-thick brick wall might have a mid-frequency sound transmission loss of about 40 dB. This means that only $\frac{1}{10,000}$ of the incident sound energy is transmitted. Recall from Figure 1.13 that a 10-dB change in sound level represents a significant reduction (i.e., a halving of the subjective loudness of a sound). Accordingly, a 40-dB change represents an even more dramatic reduction. Reductions in room-to-room sound level of 20 to 50 or more are generally needed to effectively isolate typical building activities from one another.

Single Homogeneous Walls

Figure 1.18 helps illustrate the general effects of mass in the sound transmission loss performance of constructions. For single homogeneous constructions the average sound

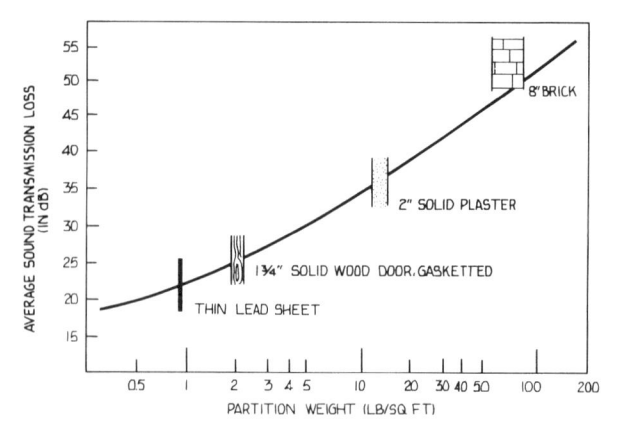

Figure 1.18 Average airborne sound transmission loss for single homogeneous partitions. (From W. J. Cavanaugh, Building Construction: Materials and Types of Construction, 5th ed. by Whitney Huntington and Robert Mickadeit. Copyright © 1981 John Wiley & Sons. Reprinted by permission of John Wiley & Sons.)

transmission loss (average from 125 to 4000 Hz) increases with increasing weight. For example, a 15-psf (pounds per square foot) plaster wall (2-in. solid plaster) would have an average sound transmission loss of about 35 dB. Doubling the partition weight (and thickness) to 30 psf (4 in.) would increase the average sound transmission loss to 40 dB. Another doubling to 60 psf (6 in.) would yield a TL of 45 dB. Clearly, single homogeneous constructions quickly reach a point of diminishing returns where increased weight and thickness are no longer practical.

Double Walls

Figure 1.19 shows the advantage of complexity rather than just increasing the weight of the sound-isolating construction. The 2-in. solid plaster partition discussed above

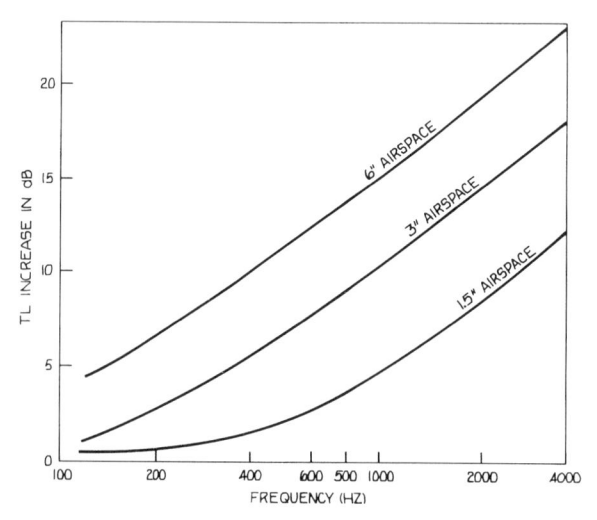

Figure 1.19 Increase in airborne sound transmission loss using double-layer construction with airspace (weight of two leaves equal). (From W. J. Cavanaugh, Building Construction: Materials and Types of Construction, 5th ed. by Whitney Huntington and Robert Mickadeit. Copyright © 1981 John Wiley & Sons. Reprinted by permission of John Wiley & Sons.)

yielded an average TL of 35 dB. Without any overall increase in weight, if the 2-in. plaster were split into two independent 1-in. leaves and separated by a 3-in. airspace, an average increase of about 8 dB would result (from Figure 1.19 the increase would be 2 dB at 125 Hz and 12 dB at 4000 Hz). In other words, double-layer construction is one way to beat the "mass law" limits of single homogeneous partition materials. This explains why so many prefabricated operable or demountable wall systems as well as other constructions that use double-layered elements can achieve relatively high sound transmission loss values with relatively low overall surface weights.

Many constructions, such as sheet metal or gypsum board stud wall systems or wood joist floor systems with gypsum board ceilings, fall somewhere between ideal mass law performance and ideal double-construction performance. In addition to optimizing the sound isolation performance of various building elements, consideration is also given to improved methods of connecting and sealing the individual elements so that maximum performance can be realized in the field.

Note also from Figure 1.19 that relatively large airspaces between the two elements of a double construction are required to achieve significant improvement over single-layer performance. Airspaces of less than about $1\frac{1}{2}$ in. do not really yield very much improvement over the equivalent single mass performance. This explains why some thin glazed double thermal insulating window systems with airspaces of $\frac{1}{4}$ to $\frac{1}{2}$ in. have disappointing sound transmission loss performance even though adequate thermal insulation may be achieved.

Cavity Absorption in Double Constructions

When the full advantages of both mass and complexity have been utilized in double-leaf constructions, a further improvement in performance can be realized by sound-absorbing material within the cavity of the construction. Fibrous, glass, or mineral wool-type insulation materials can reduce the sound energy within the cavity volume and thus increase the overall sound energy loss through the construction. If the construction is such that the two sides are extensively coupled together by internal supporting elements, the cavity absorption will have significantly less effect than if the two sides are well isolated from one another. For example, typical masonry block constructions where the two sides of the block are intimately joined by the very rigid core elements gain very little by cavity absorption. On the other hand, a double-layer gypsum board or steel panel system on widely spaced framing members can gain as much as 3 to 5 dB in improved sound isolation performance over the same construction without cavity absorption.

Cavity absorption also provides a "mufflerlike" effect to reduce sound transmission at penetrations of a double-leaf construction for electrical conduit or at other locations where shrinkage cracks may develop. On balance, whenever maximum sound-isolating performance is desired with most double-leaf structures, cavity absorption can contribute to improved results.

Composite Constructions

Often the common walls between rooms are made up of more than one component (doors, windows, two different partition elements, etc.). Unless the sound transmission loss performance of each of the components is identical, the effective sound transmission loss performance of the composite construction will fall below that of the most effective single component and approach that of the weaker element. Figure 1.20 permits the determination of the effective sound transmission loss of a two-element composite assembly. For example, assume a 200-ft^2 section of a 4-in. brick wall has

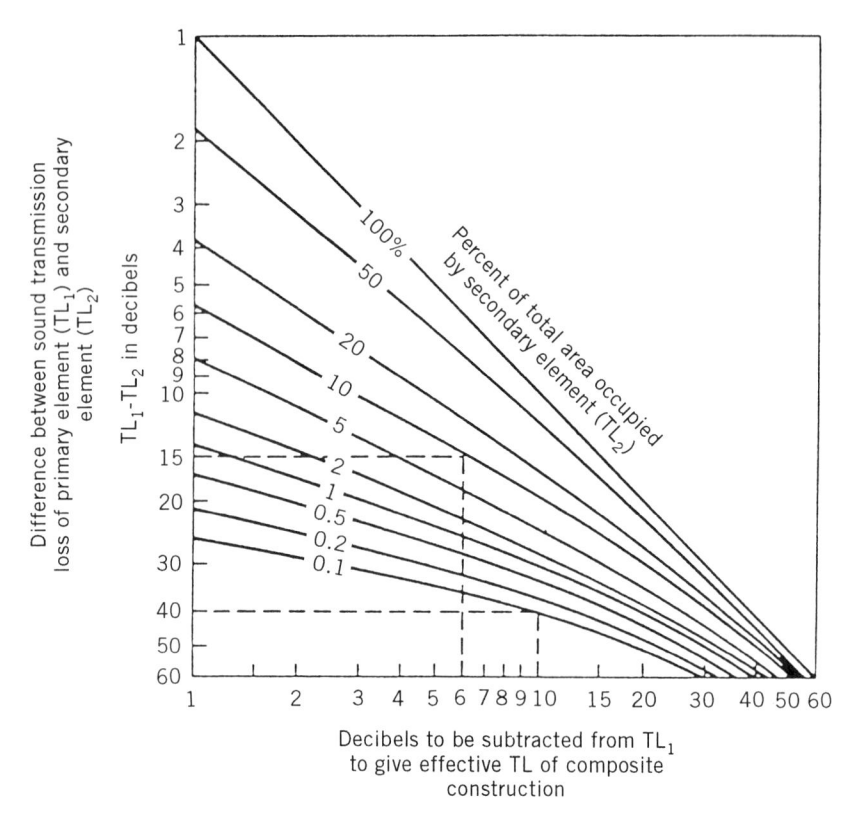

Figure 1.20 *Effective transmission loss (TL) of composite acoustic barriers made up of two elements. (From W. J. Cavanaugh,* Building Construction: Materials and Types of Construction, *5th ed. by Whitney Huntington and Robert Mickadeit. Copyright © 1981 John Wiley & Sons. Reprinted by permission of John Wiley & Sons.)*

an average TL of 40 dB. If a 7 × 3 ft pass door having an average TL of 25 dB is cut into the brick wall, the effective TL of the composite wall can be found as follows:

$$TL_{wall} - TL_{door} = 40 - 25 = 15 \text{ dB}$$

Percent of wall occupied by door = $\frac{21}{200} \approx 10\%$

From Figure 1.20, amount to be subtracted from TL_{wall} = 6 dB. Therefore, the effective sound transmission loss of the wall with door = 40 − 6 or 34 dB.

Similarly, the effect of sound leaks in a partition system can be evaluated. For example, 0.1% of crack area in a 100-ft² section of 4-in. brick wall would lower the sound transmission loss from 40 to 30 dB:

$$TL_{wall} - TL_{crack} = 40 - 0 = 40 \text{ dB}$$

Percent of crack area = 0.1%

From Figure 1.20, amount to be subtracted from TL_{wall} = 10 dB. Therefore, the effective sound transmission loss of the wall with 0.1% crack is 40 − 10 or 30 dB. One-tenth of one percent a 100-ft² wall would represent a total of only about 14 in.² of accumulated crack area!

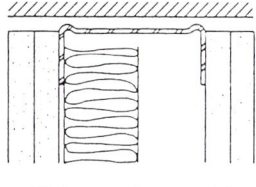

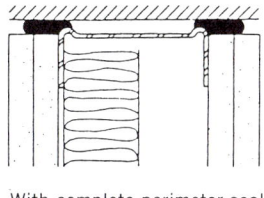

3 5/8" steel studs 24"oc. with 2 layers gypsum board each side and 2" cavity insulation

Without perimeter seal (approx. 1% of total wall area-estimated crack area)

With complete perimeter seal with hardening mastic chaulk

Measured sound transmission class (STC)   STC 27   STC 50

Figure 1.21 Typical sound leak encountered in building construction. (From W. J. Cavanaugh, Building Construction: Materials and Types of Construction, 5th ed. by Whitney Huntington and Robert Mickadeit. Copyright © 1981 John Wiley & Sons. Reprinted by permission of John Wiley & Sons.)

Sound Leaks

From the preceding discussion, the relative significance of any direct sound leakage through a construction is quite apparent. In fixed permanent wall or floor/ceiling systems, the leakage at intersections with floors, side walls, and so forth can often be avoided by "one-time" applied sealants during the initial installation. Figure 1.21 shows the reported laboratory results for a gypsum partition system with and without a typical perimeter crack. Degradation in performance by some 23 dB from a partition system potential performance of over 50 dB can be a very real concern in typical field situations. Hidden sound leaks can occur in spite of even the best field supervision of the installation (above suspended ceilings, behind convector covers, etc.).

With lightweight operable or demountable partition systems, the problem of sound leakage at the numerous panel joints, floor, ceiling tracks, and side wall intersections is even more demanding. Materials and systems must be detailed and specified to assure positive panel joint seals that will perform effectively over the expected life of the partition installation. A fixed partition is relatively easy to seal. In operable or demountable partitions, the seals themselves must also be operable and durable over the life cycle of the installation with minimum maintenance required.

Flanking

Besides direct sound leaks within the perimeter of a given common wall or floor/ceiling system, significant sound transmission may occur between adjacent rooms via so-called flanking paths. There are literally hundreds of possibilities for sound to "bypass" the obvious common partition path, and their relative importance is directly proportional to the sound isolation performance desired. Flanking becomes increasingly important with higher sound isolation performance. The sound transmission path over the partition via a suspended ceiling, against which the partition terminates, is a common condition. Others include interconnecting air-conditioning ducts or plenums, doors opening to adjacent rooms via a common corridor, and adjacent exterior windows. The list could go on, but Figure 1.22 gives an idea of the many possible flanking paths that must be considered in addition to the potential leaks occurring within the perimeter of the partition or floor/ceiling assembly itself.

Where partitions terminate at a common suspended ceiling over two adjacent rooms, a serious room-to-room sound-flanking problem may exist, as also shown in

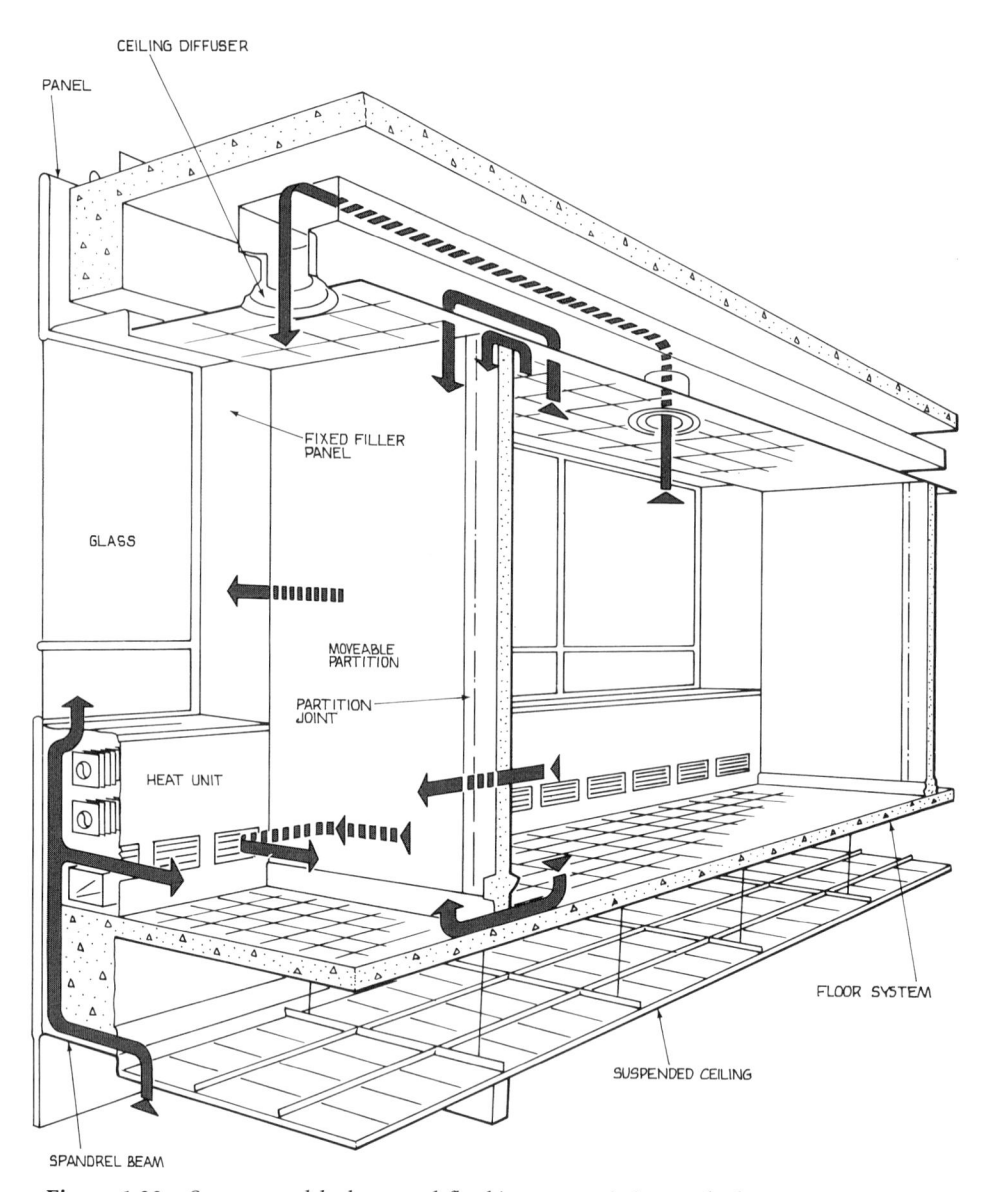

28 • *Introduction to Architectural Acoustics and Basic Principles*

Figure 1.22 Some sound leakage and flanking transmission paths between rooms. (From W. J. Cavanaugh, Building Construction: Materials and Types of Construction, 5th ed. by Whitney Huntington and Robert Mickadeit. Copyright © 1981 John Wiley & Sons. Reprinted by permission of John Wiley & Sons.)

Figure 1.22. The room-to-room sound isolation performance of suspended ceiling systems, ceiling attenuation class (CAC), is determined in the laboratory by standard test procedures (see laboratory test methods discussed below). The test measures the overall room-to-room sound attenuation considering that the source room sound follows a complex path to a receiving room (i.e., it passes through the ceiling in the source room along the common plenum over the partition and down throughout the suspended ceiling in the receiving room). Measured sound attenuation data for this rather complex room-to-room sound path can vary from about 20 to 40 dB or more, depending on the type of ceiling assembly used. Some lightweight acoustical ceiling boards may fall at the low end of this range, and other dense mineral-fiber "back-sealed" ceiling systems with tightly splined joints may perform at the upper end of this range or greater. Where high values of room-to-room isolation are required, it is often necessary to design sound barriers to increase the room-to-room sound loss via the ceiling path with additional impervious elements, either vertically above the partition line or horizontally on the back side of the suspended ceiling system. In general, the sound isolation via the ceiling path must be at least equivalent to that via the common partition for balanced engineering design between the adjacent rooms.

Similarly, acoustical data for evaluating room-room sound transmission via air plenum distribution systems serving two adjacent rooms are available from products or systems tested in accordance with available standards (e.g., Air Diffusion Council Method of Test AD-63). All such paths via air-conditioning ducts, interconnecting secondary spaces, and so forth must be investigated to avoid serious reduction in field performance of an otherwise effective sound isolation common wall system. As a general rule, the higher the expected sound isolation performance of a construction, the greater the concern need be about flanking. Indeed the achievement of room-to-room sound isolation values of greater than about 50 dB in typical field situations require extraordinary care in handling all possible flanking paths.

Laboratory Measurement of Airborne Sound Transmission Loss

There are standardized methods used throughout the world for the laboratory measurement of airborne sound transmission loss of building partitions, floor/ceiling assemblies, doors, windows, and so forth (e.g., ASTM E 90). Obviously, in laboratory testing, all of the variables normally encountered in the field can be controlled to the extent necessary to measure the sound-isolating properties of just a partition or floor/ceiling assembly alone. The requirements of current testing standards have been established to yield as realistic and practical result as is possible. For example, a sufficiently large specimen must be tested (not less than 8×6 ft). Similarly, the test specimens must be representative of the assembly to be used in field situations and must be installed in a manner that duplicates normal field conditions to the extent possible. All of these requirements, as well as the special requirements of the laboratory facility and the measurement procedure, are very precisely defined in the standards.

The test specimen is mounted in an opening between two acoustically isolated test rooms and the sound transmission loss is determined from measurement of the reduction of sound between the rooms with a high-level sound source operating in one of the rooms. The measurements are carried out in continuous $\frac{1}{3}$-octave bands with center frequencies from 125 to 4000 Hz. This is a wide-enough frequency range to cover the low, middle, and high ranges of the audible spectrum. The laboratory test rooms are carefully isolated from one another to avoid "flanking" of sound between rooms by paths other than through the test specimen itself. Measurement of the sound transmission loss of a test specimen under the procedures of ASTM E 90 will yield values that are representative of the maximum sound-isolating capability of a partition system. Figure 1.23 indicates typical laboratory test results for two partitions: (1) a simple fixed partition $4\frac{5}{8}$ in. thick, constructed of 2×4 wood studs with $\frac{1}{2}$-in. gypsum board outer surfaces and (2) a 6-in.-thick reinforced concrete floor slab. Note that, in general, the concrete floor system has higher sound transmission loss values over the entire frequency range from 125 to 4000 Hz. This is largely due to the substantial mass of the construction compared to the lightweight partition. Furthermore, the stud partition exhibits significant "dips" in the transmission loss performance curve (e.g., at 3150 and 125 Hz) compared to that for the concrete slab.

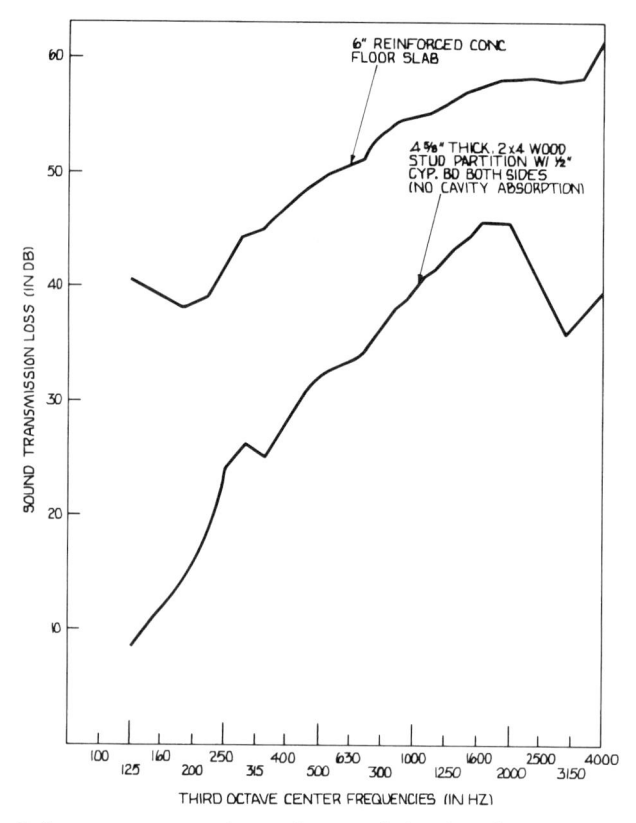

Figure 1.23 Laboratory-measured sound transmission loss for two typical constructions. (From W. J. Cavanaugh, Building Construction: Materials and Types of Construction, 5th ed. by Whitney Huntington and Robert Mickadeit. Copyright © 1981 John Wiley & Sons. Reprinted by permission of John Wiley & Sons.)

Sound Transmission Class

For design and specification purposes a "single-number" descriptor is usually desirable to indicate the sound isolation capability of a partition system. It can be seen with the test data for the constructions of Figure 1.23 that any simple arithmetic averaging of the test results over the full frequency range might be misleading especially for partition performance where large dips occur. To overcome the limitations of simple averaging, a system of rating based on fitting the test data curves (sound transmission class, STC) was established. The procedures for rating the sound transmission loss performance are standardized in ASTM E 413. Figure 1.24 shows the appropriate STC curves fitted to the laboratory data curves for the partitions of Figure 1.23 in accordance with the rules of ASTM E 413. The fitting procedure allows for an average deviation of 2 dB below the STC curve in each of 16 third-octave measurement bands (or a total deviation of 32 dB). In addition there is a maximum 8-dB deficiency allowable in any single test frequency band.

In Figure 1.24 it can be seen that the 8-dB maximum deficiency governs the fit of the STC 32 curve in the case of the fixed gypsum board partition construction. With the heavy concrete floor construction, the allowable deficiencies are spread over a wider portion of the test range. The classification procedure is designed to penalize poor performance and not allow especially good performance at other frequency ranges

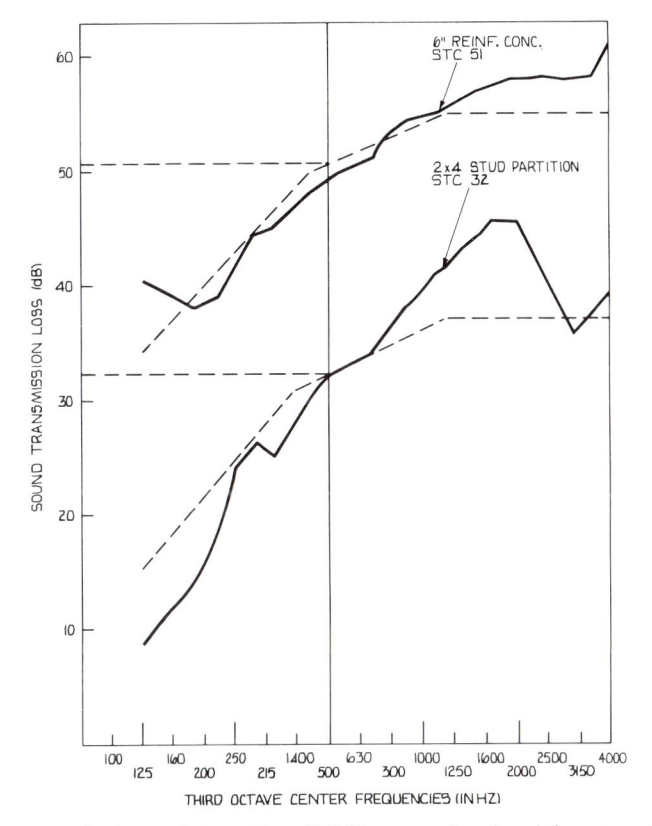

Figure 1.24 Sound Transmission Class (STC) curves fitted to laboratory data. (From W. J. Cavanaugh, Building Construction: Materials and Types of Construction, 5th ed. by Whitney Huntington and Robert Mickadeit. Copyright © 1981 John Wiley & Sons. Reprinted by permission of John Wiley & Sons.)

to "fill in" any critical performance "gaps." In general, however, the STC value does give a rough approximation of the TL value in the 500-Hz, or mid-frequency, range.

Chapter 2 includes performance tables of STC values for some common constructions. These may be compared with currently available laboratory test data in manufacturer's specification information for various specific products and systems being considered for a building design.

Ceiling Attenuation Class

For the case of sound transmission between adjacent rooms via the suspended ceilings over the room which share a common plenum (see Figure 1.22), a special laboratory test method is used. The method measures the room-to-room sound attenuation with the source room sound passing through the ceiling, along the common ceiling plenum and then down through the ceiling again into the receiving room. The method of test is ASTM E 1414-91a (1996), "Airborne Sound Attenuation Between Rooms Sharing a Common Ceiling Plenum." This method has replaced an earlier method of test, AMA 1-II (1967). The attenuation data are rated in accordance with the procedures of ASTM E 413-87 (1994), "Rating Sound Insulation," to yield a similar single number, ceiling attenuation class (CAC) for the common suspended ceiling system over the two adjacent rooms.

Field Measurement of Airborne Sound Transmission Loss

In actual building installations, it must be remembered that airborne sound is transmitted from room to room not only via the common constructions separating the rooms but by many other potential flanking paths. Also, in the field there are usually other conditions—pipe penetrations of the construction system, air-conditioning duct penetrations—that are not present in the basic partition or floor/ceiling system tested in the laboratory. Thus, as would be expected, the sound isolation between rooms in an actual building is often somewhat less than that which can be realized in an idealized laboratory situation. Notwithstanding the complexity of the real-world situation, however, there is a need to measure and to classify the sound isolation capabilities of construction systems under typical field conditions. In fact, field performance is the ultimate interest of building professionals as well as the occupants themselves.

Field Sound Transmission Class

The current field test standard (e.g., ASTM E 336) provides recommended methods for measuring sound insulation in buildings for nearly all cases likely to be encountered in the field. If the field situation is such that flanking of sound around the partition system being measured can be shown to be insignificant, meaningful field sound transmission loss values can be determined and the test data rated to yield a field sound transmission class (FSTC) value. The classification procedures discussed above for rating laboratory-derived sound transmission loss data are also used, but the added letter designation F indicates that the rating value is based on field-derived rather than laboratory test data.

Noise Isolation Class

In many complex field installations, the absence of significant flanking paths by the procedures of ASTM E 336 cannot be demonstrated. In such cases, the standard provides for simply measuring the noise reduction between the rooms in question without taking into account appropriate corrections for partition or floor/ceiling receiving room absorption and so forth. Thus the test provides only a measurement of the overall noise reduction achieved by the construction in situ and includes the effects of flanking paths between rooms in the particular environment. The noise reduction data can be rated in accordance with the procedures of ASTM 413 to yield a single-number noise isolation class (NIC) value. However, this value represents the sound isolation performance of the construction only in the particular environment tested. An NIC value attained at one building may not apply to an installation in another building where the acoustical environment is significantly different.

Therefore, several different field test procedures are standardized in ASTM E 336 to cover the various field conditions encountered. Naturally, the complexity in different measurement conditions and the various ratings appropriate for these different conditions lead to some confusion and even misuse of the data. Suffice to say that when field measurements are involved, the users of the resulting test data must understand just what was measured and all pertinent details of the field test conditions. Finally, and above all, the ratings derived from field tests (FSTC or NIC) are neither interchangeable nor directly comparable with laboratory-derived ratings (STC).

Control of Direct Structure-Borne Sound

When sound energy is directly induced into a structure (by the impact of footsteps, falling objects, hammering, or by rigidly attached vibrating mechanical equipment, etc.), the energy will travel relatively easily throughout the structure and reradiate as airborne sound in adjacent spaces. This type of direct structure-borne sound can be controlled at the source by resilient mounting of mechanical equipment, by the use of resilient or "cushioning" materials at the point of impact as with flooring materials such as carpeting, and by special isolated constructions.

Impact Insulation Class

Direct impact sound of footsteps and falling objects may be a concern with many building types (apartments, offices, etc.) where occupied spaces occur over one another. Acoustical testing of a floor/ceiling assembly's ability to reduce direct impact sound transmission is similar to that for airborne sound transmission loss testing. A test specimen construction is placed in a floor opening between two test rooms, and microphones are positioned in the lower test room to record the sound levels transmitted with a standard taping machine operating on the test specimen. The complete details of the test procedure and method of reporting the test results are included in the current standard method of test (ASTM E 492).

Figure 1.25 shows typical laboratory test results for a reinforced concrete floor slab with and without floor carpeting. The impact insulation class (IIC) increases significantly (IIC 47 vs. IIC 81), indicating that the transmitted impact sound levels due to the carpeting are considerably reduced. The improvement with a "cushioning"

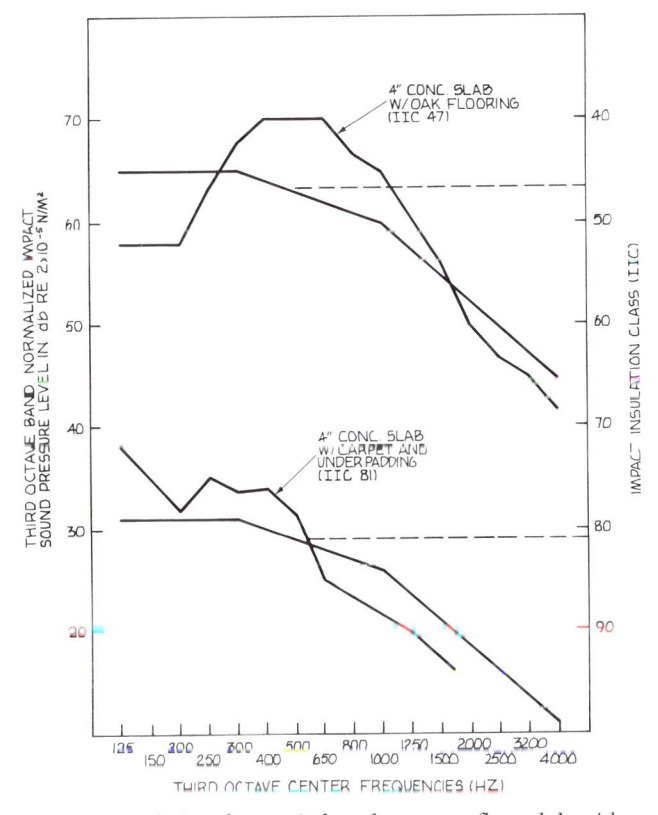

Figure 1.25 Impact insulation for a reinforced concrete floor slab with and without carpeting. (From W. J. Cavanaugh, Building Construction: Materials and Types of Construction, 5th ed. by Whitney Huntington and Robert Mickadeit. Copyright © 1981 John Wiley & Sons. Reprinted by permission of John Wiley & Sons.)

material such as carpeting at the point of impact is enormous as any apartment dweller who has experienced this situation with neighbors overhead can attest.

Chapter 2 also includes a performance table of IIC values for a number of common floor/ceiling assemblies. Also shown, for comparison, are the corresponding airborne STC values for the constructions. Wherever there is a potential for disturbance from both airborne and direct impact sounds, both the STC and IIC ratings of the separating floor/ceiling construction should be considered.

A construction's ability to effectively isolate direct impact sounds (i.e., higher IIC values) is dependent largely on the "softness" of the floor surface and/or the degree to which the directly impacted floor surface is decoupled from the radiating surfaces below. Unlike the case of airborne sound transmission, the mass of the structure plays a secondary role in a structure's ability to isolate direct impact sound except at low frequencies. Heavy concrete slab constructions are only slightly more efficient in isolating impacts than much lighter wood constructions over a significant range of audible frequencies. Soft, cushionlike flooring surfaces, however, significantly improve the IIC values for both light and more massive floor/ceiling assemblies. The low-frequency "boominess" of transmitted impact sounds associated with lightweight wood frame floors even with carpeting can be improved by carefully detailed and constructed measures such as resiliently supported or independent gypsum board ceilings below the floor surface structure and the like.

As with airborne sound transmission, the adequate performance of impact-isolating structures in the field is the ultimate objective. In actual buildings, there are numerous flanking paths for impact sound transmission as well many instances where a system that was effectively decoupled in the laboratory test specimen becomes seriously "short circuited" in the real building context. Again, detailing of the construction and field supervision is essential to be sure that special impact-isolating features of a particular assembly are retained in the actual installation. This becomes increasingly important as higher acoustical performance values are sought for both airborne and structure-borne sound transmission.

Isolation of Mechanical/Electrical Equipment

Most building services [heating, ventilation, and air-conditioning (HVAC) systems, electrical power generators or transformers, elevators, automatic delivery systems, etc.] involve rotating, reciprocating, or otherwise vibrating equipment. When such sources are located near critical occupied spaces and are directly attached to the supporting structure, they can cause serious problems due to both airborne and direct structure-borne sound transmission. The complete treatment of mechanical/electrical equipment noise and vibration control is beyond the scope of this general review. However, some fundamental principles are common to nearly all types of mechanical/electrical equipment sources. Obviously, the larger the machine capacity and its electrical power consumption, the greater the potential for noise and vibration output. A first guideline then is to select and specify (in quantitative terms if possible) the quietest available equipment for the task at hand. Second, the major noise sources should be located as far from critical areas as possible (e.g., basement mechanical equipment rooms rather than an upper level). Third, the vibrating equipment must be effectively decoupled from the building structure (i.e., vibration isolation mounts and bases, resilient connections at connecting ducts and pipes, etc.). Fourth, the enclosing mechanical room structure (floors, walls, and ceilings) must reduce airborne and structure-borne sound to adequately low levels in adjacent spaces. The latter includes careful attention to all possible flanking paths for airborne and structure-borne sound that might short circuit the

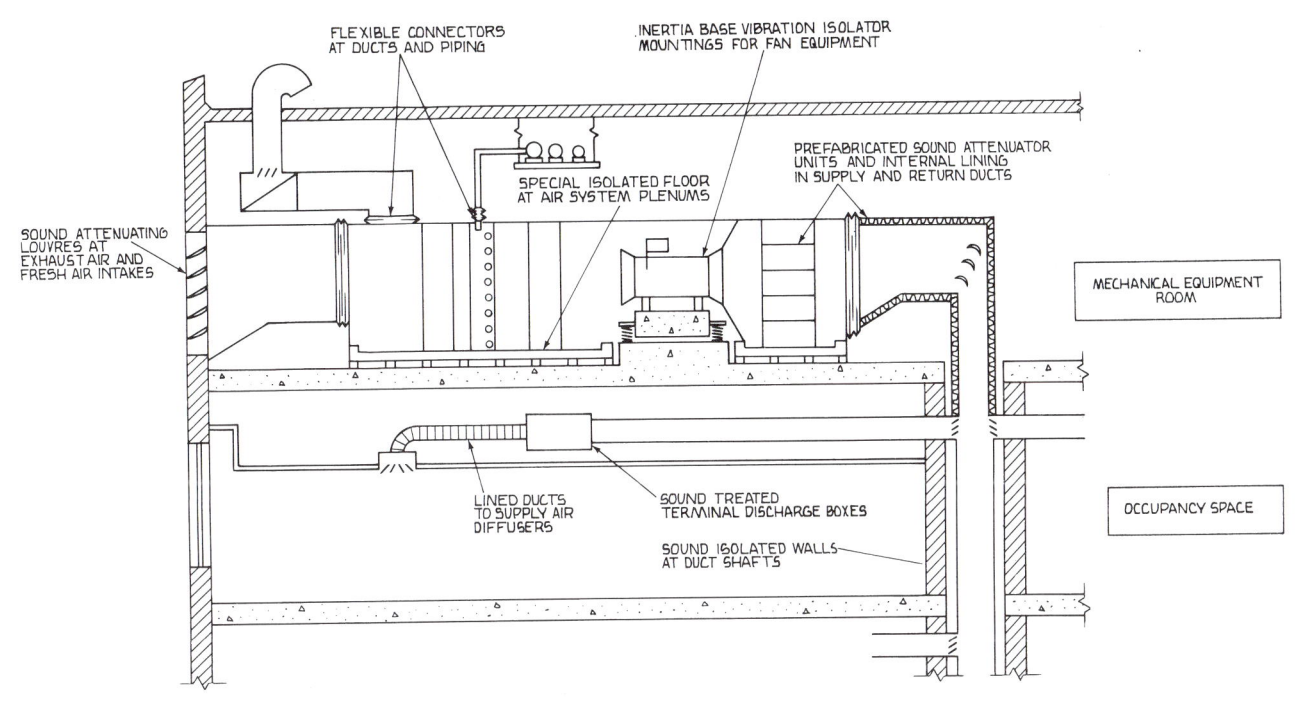

Figure 1.26 Typical mechanical equipment noise and vibration control measures. (*From W. J. Cavanaugh,* Building Construction: Materials and Types of Construction, *5th ed. by Whitney Huntington and Robert Mickadeit. Copyright © 1981 John Wiley & Sons. Reprinted by permission of John Wiley & Sons.*)

designed noise and vibration controlling system. Figure 1.26 illustrates the general approaches to noise and vibration control measures for a typical building mechanical equipment installation. Chapter 3 addresses this and other common building noise control applications as well. In addition, excellent guidance in the selection of design criteria as well as in the analysis of HVAC system noise and vibration control measures is available in technical society guide books [e.g., chapters on sound and vibration control in current editions of the *American Society of Heating, Refrigerating, and Air Conditioning Engineers (ASHRAE) Guide*].

1.3 DESIGN CRITERIA

Architectural and engineering applications of the basic concepts of sound control in buildings require establishment of reasonable design criteria guidelines and standards. A typical building may contain literally hundreds of kinds of spaces intended to house a tremendously wide variety of activities. Many of the spaces are multiuse in that they must accommodate more than one type of activity, occasionally, simultaneously. The optimum acoustical environment for one activity may be impossible for another. Even some apparently unitary-use rooms such as a music recital hall may need an adjustable acoustical environment to handle the needs of various sizes and types of performing groups, as well as variations in the acoustical environment for the music of different periods (classical or romantic versus contemporary, etc.).

Criteria have been or can be developed for every source–path–receiver situation in and around buildings. Criteria for acceptable background sound levels in various kinds of rooms as well as criteria for acceptable degrees of sound isolation from exterior sources as well as from sources within a building have become part of the building technology literature. With time many of the developing criteria form the basis for standards and codes. Keep in mind that in most building situations, people are the ultimate receptors of the sounds in question. The somewhat variable and often confusing responses observed to various situations should not be too surprising where the end result is so often a largely subjective one. In other words "one man's music can be another's noise." However, criteria that have withstood the test of time and the intelligent application thereof can minimize the risks in the engineering design decisions involved. The building designer's task is more comfortable, of course, when the criteria have found their way into hard-and-fast standard values that must be met in a particular building code. The responsibility then may be shifted to the later stage of the building project where those responsible for the field execution of a specified acoustical construction will provide the assurance that code requirements are satisfied.

This chapter briefly reviews some of the available state-of-the-art criteria for building acoustics applications, and Chapter 3 addresses certain applications in more detail. As with most aspects of the building technologies, new criteria and standards are being developed continually, and there will be new and modified criteria to meet changing societal needs. Fortunately, the fundamental laws governing the behavior of sound do not change, and a basic understanding of the basic principles and concepts for sound control in buildings should permit the building professional to intelligently deal with new and modified criteria.

Criteria for Background Sound Levels

The general background or ambient sound levels in a space are an extremely important element of the acoustical environment of that space. They form the "noise floor," so to speak, against which the occupants hear the desired sounds or undesired sounds (noise) in the space. Continuous background sound can cover up or mask the minor intrusive sounds within a space or those transmitted from an adjacent space. Just as there are a wide variety of kinds of spaces in buildings, there is an equally wide range of acceptable background sound levels. For critical spaces, such as radio broadcast or recording studios, very low background sound levels must be assured to be able to pick up the faintest desired musical speech sounds. On the other hand, excessively low levels in a typical office environment might be "deafening" in that practically all of the everyday normal activity sounds would become objectionable. A higher level of "bland," inobtrusive background sound in such spaces becomes more comfortable for the occupants. The general objective is "quiet," that is, a comfortable level of background sound appropriate for the particular space involved. The objective is not "silence," the virtual absence of sound, as it might be desired in very critical recording studios or acoustical laboratory testing chambers.

Noise criteria (NC) curves that have been extensively used for engineering design and specification of building noise control elements are shown in Figure 1.27. These criteria curves specify allowable sound pressure levels in octave bands of frequency over the full audible range. The numerical value assigned (i.e., the NC number) is the arithmetic average of the levels in the 1000-, 2000-, and 4000-Hz bands (the frequency range most important to the understanding of speech). Each criterion curve generally permits higher levels of low-frequency sound compared to mid and upper frequencies and follows the general pattern of how people respond to sound over the audible range. Low-frequency sounds are generally less annoying than high-frequency sounds within the limits expressed by the various NC curves. (Note: Since NC curves were first introduced in the late 1950s they have found wide applications in building noise

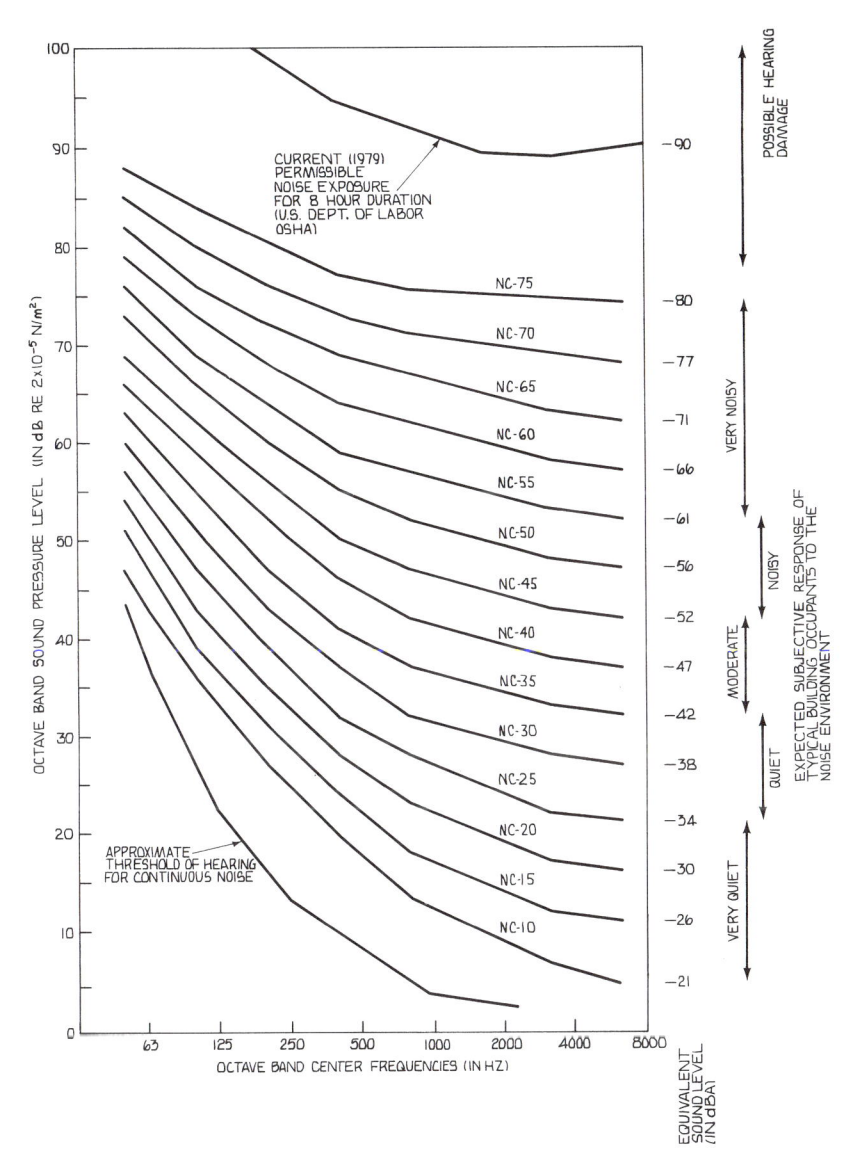

Figure 1.27 Background noise criteria (NC) curves. (From W. J. Cavanaugh, Building Construction: Materials and Types of Construction, 5th ed. by Whitney Huntington and Robert Mickadeit. Copyright © 1981 John Wiley & Sons. Reprinted by permission of John Wiley & Sons.)

control. In the 1980s and 1990s further refinements and improvements to account for how people respond to the frequency content of HVAC system sounds have evolved. Two such examples are room criteria (RC) and balanced noise criteria (NCB) methods. These newer methods are discussed further in Chapter 3 and may be found in current ANSI Standard S12.2-1995, "Criteria for Evaluation Room Noise."

Also indicated in Figure 1.27 is the equivalent single-number A scale frequency-weighted equivalent values in dBA for the individual NC curves. For example, a background sound environment that just matched the NC 35 curve would measure 42 dBA with a simple sound level meter. For many building applications, the more detailed octave-band analysis is necessary for engineering design or specification. However, the use of simple A-weighted sound levels for analysis and evaluation of final field results may be appropriate. Figure 1.27 also shows the general subjective judgment a typical building occupant might express relative to the background sound environment represented by the various NC curves and their equivalent dBA values.

Table 1.2 lists recommended criteria ranges for background sound levels in typical building spaces in terms of both NC curves and dBA values. Design criteria from the table may be selected for design purposes in developing HVAC system noise control measures and in specifying system components such as air-conditioning diffusers,

TABLE 1.2 Recommended Criteria for Steady Background Sound in Typical Building Spaces

TYPE OF SPACE OR ACTIVITY	Recommended NC Curve	Sound Level, dBA
Workspaces in which continuous speech communication and telephone use are not required	60–70	65–75
Shops, garages, contract equipment rooms	45–60	52–65
Kitchens, laundries	45–60	52–65
Light maintenance shops, computer rooms	45–55	52–61
Drafting rooms, shop classrooms	40–50	47–56
General business and secretarial offices	40–50	47–56
Laboratories, clinics, patient waiting spaces	40–50	47–56
Public lobbies, corridors, circulation spaces	40–50	47–56
Retail shops, stores, restaurants, cafeterias	35–45	42–52
Large offices, secretarial, relaxation areas	35–45	42–52
Residential living, dining rooms	30–40	38–47
General classrooms, libraries	30–40	38–47
Private, semiprivate offices	30–40	38–47
Bedrooms, hotels, apartments with air conditioning	30–40	38–47
Bedrooms, private residences, hospitals	25–35	34–42
Executive offices, conference spaces	25–35	34–42
Small general-purpose auditoriums (less than about 500 seats), conference rooms, function rooms	35 (max)	42 (max)
Small churches and synagogues	25 (max)	38 (max)
Radio, TV, recording studios (close microphone pickup)	25 (max)	38 (max)
Churches, synagogues (for serious liturgical music)	25 (max)	38 (max)
Large auditoriums for unamplified music and drama	25 (max)	38 (max)
Radio, recording studios (remote microphone pickup)	20 (max)	30 (max)
Opera performance halls	20 (max)	30 (max)
Music performance and recital halls	20 (max)	30 (max)

From William J. Cavanaugh, "Acoustics—General Principles" in *Encyclopedia of Architecture: Design, Engineering & Construction,* Joseph A. Wilkes, Ed. Copyright © 1988 John Wiley & Sons. Reprinted by permission of John Wiley & Sons.

fluorescent light ballasts, and the like. Note that the 10-dB recommended design criteria range (e.g., classrooms NC 30–40) indicates the wide range of acceptability found among typical building occupants. Generally, the middle of the recommended design range is chosen (e.g., NC 35 for classrooms). However, often a building owner may indicate a desire that a conservatively low design criterion be used because, for example, the building must accommodate hearing impaired occupants (e.g., NC 25 for classrooms). Since the lower the design criteria value, the more costly the noise control measures are likely to be to meet that design criteria, selection is an extremely important part of the building planning decision process in the context of the overall building budget. Spaces for some activities such as music performance or recording can rarely be "too quiet." For the latter, a recommended maximum limit rather than a design range is desirable as can be seen in Table 1.2. The background sound level criteria and the extent to which they are realized in a finished building has important implications for other related acoustical design aspects in particular situations (e.g., those involving acoustical privacy within and between rooms).

Criteria for High Noise Level Areas

In most industrial plants, building mechanical service areas, and other such areas, the production process or system equipment noise cannot be controlled to reasonably low levels for optimal acoustical comfort from a practical standpoint. In these spaces, it often is a matter of simply providing the best possible environment for speech communication or telephone usage. Or, if very high noise levels are likely, it may be a matter of protection of the exposed workers' hearing.

Table 1.3 indicates the nature of speech reception possible in various noise environments as well as a person's ability to carry on telephone communications in that environment.

Table 1.4 indicates the current Occupational Health and Safety Administration (OSHA) permissible noise exposures for various exposure durations. Note that these are upper-limit criteria for exposure and are not recommended design values. Even if an industrial noise environment falls below these limits (say a typical worker's exposure

TABLE 1.3 Nature of Speech Communication Possible in Various Background Sound Levels

BACKGROUND SOUND LEVEL, dBA	VOICE EFFORT REQUIRED AND DISTANCE	NATURE OF COMMUNICATION POSSIBLE	TELEPHONE USE
55	Normal voice at 10 ft	Relaxed communication	Satisfactory
65	Normal voice at 3 ft Raised voice at 6 ft Very loud voice at 12 ft	Continuous communication	Satisfactory
75	Raised voice at 3 ft Very loud voice at 12 ft Shouting at 8 ft	Intermittent communication	Marginal
85	Very loud voice at 1 ft Shouting at 2–3 ft	Minimal communication (restricted prearranged vocabulary desirable)	Impossible

From William J. Cavanaugh, "Acoustics—General Principles" in *Encyclopedia of Architecture: Design, Engineering & Construction,* Joseph A. Wilkes, Ed. Copyright © 1988 John Wiley & Sons. Reprinted by permission of John Wiley & Sons.

TABLE 1.4 Permissible Noise Exposure in Industrial Environments

DURATION PER DAY, HR	PERMISSIBLE SOUND LEVEL, SLOW METER RESPONSE, dBA
0.25 or less	115
0.5	110
1	105
1.5	102
2	100
3	97
4	95
6	92
8	90

From Paragraph 1910.95, Occupational Safety and Health Act, U.S. Department of Labor (1979).

is less than 90 dBA for an 8-hr work day), there is still a potential hearing damage hazard. There is current legislative action in progress toward lowering the exposure limit to 85 dBA from 90 dBA for 8-hr exposure with corresponding reductions for shorter exposure durations. A building professional may well be involved in critical decisions concerning an industrial building design since the building or equipment enclosures can influence the mechanical equipment room noise to which personnel are exposed.

Sound Isolation between Dwelling Units

The Federal Housing Administration (FHA) of the U.S. Department of Housing and Urban Development (HUD) recommends criteria for all federally subsidized housing to ensure that both airborne and impact sound transmission between dwelling units will be controlled. Constructions that meet the criteria and are properly installed in the field will provide good sound insulation between dwelling units and should satisfy most occupants. Since the level of background sound varies in different building site environments, 3 criteria grades are established:

Grade I. Generally "quiet" suburban and peripheral suburban areas where the nighttime exterior background noise levels might be about 35 to 40 dBA or lower. In addition, Grade I is applicable to dwelling units in high-rise buildings above about the eighth-floor level and to apartment buildings desiring maximum sound insulation regardless of location.

Grade II. Generally "average" suburban and urban residential areas where the nighttime exterior background noise levels fall in the 40- to 45-dBA range.

Grade III. Generally "noisy" suburban or urban areas where the nighttime exterior background noise levels exceed 55 dBA. This category is considered as minimum desired sound isolation between dwelling units.

Figure 1.28 and Table 1.5 indicate key FHA recommended criteria for airborne and impact sound isolation criteria in terms of minimum STC and IIC values for each of the 3 grades. Table 1.6 indicates key criteria for airborne sound isolation within a dwelling unit. As expected, FHA interior criteria are less demanding than for neighbor-

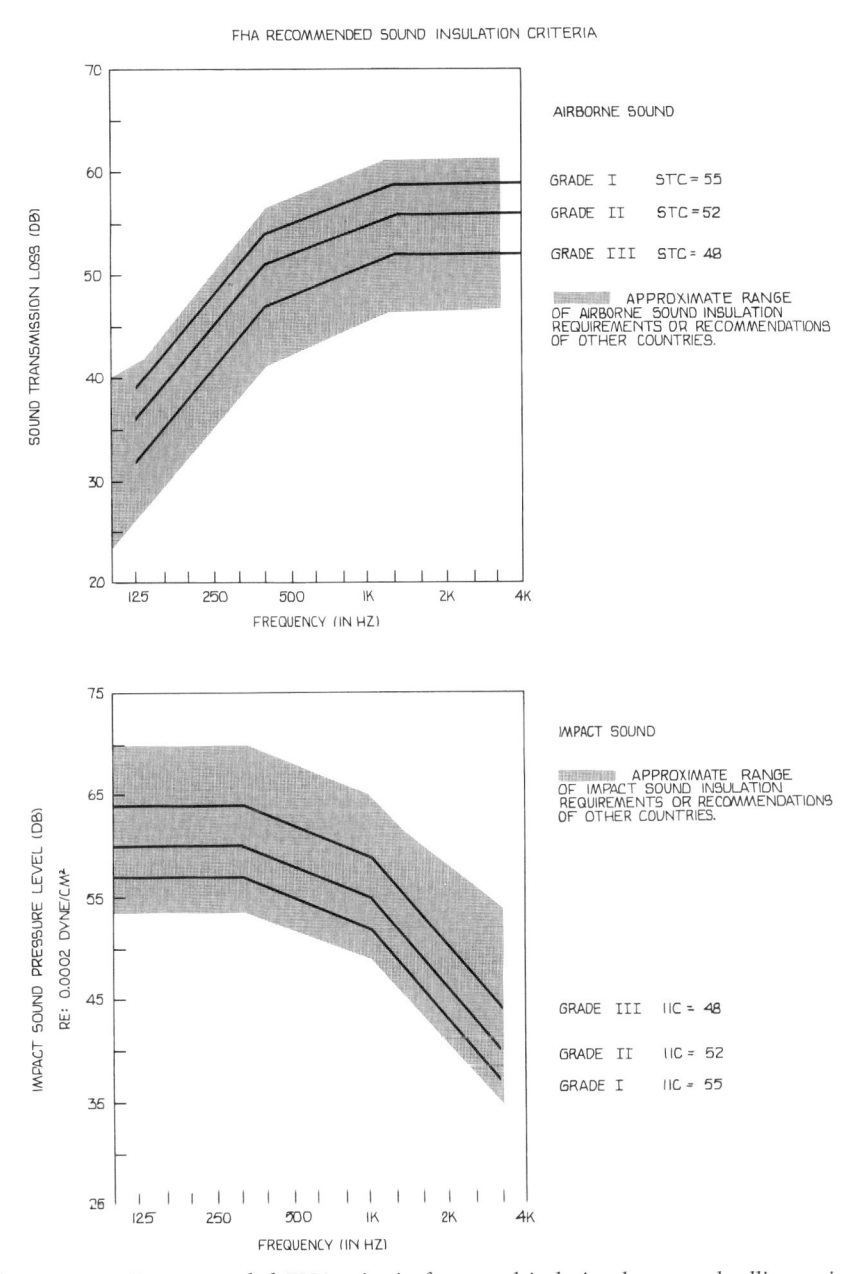

Figure 1.28 Recommended FHA criteria for sound isolation between dwelling units. *(From W. J. Cavanaugh,* Building Construction: Materials and Types of Construction, *5th ed. by Whitney Huntington and Robert Mickadeit. Copyright © 1981 John Wiley & Sons. Reprinted by permission of John Wiley & Sons.)*

ing occupancies. Also, no criteria for impact sound are suggested since it is assumed that activities within a dwelling unit may be controlled by the occupants themselves. It is usually a neighbor who makes noise not our own family!

Criteria for Mechanical Systems

The major building air-handling and electrical power systems, for example, must be located for efficient distribution and service to various parts of the building. Most

TABLE 1.5 FHA Criteria for Sound Insulation between Dwelling Units

	QUALITY AND LOCATION GRADE		
	Grade I	Grade II	Grade III
Party walls	STC 55	STC 52	STC 48
Party floor/ceilings	STC 55	STC 52	STC 48
	IIC 55	IIC 52	IIC 48
Mechanical equipment room to dwelling unit	STC 65[a]	STC 62[a]	STC 58[a]
Commercial space to dwelling unit	STC 60	STC 58	STC 56
	IIC 65	IIC 63	IIC 61

[a] Special vibration isolation of all mechanical equipment is required.

From William J. Cavanaugh, *Building Construction: Materials and Types of Construction*, 5th ed. W. Huntington and R. Mickadeit, Eds. Copyright © 1981 John Wiley & Sons. Reprinted by permission.

TABLE 1.6 FHA Criteria for Sound Insulation within Dwelling Units

	QUALITY AND LOCATION GRADE		
	Grade I	Grade II	Grade III
Bedroom to bedroom	STC 48	STC 44	STC 40
Livingroom to bedroom	STC 50	STC 46	STC 42
Bathroom to bedroom	STC 52	STC 48	STC 45
Kitchen to bedroom	STC 52	STC 48	STC 45
Bathroom to livingroom	STC 52	STC 48	STC 45

From William J. Cavanaugh, *Building Construction: Materials and Types of Construction*, 5th ed. W. Huntington and R. Mickadeit, Eds. Copyright © 1981 John Wiley & Sons. Reprinted by permission.

often, especially in high-rise structures, all mechanical equipment spaces cannot be located in remote basement areas. Above-grade locations at intermediate levels and at the upper floors of a building are almost always necessary for fans, pumps, cooling towers, emergency generators, and the like. Accordingly, these spaces must be adequately isolated from occupied spaces above and below. This involves the specification and detailing of adequate floor/ceiling and enclosing wall systems to reduce airborne sound transmission. As is apparent from the FHA criteria of Table 1.5, in apartment buildings, generally high orders of sound isolation are required where sensitive occupants are immediately adjacent to major mechanical/electrical noise sources. When the noise emission levels for the various potential sources are known or can be estimated, it is relatively easy to determine the required isolation to meet a specified background noise level in an adjacent occupied area. Then suitable constructions can be designed and specified.

A special problem with such noise sources is the possible direct excitation of the building structures by the vibrating mechanical equipment through the many ducts, pipes, or electrical conduit that must also be connected to the equipment. Decoupling of all sources of vibration from the building structure is axiomatic. This is accomplished by means of special resiliently supported isolation bases and mounts, resilient hangers, flexible couplings, and so forth, all intended to avoid direct contact of the sources of vibration with the structure that would otherwise reradiate the sound energy in other spaces throughout the building. Chapter 3 addresses in more detail the criteria and selection guidelines for typical mechanical equipment installations in buildings.

Criteria for Rooms for Listening and Performance

Auditoria, music and drama performance halls, conference rooms, sports stadia, classrooms and, for that matter, all spaces large and small where audiences listen to some desired sound source or sources, must satisfy certain fundamental acoustical requirements in order to permit satisfactory listening conditions. Chapters 4 5, and 6 address in great detail the acoustical design issues in spaces of all types and capacities for listening and performance. However, there are some design criteria common to all listening rooms large and small. The basic objectives for any are simply stated in terms of two aspects of the basic building design: (1) the control of all undesired sounds, from exterior sources, adjacent spaces within the building, the HVAC systems serving the space, and so forth, and (2) the control of all desired sounds the audience has come to hear so that they are adequately loud and properly distributed without echo or distortion throughout the space.

The first is rather obvious. However, often serious oversights, such as inadequate control of HVAC system noise, can mask significant parts of the desired speech or music program. The poor acoustics of a church or school auditorium can often be corrected by simply turning off the air-conditioning fans. Such situations should not occur. Properly chosen background sound levels (refer to Table 1.2) and then the design of sound-attenuating constructions to exclude all potential intrusive sounds will satisfy this extremely important first requirement. Table 1.2 also suggests the range of acceptable background sound levels for various types of auditoria spaces depending on their size and type of program material. Small conference or classrooms, where speaker-to-listener distances are small, are less demanding than larger performance auditoria with critical program material.

The control of the desired sounds is a much more complex matter since in most spaces the full range of sources (speech through music) must be accommodated and the audience itself is dispersed over a major part of the enclosing volume of the room. With systematic study of typical source-to-receiver paths, the complexities can be overcome in most rooms, especially with today's readily available computer-aided design and analysis procedures.

To begin with, the source must be made adequately loud at all possible listener locations. This is accomplished by taking advantage of the natural reinforcement of the major room surfaces that can direct reflected sound in mirrorlike fashion from the source to the listener (see Chapter 4). In larger rooms, or for some sources that are weak to begin with, electronic reinforcement systems must supplement the natural loudness of the desired sounds (see Chapter 5). The coordination and integration of sound amplification system equipment with the basic room acoustics design is often an important part of the overall acoustical design. In very large auditoria and sports arenas, electronic amplification systems, as described in Chapter 5, do the entire job of providing adequate loudness.

Another corollary requirement associated with the loudness requirement is that the desired sound must be distributed uniformly throughout the listening space without long delayed discrete reflections (echo), focused reflections, repetitive reflections (flutter echoes), or other undesirable colorations of the original source. These detailed design considerations are important but rarely amenable to simple criteria. Simplified ray diagram analyses for the various principal source locations can reveal the general pattern of sound distribution throughout the space and the presence of possible deleterious reflections. In general, reflected signals that arrive within about 40 msec after the direct sound has arrived (i.e., a path difference of 40 ft or less between the direct and reflected sound) contribute to the apparent loudness of the sound. Reflected sounds of sufficient level arriving after about 60 msec may be distinguished as discrete separate signals (or echoes). Intermediate delays between about 40 and 60 msec may simply result in "fuzziness" of the sound received with no real contribution to its loudness or intelligibility.

A final requirement for good listening conditions is adequate reverberation control. Excessive reverberation can destroy speech intelligibility, yet inadequate persistence of sound can make the music sound dead and lifeless. In most rooms, the selection of criteria for reverberation is largely a matter of judgment and, in some cases, compromise between the ideal for either extreme, music or speech uses. Chapter 4 includes design criteria for reverberation time, (RT), early decay time (EDT), clarity (C), early-to-total energy ratio (D), loudness (L), and other factors that are crucial in the analysis and design of new buildings for the performing arts as well as for understanding the acoustical environments in existing facilities. The past several decades have seen extensive research in the acoustics of listening spaces and in the psychoacoustical responses of typical listeners themselves, as discussed in Chapter 6, all of which contributes to better spaces for listening and performance and ultimately evolves into acoustical design criteria and design guidelines for all spaces large and small in which people are to hear and enjoy desired music or speech sounds.

1.4 SELECTED STANDARDS IN BUILDING ACOUSTICS

There are literally hundreds of standards promulgated by national and international standards bodies, industrial and trade organizations, and technical and scientific societies concerned with acoustics. The following list of selected standards includes those most likely to be of interest in building construction. The U.S. Department of Commerce National Institute of Standards and Technology (NIST) [formerly the National Bureau of Standards (NBS)] publishes a comprehensive listing with annotated summaries of practically all acoustical standards currently in force as well as those in draft and proposal stage (NBS Special Publication 386 "Standards on Noise Measurements, Rating Schemes, and Definitions: A Compilation," Order SD Catalogue No. C13.10: 386, Superintendent of Public Documents, U.S. Government Printing Office, Washington, D.C. 20402. Price: $1.10). In addition, complete copies of any standard may be ordered directly from the standards producing organization involved, from the partial listing below, along with up-to-date lists of standards in force as well as those in "draft" or "working document" stage.

- *Acoustical and Board Products Association (APBA)*
  205 West Touhy Avenue
  Park Ridge, IL 60068
  (formerly Acoustical and Insulating Materials Association)

  AMA-1-II-1967, Method of Test. Ceiling Sound Transmission by Two-Room Method
    [this method of test has been replaced by ASTM E 1414-91a (1996), see American Society for Testing and Materials below]
  AIMA Building Code Report. AIMA Model Noise Control Ordinance. 1971

- *Air Diffusion Council (ADC)*
  435 North Michigan Avenue
  Chicago, IL 60611

ADC Test Code 1062R3. Equipment test code (1972). (Replaces 1062R2)
AD-63 (1963). Measurement of Room to Room Sound Transmission through Plenum Air Distribution Systems

- *American National Standards Institute (ANSI)*
Standards Secretariat
Acoustical Society of America
120 Wall Street, 32nd Floor
New York, NY 10005–3993

ANSI S1.1-1994 American National Standard Acoustical Terminology
ANSI S1.4-1983 (R 1990) American National Standard Specification for Sound Level Meters
ANSI S1.6-1984 (R 1994) American National Standard Preferred Frequencies, Frequency Levels, and Band Numbers for Acoustical Measurements
ANSI S1.8-1989 American National Standard Reference Quantities for Acoustical Levels
ANSI S1.9-1996 American National Standard Instruments for the Measurement of Sound Intensity
ANSI S1.10-1966 (R 1986) American National Standard Method for the Calibration of Microphones
ANSI S1.11-1986 (R 1993) American National Standard Specification for Octave-Band and Fractional-Octave-Band Analog and Digital Filters
ANSI S1.13-1995 American National Standard Measurement of Sound Pressure Levels in Air
ANSI S1.40-1984 (R 1994) American National Standard Specification for Acoustical Calibrators
ANSI S1.42-1986 (R 1992) American National Standard Design Response of Weighting Networks for Acoustical Measurements

- *American Society of Heating, Refrigerating and Air Conditioning Engineers (ASHRAE)*
1791 Tullie Circle NE
Atlanta, GA 30329

ASHRAE 1986, Laboratory Method of Testing in Duct Sound Power Measurement Procedure for Fans

- *American Society for Testing and Materials (ASTM)*
100 Barr Harbor Drive
West Conshohocken, PA 19428

ANSI/ASTM Designation: C 423-90a. Test Method for Sound Absorption and Sound Absorption Coefficients by the Reverberation Room Method
ANSI/ASTM Designation: C 634-96. Terminology Relating to Environmental Acoustics
ASTM Designation: E 90-97. Test Method for Laboratory Measurement of Airborne Sound Transmission Loss of Building Partitions
ANSI/ASTM Designation: E 336-97. Test Method for Measurement of Airborne Sound Insulation in Buildings
ASTM Designation: E 413-87 (1994). Classification for Rating of Sound Insulation
ANSI/ASTM Designation: E 492-90 (1996). Test Method for Laboratory Measurement of Impact Sound Transmission Through Floor-Ceiling Assemblies Using the Tapping Machine

ASTM Designation: E 557-93. Practice for Architectural Application and Installation of Operable Partitions

ASTM Designation: E 1374-93. Guide for Open Office Acoustics and Applicable ASTM Standards

ASTM Designation: E 1414-91a (1996). Airborne Sound Attenuation Between Rooms Sharing a Common Ceiling Plenum

- *General Services Administration (GSA)*
  Public Buildings Service
  Office of Construction Management
  19th and F Street, N.W.
  Washington, D.C. 20405

PBS-C.1. Test Method for the Direct Measurement of Speech-Privacy Potential (SPP) Based on Subjective Judgments (1972)

PBS-C.2. Test Method for the Sufficient Verification of Speech-Privacy Potential (SPP) Based on Objective Measurements Including Methods for the Rating of Functional Interzone Attenuation and NC-Background (1972)

Guide for Acoustical Performance Specification of an Integrated Ceiling and Background System (1972)

- *International Standardization Organization (ISO)*
  Central Secretariat
  Geneva, Switzerland
  Attention: Technical Committee on Building Acoustics ISO/TC 43/SC2

ISO Recommendation R140. Field and Laboratory Measurements of Airborne and Impact Sound Transmission (1960)

ISO Recommendation R717. Rating of Sound Insulation for Dwellings (1968)

- *National School Supply and Equipment Association (NSSEA)*
  1500 Wilson Boulevard
  Arlington, VA 22209

NSSEA Test Procedure. Testing Procedures for Measuring Sound Transmission Loss through Movable and Folding Walls (R1972)

REFERENCES AND FURTHER READING

Apfel, R. F., *Deaf Architects and Blind Acousticians . . . A Guide to the Principles of Sound Design,* Apple Enterprises Press, New Haven, CT, 1998.

Barron, M., *Auditorium Acoustics and Architectural Design,* E & FN Spoon, an imprint of Chapman & Hall, London and New York, 1993.

Beranek, L. L., *Concert and Opera Halls . . . How They Sound,* American Institute of Physics, Woodbury, NY, 1996; originally published as *Music, Acoustics and Architecture,* Wiley, New York, 1962.

Beranek, L. L., *Noise and Vibration Control,* rev. ed., Institute of Noise Control Engineering (INCE), Poughkeepsie, NY, 1988.

Cavanaugh, W. J., "Acoustics—General Principles," in *Encyclopedia of Architecture: Design, Engineering and Construction,* J. A. Wilkes, Ed., Wiley, New York, 1988.

Cavanaugh, W. J., "Acoustical Control in Buildings" (Chapter 10) in *Building Construction: Materials and Types of Construction,* 5th ed., W. Huntington and R. Mickadeit, Eds., Wiley, New York, 1981.

Cremer, L., and Muller, H. A., *Principals and Applications of Room Acoustics,* Vols. I and II, Applied Science, London and New York, 1978 (translated by T. J. Schultz).

Crocker, M. J., Ed., *Encyclopedia of Acoustics,* Vol. 3, Part IV, *Architectural Acoustics* (Chapters 90–98), Wiley, New York, 1997., and *Handbook of Acoustics* (Derived from *Encyclopedia of Acoustics,* Vols 1-4); Wiley, New York, 1998.

Egan, M. D., *Architectural Acoustics,* McGraw-Hill, New York, 1988.

Hunt, F. V., *Origins in Acoustics* (with foreword by R. E. Apfel), American Institute of Physics, Woodbury, NY, 1992; originally published 1978, Yale University Press, New Haven, CT.

Irvine, L. K., and Richards, R. L., *Acoustics and Noise Control Handbook . . . For Architects and Builders,* Kreiger, Melbourne, FL, 1998.

Knudsen, V. O., and Harris, C. M., *Acoustical Designing in Architecture,* American Institute of Physics, Woodbury, NY, 1980; originally published 1950, Wiley, New York.

Kopec, J. W., *The Sabines of Riverbank,* Acoustical Society of America, Peninsular Press, Woodbury, NY, 1997.

Lubman, D., and Wetherill, E. A., Eds., *Acoustics of Worship Spaces,* Acoustical Society of America, Woodbury, NY, 1985.

McCue, E. R., and Talaske, R. H., *Acoustical Design of Musical Education Facilities,* American Institute of Physics, Woodbury, NY, 1990.

Northwood, T. D., Ed., *Architectural Acoustics,* Vol. 10, Benchmark Papers in Acoustics Series, Dowden, Hutchinson and Ross, Stroudsburg, PA, 1977.

Sabine, W. C., *Collected Papers on Acoustics,* Acoustical Society of America, Peninsular Press, Woodbury, NY, 1994; originally published 1921, Harvard University Press, Cambridge, MA, and 1964, Dover, New York.

Salter, C. M., *Acoustics: Architecture, Engineering, the Environment,* William Stout, San Francisco, 1998.

Talaske, R. H., and Boner, R. E., *Theatres for Drama Performance: Recent Experience in Acoustical Design,* Acoustical Society of America, Woodbury, NY, 1987.

Talaske, R. H., Wetherill, E. A. and Cavanaugh, W. J., Eds., *Halls for Music Performance: Two Decades of Experience, 1962–1982,* Acoustical Society of America, Woodbury, NY, 1982.

Case Study

FOGG ART MUSEUM LECTURE HALL, HARVARD UNIVERSITY (1895–1973): THE BEGINNINGS OF MODERN ARCHITECTURAL ACOUSTICS

In 1895, Harvard dedicated the new building of the Fogg Art Museum (no longer standing). The acoustics of the building's main feature, the Lecture Hall, were a disaster, such that its use had to be immediately abandoned. Harvard's President, Charles W. Eliot approached the Physics Department for help, and the chairman referred him to a 27-year-old new assistant professor, Wallace Clement Sabine. Sabine's colleagues looked upon his new assignment as a "no-win," fruitless task. However, Sabine undertook the challenge with apparent relish and totally dedicated himself to coaxing from science a logical "quantifiable" answer to the age-old problem of why the acoustics of some rooms are good and in others mediocre or really impossible as were speech intelligibility conditions at the Fogg Lecture Hall.

The acoustics of Fogg obsessed Sabine for 3 years and kept the hall out of service until 1898. Except for teaching physics classes assigned him, Sabine isolated himself from colleagues and usually with the help of two laboratory assistants worked between late evenings and early dawn to be free of street noise and vibrations from the new Harvard Square subway line nearby. With a promise to university authorities to return everything by class-time each morning, Sabine and his assistants dragged hundreds of upholstered seat cushions from the nearby Sanders Theatre to the Fogg after midnight and returned them each morning. Sabine, like many careful observers up to that time, knew that too much reverberation (i.e., excessive persistence or sound) makes speech communication difficult if not impossible. He also observed that carpets, draperies, seat cushions, and the like could reduce reverberation to acceptable levels.

With Sanders Theatre (excellent acoustics), the Jefferson Hall Lecture Room (generally mediocre but favorable acoustics), and the Fogg Lecture Hall (atrocious acoustics) as his physical real-world models, Sabine set out to study and measure each of these halls as well as to conduct experiments in the Jefferson Hall basement, which he fitted up as a reverberation test chamber. Using organ pipes as the source of sound Sabine would excite the rooms at the mid-frequency tone of 512 Hz and, from the instant the source was shut off, he would measure the time it took for the sound to decay to inaudibility. It should be remembered that these remarkable experiments were long before the development of precise electronic measurement equipment. Sabine's measuring instruments were his ears and those of his laboratory assistants and a stopwatch! From his studies and measurements Sabine derived the famous Sabine reverberation equation and useful absorption coefficients for many common building materials. The Sabine equation states that the reverberation time of a room is directly proportional to the cubic volume of the room and inversely proportional to the sound absorption provided at the room boundary surfaces and by the room's furnishings. This relationship was immediately very appealing to practicing architects and others since it required little knowledge of its theoretical underpinnings and could be calculated simply by knowing the architectural dimensions of a room being designed and the absorption coefficients provided by Sabine. His studies led to recommendations to correct the Fogg Lecture Hall acoustics permitting the Fogg to reopen for use in 1898.

Based on his success at the Fogg, Harvard president recommended Sabine to the prestigious architectural firm McKim Mead and White of New York who engaged him as acoustical consultant for the new Boston Symphony Hall then in design. Sabine's enormous contribution was to prevent the architects from adopting a design, already well underway, that would have been a disaster. Instead he persuaded McKim Mead and White to adopt a "shoe box" design duplicating much of the Boston Symphony Orchestra's old Music Hall's dimensions as well as those of many European halls that Sabine had studied and observed. Sabine's equation validated the new Boston Symphony Hall design and history has proven it one of the best concert halls in the world.

Ewart A. Wetherill, FASA, RBIA, traces the historical details of the Fogg Lecture Hall in his invited paper at the *Wallace Clement Sabine Centennial Symposium* at MIT, 5–7 June 1994 (see Paper 1AAAa5, Proceedings available from the Acoustical Society of America, Woodbury, NY):

> The fact that it [the Fogg Lecture Hall] ever yielded such profound results can be attributed to a remarkable confluence of circumstances. First, the controversy and criticism that began well before its completion had brought the new building to everyone's attention. Second, the lecture room had been designed expressly for Professor Charles Eliot Norton, whose prestige was largely responsible for the creation of the new museum. He was also a cousin of the president of Harvard University and wielded considerable power. When he found that his new lecture room was unusable he demanded that it be corrected in a blistering Resolution to the Board of Overseers. "The building is quite as unsatisfactory as every other lecture room in the Yard. It is the duty of the graduates to remove this impediment." Third, it is entirely possible that in the hands of anyone other than Wallace Clement Sabine the difficulties could have been mitigated without looking at the fundamental problems of acoustics. Finally, the subsequent design studies for Boston's Symphony Hall presented Sabine with another challenge appropriate to his research and brought him to the attention of the world beyond the University.

Defining the Problem

In his paper entitled Architectural Acoustics, presented at the annual convention of the American Institute of Architects on 2 November 1898, Sabine opened with a concise summation of the conditions for good hearing. We can only guess how much time he spent in research of reference material, and in carrying out his own investigations to derive 3 simple yet precise criteria for good hearing, "the greatest loudness . . . , proportional loudness of all component notes, and the greatest distinction of successive enunciations, and the disarmingly simple observation that the only two acoustical variables in a room are shape including size, and materials including furnishings."

Preliminary evaluation of the Fogg lecture room began with an assessment of the length of time that a sound persisted in the unoccupied space after the source had stopped. Sabine's description of the relation of sound decay to intelligibility of speech is profound in its simplicity: "a word spoken in an ordinary tone of voice was audible for five and a half seconds afterwards . . . the successive enunciations blended into a loud sound, through which and above which it was necessary to hear and distinguish the orderly progression of the speech." Many other spaces were studied and compared to the Fogg lecture room, and the effect of an audience on the decay of sound was established in qualitative terms. This set the stage for quantitative evaluation of the "sound absorbing power" of various materials.

Acoustical Tests

Having established a method by which sound decay could be measured, Sabine made measurements in a wide variety of other spaces, some to ensure that his conclusions were applicable to spaces other than the Fogg lecture room and some because the spaces were ideally suited for specific measurements, as noted in the following examples.

- The duration of audibility of the residual sound is nearly the same in all parts of an auditorium—Steinert Hall, Boston. Sabine noted that later, more accurate experiments verified this conclusion. To date we have little more information than is shown in the plan and section.
- The duration of audibility is nearly independent of the position of the source—Jefferson Physical Laboratory, large lecture room.
- The efficiency of an absorbent in reducing the duration of the residual sound is, under ordinary circum-

stances, nearly independent of its position—Fogg lecture room. This was determined by measurements with the same area of cretonne cloth, first under the low ceiling and then near the high domed ceiling.
- Distribution of sound intensity in a room due to interference effects is frequency dependent—Constant-temperature room, Jefferson Physical Laboratory.
- The importance of consistency in making measurements—Fogg lecture room. Anomalies in some measurements of seat cushion absorption led to the discovery that seat cushions when spaced apart had greater absorptivity than with edges touching.
- Measurement of the total absorptivity of the empty room in terms of running meters of seat cushion—Fogg lecture room. This series of measurements prompted two subsequent investigations—to find a universal unit of absorption, and to determine the absorptivity of each building material.
- Comparison of absorption of seat cushions to that of an open window—several rooms including lobby of Fogg Art Museum.
- Applicability of the reverberation equation regardless of room size—a wide variety of spaces ranging from a committee room (65 cu. meters) to Sanders Theatre (9,300 cu. meters).
- Measurement of absolute decay of residual sound, and determination of mean free path between reflections—Boston Public Library, lecture room and small room.
- Reduction of the initial intensity of sound as a function of the amount of sound absorption, and establishment of a standard for initial intensity—Boston Public Library, lecture room.
- Comparison of sound absorption due to air viscosity with absorbing power of building materials—Boston Public Library, lecture room and small room.
- Absorbing power of an audience—Fogg lecture room and Jefferson Physical Laboratory, large lecture room. This particular series of tests clearly exhibited the meticulousness of the researcher. Because of the increasing sophistication of the tests, Sabine concluded that an earlier set of data, comprising 2 months of work and over 3000 observations, had to be discarded because of failure to record the kind of clothing worn by the observer.

The Fogg lecture study was officially closed in September 1898 when corrective treatment consisting of hair felt was installed at the upper part of the rear wall and in the recesses in the domed ceiling, " . . . the room was rendered not excellent, but entirely serviceable."

In the course of this investigation and its related studies, Sabine made several observations that clearly showed his awareness of unexplored dimensions of acoustics and indicated that he looked upon these as a source of future research. A typical example was the conclusion that sound dissipation as heat in an absorber was essentially the same to the acoustics of the space as sound transmission that he would later explore in the New England Conservatory of Music. Along the way he found the time to examine the sound absorbing characteristics of a wide range of building materials and to recognize that even slight variations in construction could mean changes in the sound absorbing properties of the materials. This information was of particular importance for his consultation on the design of Boston Symphony Hall—the first example of calculating reverberation in advance of construction.

Key elements of Sabine's success were his ability to arrive at firm conclusions and to translate his findings in clear, precise terms that could be easily understood by people with a limited technical background. He also had an ability to express profound ideas without fanfare. Probably the outstanding example is his comment that the standard of initial intensity for measuring the duration of audible sound would be one million times the minimum audible intensity "as this is the nearest round number to the average intensity prevailing during these experiments." On deriving the equation for relationship between absorption, volume and duration of audible sound he comes close to humor when he notes "It also furnishes a more pleasing prospect, for the laborious handling of cushions will be unnecessary." However, the one truly amusing episode occurred in measurements in the dining room of Memorial Hall, which were carried out late at night. "There was no opportunity to carry the experiment further than to observe the fact that the duration was surprisingly short, for the frightened appearance of the women from the sleeping-rooms at the top of the hall put an end to the experiment."

Fogg Lecture Room after 1898

Despite Sabine's apparent satisfaction with the corrections done in 1898, the lecture room seems to have been very unpopular with the Fine Arts department and its subsequent users. In 1912 it was reduced in size from around 400 seats to 200 seats by the addition of a semicircular wall at the column line and a flat floor was added. In the words of the museum director, "We hope for . . . a roof that does not leak; [and] a medium-sized lecture hall instead of a large one in which you cannot hear." In 1927 the new Fogg Art Museum was opened and the old building, renamed Hunt Hall after its architect, was occupied by the Architecture, Planning and Landscape departments. Of the hundreds of architects exposed to this building, it would be interesting to know how many learned its lessons.

There are few records of the subsequent changes in the lecture room. Around the time of the Second World War much of the semi-circular wall was

Figure 1.29 Lecture Hall and detail of the acoustical treatment Hunt Hall (original Fogg Art Museum, demolished 1973). (Courtesy of the Fogg Art Museum Harvard University Art Museums.)

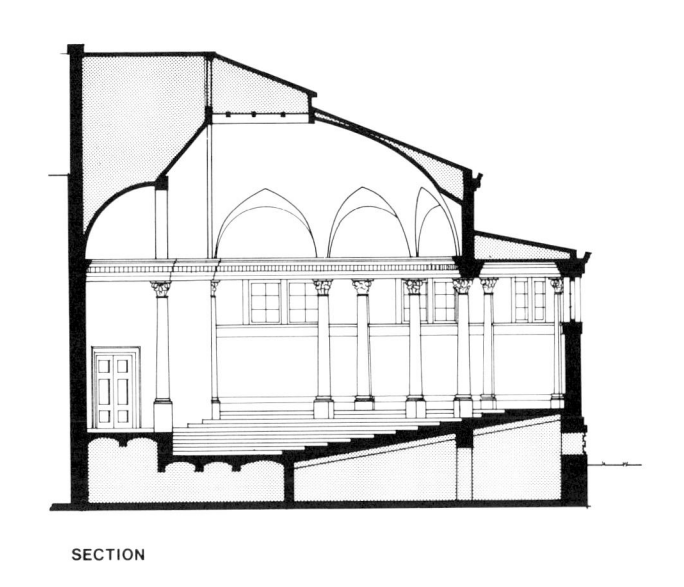

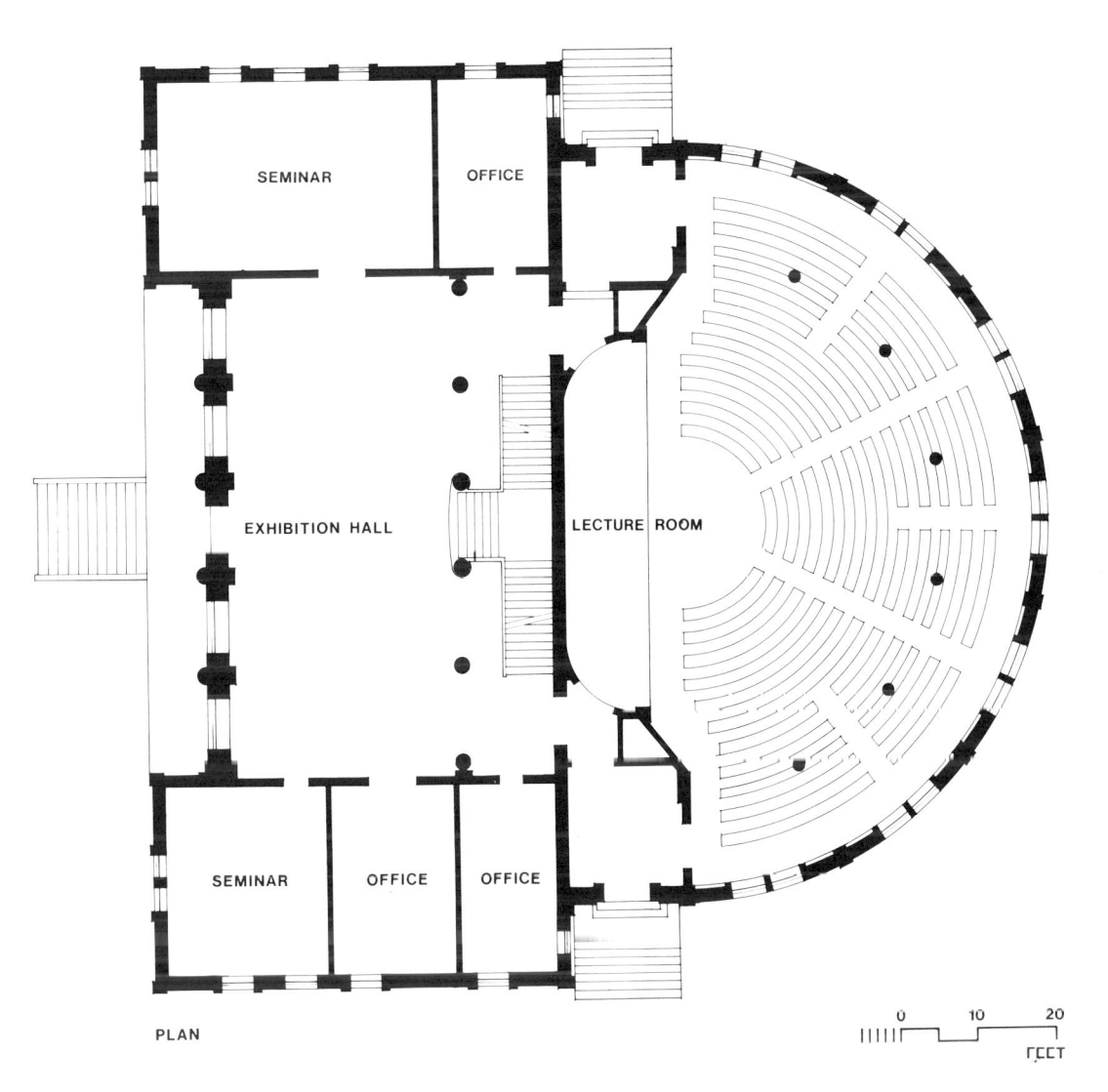

FOGG ART MUSEUM   c. **1895**

Figure 1.30 Plan and section of Fogg circa 1895 (courtesy E. A. Wetherill, FASA).

SECTION

PLAN

HUNT HALL
FORMERLY FOGG ART MUSEUM

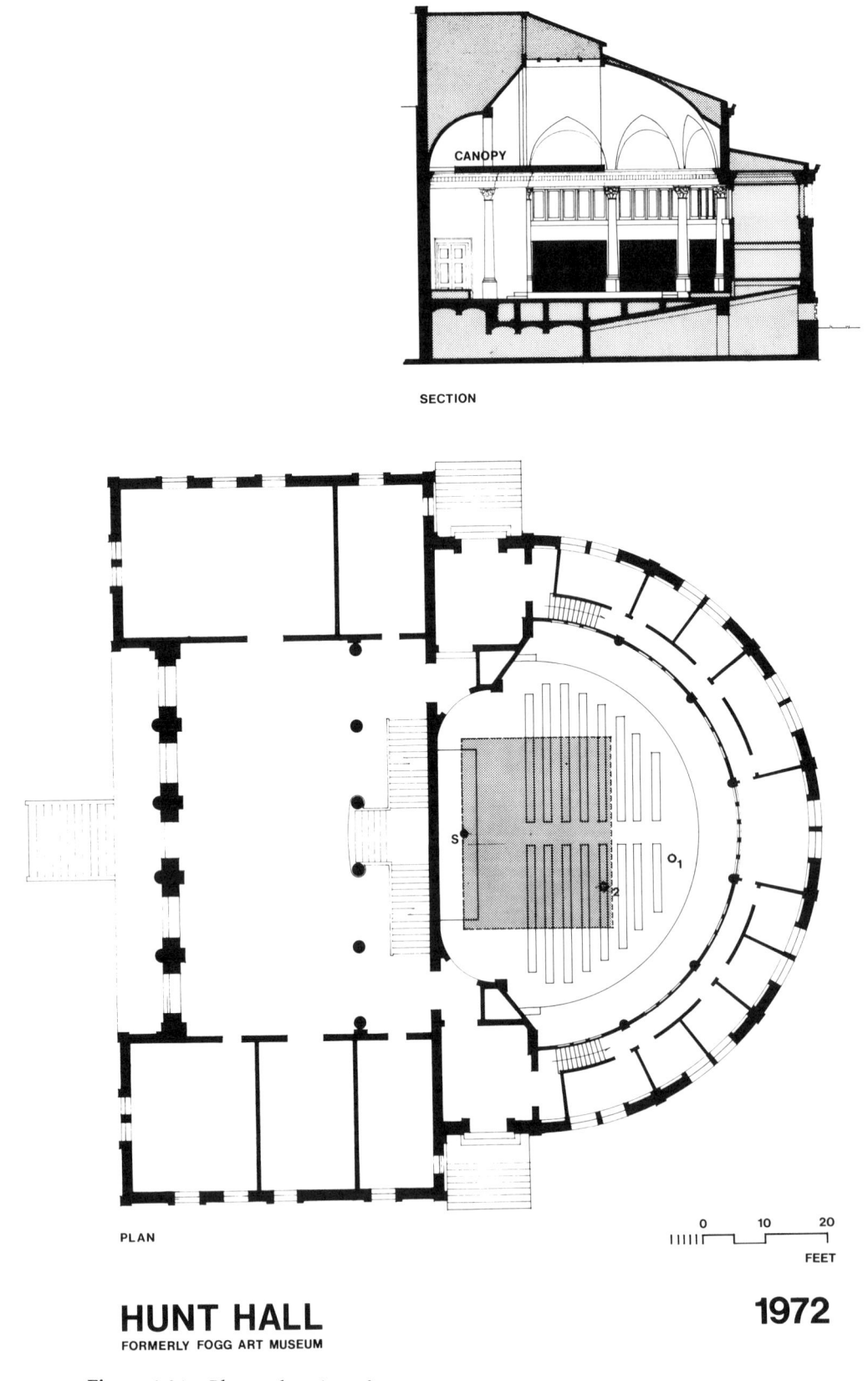

1972

Figure 1.31 Plan and section of Fogg circa 1972 (courtesy E. A. Wetherill, FASA).

Figure 1.32 1973 Acoustical impulse measurements of Hunt Hall, formerly Fogg Lecture Hall (courtesy E. A. Wetherill, FASA).

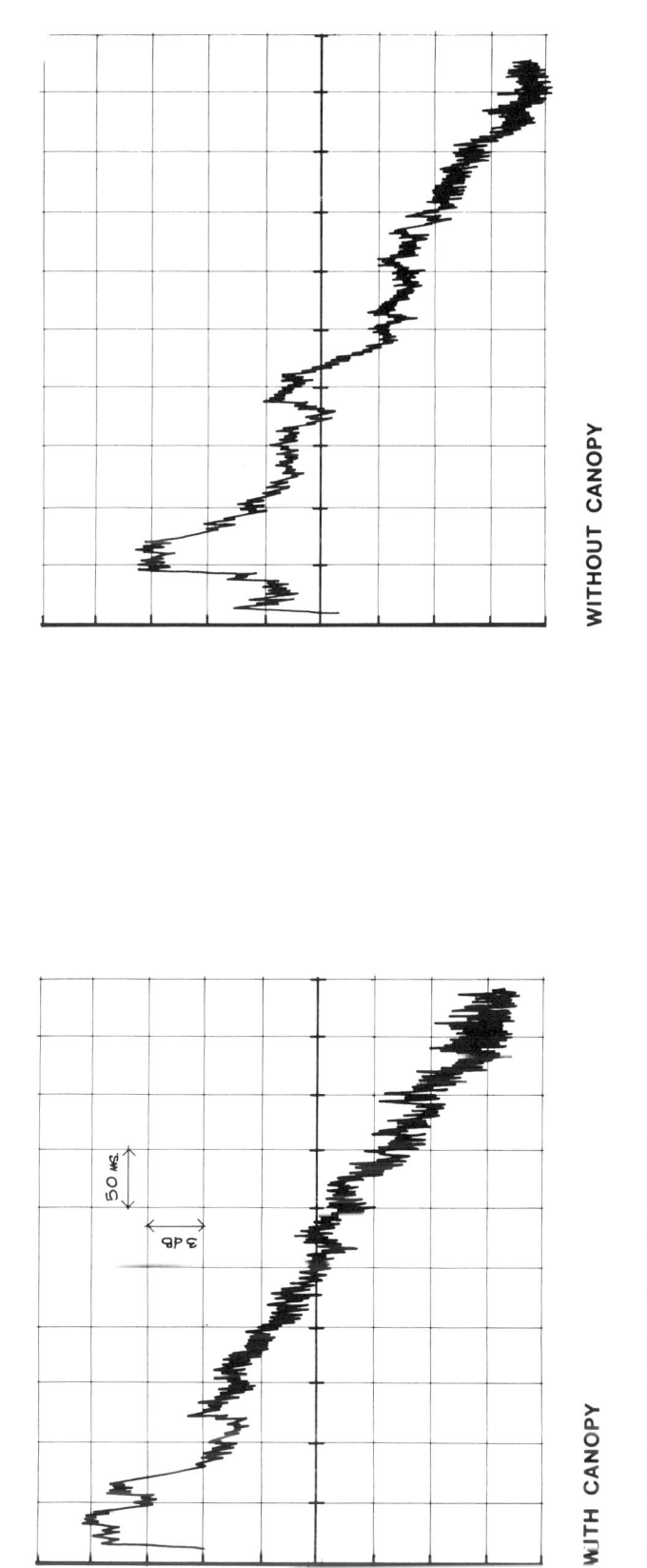

covered with hair felt and perforated asbestos board, and in the 1960s the floor was carpeted. Student comments refer to the whispering gallery effect and to the difficulty of hearing at some locations. In 1972, a year before the building was demolished, the semicircular wall was covered to a height of 8 feet with highly absorptive glass fibre board and a large horizontal canopy was suspended over the platform and front seating. This created a significant change in intelligibility, so for one last academic year students heard well in this space.

In the two weeks prior to its demolition in 1973, reverberation and impulse response measurements were made in the reduced lecture room, both with and without the canopy and glass fibre board. Measurements with the canopy show an increase in Articulation Index and indicate that both the semicircular wall and the domed ceiling contributed to the acoustical difficulties.

Figures 1.29 through 1.32 were prepared for an Acoustical Society of America technical paper in the mid-1970s by Wetherill and, as his enlightening discussion above clearly shows, demonstrate the profound significance of the Fogg Museum Lecture Hall and Sabine's further work in the field of architectural acoustics. Indeed, Sabine has earned the title of Father of Modern Architectural Acoustics and the Fogg Lecture Hall was its incubator. The century since the beginning of Sabine's work was duly celebrated at the *Wallace Clement Sabine Centennial* in conjunction with the 127th meeting of the Acoustical Society of America in June 1994 at the Massachusetts Institute of Technology in Cambridge not far from the original Fogg Museum site and the home and workspace of Sabine at Harvard University. Dr. Leo L. Beranek, internationally renowned acoustician wrote, in connection with the Sabine Centennial, "though he [Sabine] had become the principal acoustical consultant in the United States, Sabine never received an earned doctorate. When asked why not, he responded: "When the proper time came for me to do so, I should have been my own examiner." The field of architectural acoustics is fortunate that Sabine took seriously the task of "fixing up" the acoustics of the Fogg Lecture Hall. The authors and the editor of this book, *Architectural Acoustics: Principles and Practice* have sought to live up to the standards set by Sabine over 100 years ago and to record the state-of-the-art and the enormous progress that has occurred in architectural acoustics throughout the 20th century. Thanks to Sabine and countless acousticians since, the field is well equipped to continue the research, development, and practical applications in architectural acoustics for the next century and beyond.

Acoustical Materials and Methods

Rein Pirn

2.1 INTRODUCTION

All building materials are in a way acoustical because all affect the manner in which sound is reflected, absorbed, or transmitted. This includes both the most common materials and a great many products—some common, others not so common—that are or are perceived to be acoustical. It also includes a number of special devices, whose purpose is strictly acoustical.

Since the term *acoustical* can have many different meanings, one must differentiate among the processes to which sound in buildings is subjected. Also, to avoid possible confusion, it is necessary to define the words "insulation" and "isolation" as used in an acoustical context.

As explained elsewhere in this book and restated later in this chapter, sound, upon striking the boundary of a room, is partially reflected, partially absorbed, and partially transmitted to the next room. All 3 events are acoustical, yet their effect on the built environment is very different. The first two processes—reflection and absorption—are mutually exclusive, and they both fall in the realm of room acoustics. By selecting the appropriate materials, sound can be sustained (reflected) or made to disappear (absorbed). Of course, this happens in degrees, as no material is totally reflective nor totally absorptive. The third process—transmission (or its counterpart, attenuation)—falls under the broad heading of sound isolation. The issue is not how much sound is reflected or absorbed, but how much of it is allowed to pass through the material or materials of which the walls or other room boundaries are made.

Acoustical insulation, as used in this book, means a material that absorbs sound. Isolation, on the other hand, describes a process—the process of reducing or attenuating noise or vibration. (One should be cautioned that such usage of these words is not universal. In Britain, for instance, it is common to use insulation in both cases.)

Having defined the terms, it behooves us to review the two acoustical processes, sound attenuation and sound absorption, that are basic to an understanding of how materials affect the acoustical environment.

2.2 SOUND ATTENUATION

Airborne Attenuation

The basic concept, as it applies to buildings, is best described as follows. Consider a pair of adjacent rooms, as illustrated in Figure 2.1. One, the source room, contains noise, whose transmission into the other, the receiving room, is to be reduced or prevented. The degree to which this is possible depends mainly on the attenuation offered by the wall (or floor/ceiling) between the two rooms. It also depends on so-called flanking paths that may allow sound to bypass the principal barrier; these must offer comparable attenuation.

In this concept, the reverberant sound level in the source room is compared with the resultant reverberant sound level in the receiving room. Each of these levels, and therefore the difference between them, is expressed in decibels. Thus, attenuation also is expressed in decibels.

Attenuation is described more precisely by two well-defined terms. First, the sound level difference between two rooms that one measures or experiences is called noise reduction (NR). However, the sound level difference due to the principal barrier (and not to other factors, as explained below) is called transmission loss (TL). Transmission loss is a specific property of the barrier under consideration; NR is not. The relationship between the two is expressed by

$$NR = TL - 10 \log S + 10 \log A_2$$

where S is the surface area of the barrier and A_2 is the amount of acoustical absorption present in the receiving room. (S and A_2 must be expressed in compatible units; square meters and metric sabins, or square feet and sabins.) It is apparent that if $S = A_2$, then NR = TL. In actuality, S and A_2 are rarely equal, but they are usually close. In most cases, the noise reduction provided by a partition lies within a few decibels of its transmission loss.

It should be clearly understood that for a partition or other barrier to offer the expected attenuation it must be complete in all respects. It must extend across the full height and width of the opening between the two rooms, contain no acoustically inferior elements (such as doors), and absolutely no holes or cracks. The latter, if present, may seriously degrade performance.

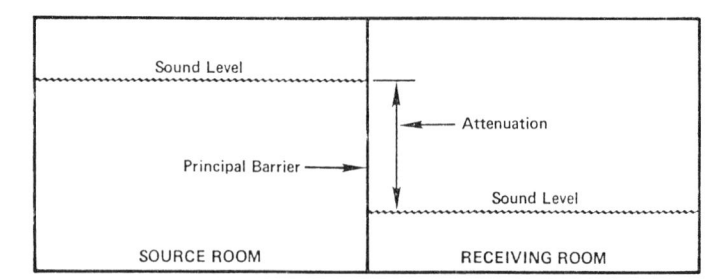

Figure 2.1 Schematic illustration of airborne sound attenuation between two rooms. (From Rein Pirn, "Acoustical Insulation and Materials," in Encyclopedia of Architecture: Design, Engineering, and Construction, Vol. 1, Joseph A. Wilkes and Robert T. Packard, Eds. Copyright © 1988 John Wiley & Sons, Inc. Reprinted by permission of John Wiley & Sons.)

Transmission loss, and therefore noise reduction and attenuation in general, is frequency dependent. The TL of a perfectly limp material, such as lead sheet, increases 6 dB per octave. (Theoretically, TL also increases 6 dB with every doubling of the material's mass.) However, most materials are not limp, and their TL curves are not straight. This is illustrated in Figure 2.2, which compares 2.5-mm (0.1-in.) lead with 100-mm (4-in.) concrete block and a comparably thick stud wall.

Analyzing and designing for sound isolation by frequency requires familiarity with the various materials' acoustical properties. This is best left to qualified acousticians. Architects, engineers, builders, and others who are involved in the building process should, however, be familiar with the single-number method of quantifying TL. In the United States and Canada, it is the Sound Transmission Class (STC) of a material or combination of materials.[1-3] The sound transmission class value is determined as follows. A standard curve is fitted to the TL curve so that the sum of the deficiencies in the sixteen $\frac{1}{3}$-octave bands from 125 to 4000 Hz is no greater than 32 dB and no single deficiency exceeds 8 dB; upon such fitting, the STC is given by the standard curve's position at 500 Hz. The resultant number does not represent TL at any frequency. It merely permits comparison of diverse materials, but with the following qualifications:

> As noted in Figure 2.2, all 3 curves correspond to an STC 40 rating. But they are distinctly dissimilar and, therefore, unequal at any given frequency.

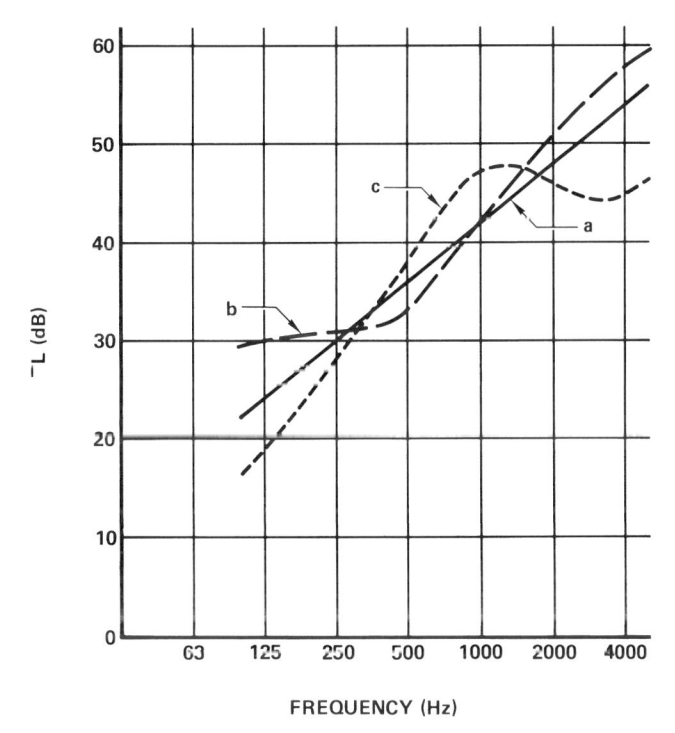

Figure 2.2 Some typical transmission loss curves, all rated STC 40: (a) lead sheet, (b) concrete block, and (c) gypsum board on studs. (From Rein Pirn, "Acoustical Insulation and Materials," in Encyclopedia of Architecture: Design, Engineering, and Construction, *Vol. 1,* Joseph A. Wilkes and Robert T. Packard, Eds. Copyright © 1988 John Wiley & Sons, Inc. Reprinted by permission of John Wiley & Sons.)

Heavy materials such as concrete and masonry perform well in the lower frequencies, are relatively inefficient in the 500-Hz range, and regain their excellence in the higher frequencies.

Lightweight materials such as gypsum board or plaster on studs (also glass) perform poorly in the low frequencies, quite well in the middle frequencies, and again less well at the highest frequencies of interest.

It follows that heavy barriers are required in order to contain low-pitched sounds, such as those of mechanical equipment or of musical instruments. Considerably lighter barriers, such as stud partitions, may be employed where nothing more than speech isolation is involved. (Speech intelligibility, and hence distraction due to unwanted speech, is almost entirely governed by sounds in the 500- to 4000-Hz range.)

Sound transmission class ratings of various materials, as well as some construction details, are further discussed later in this chapter.

Before leaving the subject of attenuation, some other forms of this process and some other terms related to it should be mentioned. Next to airborne sound attenuation, as discussed above and as illustrated in Figure 2.1, architects and other building professionals should also have a grasp of the following concepts: acoustical shielding, impact attenuation, vibration isolation, and duct attenuation.

Acoustical Shielding

Acoustical shielding is provided by incomplete barriers, for example, the partial-height partitions often used in open office spaces. Performance is almost always limited by the barrier's size and not by its TL. To be at all effective, the barrier must extend well beyond the line of sight between the source and the receiver, and sound must not be allowed to reflect over or around the barrier (hence the need for very absorptive surfaces; see Section 2.3). In practice, barrier attenuation seldom exceeds 20 dB. More usually, it is much less than that. It is therefore unnecessary to build partial-height barriers of high-TL materials.

Impact Attenuation

Impacts are generated by direct, physical contact with a surface whose other side is exposed to an adjoining room. A typical case is that of footfalls and their transmission to the floor below. Attenuation is achieved by avoiding hard contact (i.e., by covering the floor with a resilient material) and/or by resiliently separating the floor from the ceiling below. Impact noise attenuation of floor/ceilings is rated much like airborne attenuation, except that a so-called tapping machine is placed and operated on the floor, its noise is measured in the room below and then normalized.[4] Like airborne attenuation, impact attenuation is frequency dependent. Like STC, the single-number rating for impact sounds, the impact insulation class (IIC), is determined through curve fitting.

Vibration Isolation

Vibration isolation pertains specifically to mechanical equipment. No piece of rotating, reciprocating, or vibrating equipment is perfectly balanced. The imbalance causes it as well as its supporting structure (floor, beam, etc.) to vibrate. If unchecked, the vibration may be felt or heard as noise, often a considerable distance from the equipment. Attenuation is achieved through use of vibration isolators (spring and elastomeric mounts or hangers, flexible sleeves and connections) that allow the equipment to float free of the structure. There are no standards as such. To be effective, the natural frequency of vibration-isolated equipment must be substantially lower than the driving frequency. In practical terms, for mounts and hangers, this means ample static deflection.

Duct Attenuation

Noise, as from ventilation fans or due to turbulence in the airstream, propagates along ducts and enters rooms through the air supply and return grilles. (Note that sound travels almost equally with and against airflow.) Attenuation is achieved by internally lining the ducts with an absorptive material and/or by inserting commercially available silencers. Large heating, ventilation, and air-conditioning (HVAC) systems serving acoustically sensitive spaces may require 30 m (100 ft) or more of lined duct between the fan and the room.

Additional information on sound attenuation, especially as it relates to noise and vibration associated with mechanical systems and equipment, can be found in Chapter 3.

2.3 SOUND ABSORPTION

Basic Principles

As sound strikes a surface, it is transmitted, reflected, or absorbed. Usually, all 3 events occur. The percentage of energy transmitted tends to be very small, generally less than 1%, which would correspond to a transmission loss of 20 dB (the transmitted sound level, in decibels, equals 10 log of the transmitted energy fraction). All remaining energy is partially reflected and partially absorbed.

The processes that result in acoustical absorption are friction and resonance, which are discussed below. Contrary to popular belief, a coarse or broken-up surface made of an impervious material does not absorb sound, though it may diffuse (i.e., scatter) it in many different directions.

Absorption through friction occurs when sound has access to the fine pores and interstices that one finds in porous and fibrous materials. The air molecules are restrained from continuing their cycle of compression and rarefaction. The energy thus lost is converted into heat. Most acoustical materials whose purpose is to absorb sound are based on this principle.

Absorption through resonance occurs when a stiff but not totally rigid system (e.g., a plate or a confined volume of air, as in a Helmholtz resonator) is set in motion by sound. The system will absorb and dissipate the energy if its natural frequency corresponds to that of the incoming sound. Few products use this principle by design, but there are many building materials (glass, wood paneling, gypsum board, etc.) that, if not well restrained, will resonate and thus absorb sound.

Acoustical absorptivity is quantified by the sound absorption coefficient α, which represents the fraction of the incident energy that is absorbed. Highly reflective materials have coefficients near zero, highly absorptive materials near one. Virtually no material is perfectly reflective or perfectly absorptive. In general, materials with coefficients below 0.20 are rather reflective, whereas those with coefficients of 0.80 or higher are very absorptive.

60 • *Acoustical Materials and Methods*

Absorption, like attenuation, is frequency dependent. Porous-fibrous absorbers are most efficient in the higher frequencies but also perform well in the middle and low frequencies if sufficiently thick or if backed by an airspace. Resonant absorbers, on the other hand, work best in the low frequencies (specifically, at their natural frequency) and are quite reflective at other frequencies. This is illustrated in Figure 2.3.

Absorptivity, like TL, can be described by a single number. The descriptor used in the United States and Canada is called the noise reduction coefficient (NRC). It is determined by averaging the sound absorption coefficients at 250, 500, 1000, and 2000 Hz and rounding off the result to the nearest multiple of 0.05. The NRCs of the four materials in Figure 2.3 would be 0.75 for both the fiberglass and the suspended tile, 0.30 for the carpet, and 0.10 for the wood paneling.

Like any simplification of a complete set of performance data, the NRC is flawed. It suggests equality of the tile and the fiberglass, yet they are not equal. If low-frequency absorption were required, suspended tile would be far better than an inch of fiberglass (in direct contact with a solid surface). If best possible performance in the 500- to 4000-Hz speech intelligibility range were required, fiberglass would be better. It should be noted that despite their clearly unequal NRCs (0.75 and 0.30), tile and carpet can be almost equally absorptive near 4000 Hz.

Acoustical absorptivity of materials is determined by one of two methods: in a reverberation chamber[5] or in an impedance tube.[6] The latter yields normal-incidence absorption coefficients, from which random-incidence coefficients can be estimated.

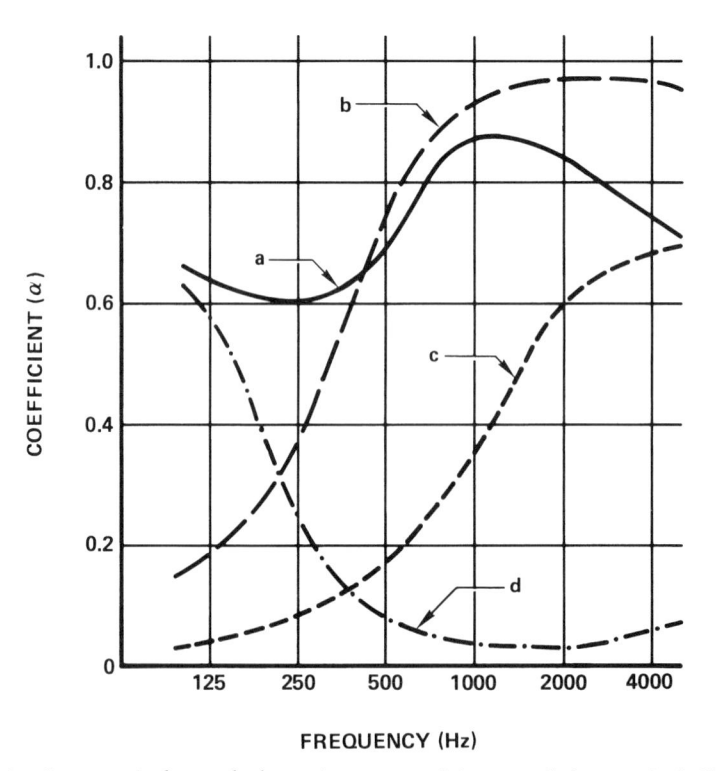

Figure 2.3 *Some typical sound absorption curves: (a) suspended acoustical tile, (b) 25-mm glass fiber board, (c) carpet, and (d) thin wood paneling. (From Rein Pirn, "Acoustical Insulation and Materials," in* Encyclopedia of Architecture: Design, Engineering, and Construction, *Vol. 1, Joseph A. Wilkes and Robert T. Packard, Eds. Copyright © 1988 John Wiley & Sons, Inc. Reprinted by permission of John Wiley & Sons.)*

The reverberation chamber method more closely simulates the conditions encountered in buildings, where sound impinges on the material from many different directions. Most of the absorption data published by manufacturers of acoustical materials are based on this method. A brief description follows.

A sample of the material, usually 244 × 274 cm (8 × 9 ft), is placed in a very reverberant, hard-surfaced room of some 250 m³ (9000 ft³). Reverberation time is then measured and compared with the reverberation time of the same room, but without the sample. Since reverberation time is inversely proportional to absorption, and since the room's absorption without the sample is known, that due to the sample can be calculated and then divided by the sample area. The resultant number per unit area is the absorption coefficient α. The absorption provided by a square foot of totally absorptive material ($\alpha = 1$) is called a sabin; that of a square meter of such material is called a metric sabin. The significance of sabins, as opposed to α, is discussed below.

A word of caution: Sometimes the reverberation chamber method yields coefficients in excess of 1 owing to diffraction effects along a sample's edges. Such numbers are unrealistic because $\alpha = 1$ means total absorption. A realistic maximum for any material is 0.99.

Absorption data are often accompanied by a mounting type that describes one of several conditions, as standardized by the American Society for Testing and Materials (ASTM).[7] Comparable designations by the Acoustical and Board Products Manufacturers Association (ABPMA), which predate the ASTM standard, may still be found in some older literature. The most common conditions are:

ASTM Type A or ABPMA No. 4   sample placed directly against a reflective test surface
ASTM Type B or ABPMA No. 1   sample cemented to gypsum board, for example, to simulate glued-on acoustical tile
ASTM Type E–400 or ABPMA No. 7   sample supported 400 mm (approximately 16 in.) off the test surface, to simulate a suspended ceiling

Every acoustical event involves a source, a path, and a receiver. As discussed earlier, attenuation usually involves two or more spaces, one of which is the source room. One speaks of attenuation *between* rooms. Absorption, on the other hand, affects the acoustics of one room, which contains both the source and the receiver. One speaks of absorption *in* a room, where it can serve any of the following purposes: reverberation control, noise control, control of discrete reflections, and improved attenuation.

Reverberation Control

The reverberation time of a room can be given by the Sabine equation, which, in a simplified form, may be written as

$$T = 0.16V/A \quad \text{(metric units)}$$

or

$$T = 0.05V/A \quad \text{(English units)}$$

where T is the reverberation time in seconds, V is the room volume in cubic meters or cubic feet, and A is the combined absorption of all finishes and furnishings (plus that of air) in metric sabins or sabins. It is obvious that rooms containing little absorption can be very reverberant. Covering some of the major surfaces with even moderately efficient absorbers will greatly reduce reverberation. Further treating an already absorptive room will not result in a further, drastic reduction in reverberation.

Efficiency (as given by α or NRC) is not as important as the total number of sabins. Larger areas of less efficient materials are just as effective as smaller quantities of more absorptive materials. Reverberation time is halved with every doubling of absorption.

Noise Control

Except in the immediate vicinity of a source, the sound level (in a room) due to that source tends to be the same regardless of distance. This so-called reverberant sound level may be expressed by

$$L = P - 10 \log A$$

where L is the sound level in decibels, P is the sound power level (here treated as a constant) of the source, and A again represents the absorption in sabins; L is thus controlled by A. A room containing few or inefficient absorbers will be noisy. Adding or upgrading the absorbers will make it less noisy, but not by very much if the room is already quite absorptive.

Again, total absorption is more important than the efficiency of individual materials. Every doubling of absorption reduces the reverberant noise level by 3 dB, which, incidentally, is not nearly as drastic a change as a halving of the reverberation time.

Control of Discrete Reflections

A typical case is that of echoes off a distant surface, such as the rear wall of an auditorium. Here, it is not the repeated interaction between sound and the material (reverberant sound will strike it many times and lose some of its energy at every contact) but a single contact that matters. In that one instant, the sound level must be appreciably reduced. The reflected level L_r, relative to the incident level L_i, both in decibels, is given by

$$L_r = L_i + 10 \log(1 - \alpha)$$

It follows that a material whose α is 0.50 will reduce the level by just 3 dB—hardly enough to eliminate an echo. Absorption coefficients of 0.80, 0.90, or higher usually are required to eliminate discrete reflections. Efficiency, and not the total number of sabins, is of the essence.

Improved Attenuation

Absorption helps attenuation in two situations: if located in a buffer space between the source and receiving rooms and if applied to the boundaries (especially the ceilings) of large but low rooms that contain both the source and the receiver.

In the first of these cases, sound that is in transit from the source to the receiver is reduced by very much the same mechanism as in noise control. Typical examples include absorptively treated sound locks, as between a lobby and a hall, and, on a smaller scale, insulation in the stud space of partitions. In the second case, one is more concerned with discrete reflections (e.g., of speech) that may distract other occupants of the room. Examples include open offices where many people share a physically contiguous space, yet expect privacy from one another.

2.4 COMMON BUILDING MATERIALS

As noted in the introduction to this chapter, every material has certain acoustical properties. It is not the purpose of this work to review these properties in detail, or to cover every conceivable material, but rather, to provide an overview. The most commonly encountered materials, regardless of their acoustical value, are briefly described, and their properties as sound attenuators and/or absorbers are discussed.

Brick. Brick is a modular building block, made of clay. It is used to build load-bearing as well as non–load-bearing walls, as a facing (brick veneer), and as a paving material.

Owing to its considerable mass [approximately 2.1 kg/dm^3 (130 lb/ft^3)], brick attenuates airborne sound very well. Exceptionally high orders of attenuation can be achieved with two side-by-side but unconnected brick walls. Joints must be fully mortared or otherwise sealed. Absorption is negligible since there is little or no porosity and the material is rigid. Consequently, brick is a good all-frequency sound reflector.

Concrete. Concrete is a mixture of Portland cement, stone and sand aggregates, and water, cured into a hard mass of superior compressive strength. It is often reinforced with steel and used for structural slabs and walls.

Normal-weight concrete [approximately 2.3 kg/dm^3 (144 lb/ft^3)] is among the best attenuators of airborne sound. Lightweight concrete is less effective, unless of equal mass per unit area. Like any hard material, concrete readily accepts and transmits impact sounds. Concrete provides virtually no absorption. There are, however, aerated concretes that are intentionally porous. These can be fairly absorptive.

Concrete Masonry Units. Concrete masonry units are modular building blocks made of concrete. They are usually manufactured with hollow cores. Normal-weight units can support considerable loads. Lightweight units are generally used to build non–load-bearing partitions.

The attenuation provided by concrete masonry units depends mainly on their weight. Lightweight units may be adequate in noncritical cases. Normal-weight units, especially if solid or if their cores are filled with sand or grout, attenuate sound very well. Two unconnected concrete masonry walls (like those of brick) can provide exceptionally high orders of sound attenuation. Since its surface is somewhat porous, concrete masonry (especially cinder block) is slightly absorptive, unless painted or otherwise sealed. If well sealed, it becomes a good all-frequency reflector.

Glass. Glass is a usually light-transparent sheet made of a mixture of silicates. It is used principally to glaze windows and other openings that need to be closed, but without excluding light.

Despite its mass [approximately 2.5 kg/dm^3 (156 lb/ft^3)], glass is a marginal sound attenuator because it is thin and the mass per unit area is quite small. Superior performance is provided by well-separated double glazing and by certain types of laminated glass (see Section 2.5). Almost totally reflective in the higher frequencies, glass resonates and, through this mechanism, can absorb appreciable amounts of low-frequency sound.

Gypsum Board. Gypsum board is a fire-resistive sheet material made of calcined gypsum and certain additives, sandwiched between sheets of special paper. Typically attached to studs, joists, or some form of furring, it is one of the most common wall and ceiling finishes in use today.

Although not very heavy [approximately 0.8 kg/dm^3 (50 lb/ft^3)] or thick, gypsum board partitions can provide a fair amount of sound attenuation. However, much depends on the way the construction is detailed. Best results are achieved with multiple layers of gypsum board, with resilient separation between the two faces of the partition, and with absorptive material in the stud space. Joints must be perfectly sealed. Gypsum board, unless attached directly (without an airspace) to a solid substrate, resonates and thus absorbs low-frequency sound. At higher frequencies, it is highly reflective.

Masonry. Masonry is any of a large variety of stonelike materials. Acoustical properties vary, but in general they are comparable to those of brick, concrete, and concrete masonry units.

Metals. Metals are any of a family of alloys, but especially steel, which is commonly used to provide structural support. See the discussion of steel decking in this section and that on lead sheet in Section 2.5.

Plaster. Plaster is a pasty substance made of sand, water, and a binder such as gypsum or perlite. It is applied as a finish to either masonry or lath that is attached to studs or ceiling joists.

Plaster skins applied to studs or joists attenuate sound much like those made of gypsum board (see above). If applied to masonry, the improvement over the unplastered masonry is small or negligible. Plaster provides very little absorption except in the low frequencies, if suspended or furred out from a solid surface. Also see the discussion of acoustical plaster in Section 2.5.

Plywood. Plywood is a laminate of several layers of wood veneer. It is used in wood construction as an underlayment for floors, as sheathing on studs or rafters, or as finished paneling on walls.

Mainly because of its modest mass [approximately 0.6 kg/dm^3 (36 lb/ft^3)], specifically its mass per unit area, plywood is relatively ineffective as a sound attenuator. However, it is often adequate in combination with other materials or where high performance is not required. Thin plywood, if furred out from a solid wall, is a potent low-frequency absorber. Specially detailed resonant absorbers, made of plywood, are sometimes used to "tune" special-purpose rooms. At the higher frequencies, plywood is quite reflective.

Resilient Tile. Resilient tile is one of a family of floor tiles (also sheets), previously made of vinyl resins and asbestos fibers, but now usually of vinyl, asphalt, rubber, cork, and other asbestos-free materials. It is used as a finish on concrete as well as other substrates.

Such tile has little effect on the attenuation provided by the substrate. However, the tile's nominal resiliency (especially if foam backed) provides some attenuation of high-frequency impact sounds. Absorption is negligible; resilient tile is almost as reflective as concrete or any other hard floor finish.

Steel Decking. Steel decking is sheet steel, usually corrugated for greater strength, installed between structural supports (beams, purlins) as a form or base for other materials. It also has other uses; for example, noise barriers along highways are often made of steel decking.

Attenuation is usually governed by the combined mass of the deck and the topping. If the topping is concrete, which is typical of floors, the additional mass of the steel tends to be negligible. If used as an outdoor barrier, attenuation invariably is limited by barrier height. Steel decking is highly reflective unless free to vibrate and hence to absorb low-frequency sound by resonance. Also see discussion of acoustical decks in Section 2.5.

Steel Joists and Trusses. Steel joists and trusses are structural members of many different configurations, including beams, designed to support floors and roofs. (Similar properties hold for joists and trusses made of other structural materials.)

On their own, these members do not attenuate sound, but their spacing and rigidity can affect attenuation, especially vibration isolation. In general, rigid structures, that is, stiff, closely spaced supports, are more favorable than long-span structures that deflect more. Joists and trusses in general do not absorb sound, but may diffuse sound if exposed.

Steel Studs. Steel studs are framing members for partitions, usually fabricated in the shape of an angular C (channel studs) and covered with gypsum board.

Light-gauge steel studs are slightly resilient, which tends to decouple the partition's two faces and thus helps attenuation. Heavy-gauge studs (also wood studs), because they are stiffer, work less well. Best performance is obtained with two unconnected rows of studs, each supporting one side of the partition. In that case, it does not matter if the studs are stiff. Studs have no significant effect on absorption.

Stone. Stone is a natural material of considerable mass that is used for load-bearing walls, as a facing (stone veneer), and as paving. In a broader sense, it includes reconstituted materials like terrazzo.

Airborne sound attenuation of stone depends, once again, mainly on its mass. If thick and well sealed, walls built of stone can be very effective. However, as a paving, stone provides no impact isolation. Stone (e.g., marble) is among the acoustically most reflective materials, though this may not apply to some stones that are naturally porous.

Wood Decking. Wood decking is one of several structural materials supported by beams or trusses to form floors and roofs. It is often exposed as a finished ceiling.

Owing to its relatively low mass (as compared to concrete), wood decking provides only nominal attenuation unless ballasted with heavier materials. Wood decks are generally reflective, but unsealed cracks between the boards have been known to contribute a fair amount of absorption.

Wood Paneling. Wood paneling means a relatively thin finish made of wooden boards or panels. These are usually attached to furring and thus kept clear of the wall behind the paneling.

Wood paneling on a wall generally results in negligible improvement over the attenuation provided by the basic wall. Wood absorbs low-frequency sound by resonance and may lead to serious bass deficiency in music rooms unless it is thick and/or well restrained; for example, attached directly (without airspace) to the solid wall.

Wood Studs and Joists. Wood studs and joists are framing members for partitions and floors, especially in wood-frame buildings. They are covered with gypsum board, plaster, or in the case of floors with plywood and a variety of finishes.

Wood studs and joists are quite rigid; the two sides of a partition remain well coupled (compare with steel studs). Attenuation can be improved considerably by resilient channels (see Section 2.6) and by absorptive material in the stud or joist space. Staggered studs that provide for complete separation of the partition's two sides are best. Studs and joists have almost no effect on absorption.

2.5 ACOUSTICAL MATERIALS

The materials reviewed below include a variety of products whose purpose is strictly acoustical or that are commonly perceived as having acoustical value. Again, the listing may not include every product that may be so classified, but familiarity with those described should allow one to draw conclusions regarding the acoustical performance of other similar materials.

Only the principal acoustical property of each material is discussed. Unless otherwise noted, those that attenuate sound do not absorb, and those that absorb do not offer much attenuation. This should be obvious since in order to stop sound, a material must be solid, and in order to absorb, it typically must be porous.

Acoustical Deck. An acoustical deck is a structural deck, usually made of perforated steel, which is backed by an absorptive material such as fiberglass (see Figure 2.4). The term also includes decks made entirely of fibrous materials (see discussion of fibrous plank).

An acoustical deck *absorbs* sound. Its NRC ranges from about 0.50 to as high as 0.90. If exposed, as is usually intended, an acoustical deck can greatly reduce noise and reverberation in spaces such as gymnasiums, factories, and workshops.

Acoustical Foam. Acoustical foam is one of a variety of cellular materials, usually made of polyurethane. Foams are manufactured either with open cells (air can be blown into and through the material) or with closed cells (each cell is sealed; the material is airtight).

Open-cell foams are excellent sound *absorbers,* provided they are sufficiently thick. The noise reduction coefficient ranges from approximately 0.25 for 6-mm (0.25-in.) foam to 0.90 and higher for 50-mm (2-in.) or thicker foams. Their uses include padding for upholstered theater seats to stabilize reverberation regardless of occupancy.

Closed-cell foams also absorb sound, but less efficiently and less predictably. They are more often applied to ringing surfaces, such as large metal plates, to provide damping.

Acoustical Plaster. Acoustical plaster is a plasterlike product, distinguished by its porosity after it dries. It was originally intended to create jointless surfaces (like those of ordinary plaster) that *absorb* sound, which ordinary plaster does not.

The performance of acoustical plaster is highly dependent on the correct mix and application technique. Noise reduction coefficients on the order of 0.60 have been obtained under controlled conditions, but field installations usually yield much less. Acoustical plaster is not a reliable sound absorber.

Acoustical Tile. Acoustical tile is a widely used ceiling material made of mineral or cellulose fibers or of fiberglass (see below). It is available in a variety of modular sizes from approximately 30 × 30 cm (12 × 12 in.) to 61 × 122 cm (24 × 48 in.) and larger. Acoustical tile is usually suspended in a metal grid, but some types of tile

Figure 2.4 Diagram of an acoustical deck. (© 1998 Epic Metals Corporation. All rights reserved.)

can be glued or otherwise attached to solid surfaces. It is prone to damage when contacted and is therefore not recommended for surfaces, especially walls, that are within human reach.

The original purpose of acoustical tile was to *absorb* sound, which to this day remains its principal function. Absorptivity ranges from approximately NRC 0.50 for the least efficient tiles to NRC 0.95 for the best (typically fiberglass) lay-in panels. Suspended tile provides more low-frequency absorption than glued-on tile. Membrane-faced tiles provide less high-frequency absorption than those whose faces are porous. In general, the thicker the tile the better it absorbs.

Many acoustical tiles also *attenuate* sound. They are distinguished by an STC rating (in addition to the NRC rating), which quantifies attenuation through the suspended ceiling, along a standardized plenum, and through the ceiling of the adjoining room.[8] This is of vital importance where the plenum is continuous, that is, where the partitions stop against or just above the ceiling. Such tiles are usually made of mineral fiber and backed by a sealed coating or foil. The two-room STC ratings range from about 30 to 45.

Carpet. Carpet is any of a variety of soft floor finishes made of synthetic materials such as nylon or natural materials such as wool. It is either glued directly to the floor or installed over an underlayment of hairfelt or foam rubber.

Carpet *absorbs* sound. It is, in fact, the only floor finish that absorbs sound. Absorptivity depends primarily on the total thickness of the pile and the porous or fibrous underlayment (if present and provided the carpet does not have an airtight backing). Noise reduction coefficients range from 0.20 to about 0.55. Absorptivity is confined mainly to the high frequencies.

Carpet also *attenuates* impact sounds because it prevents hard contact with the floor. If sufficiently thick, it can be extremely effective. However, on wood floors, it will not eliminate low-frequency thuds.

Cellulose Fiber. Cellulose fiber is one of a variety of fibers that forms the basis for materials such as acoustical tile, wood wool, fibrous spray, and so on (also see discussion of mineral fiber). Each of these materials is designed to *absorb* sound, as discussed under the relevant headings in this chapter.

Curtains and Fabrics. Curtains and fabrics include a range of textiles the are used on their own (as curtains) or as coverings for other materials that may or may not be sound absorbing. Curtains *absorb* sound if they are reasonably heavy [at least 500 g/m² (15 oz/yd²)] and, more importantly, if their flow resistance is sufficiently high to the point of severely impeding, but not stopping, airflow through the material. A light curtain may have an NRC of only 0.20; a heavy, flow-resistant fabric, draped to half area, may rate NRC 0.70 or more.

Fabrics attached directly to hard surfaces do not absorb sound. However, if stretched over materials such as fiberglass (see discussion of fibrous board), and provided they are not airtight, they make an acoustically excellent finish that fully preserves the substrate's absorptivity.

Duct Lining. Duct lining is one of a few materials that is literally an acoustical insulation. It is usually made of fiberglass and comes in thicknesses of up to 50 mm (2 in.), as shown in Figure 2.5. The lining is mechanically fastened to the interior surfaces of sheet metal ventilation ducts. In high-velocity ducts, it may be faced with perforated metal to prevent erosion.

Duct lining *absorbs* sound and thus *attenuates* noise as it propagates along ducts. Compared with an unlined metal duct, which may attenuate mid-frequency sound by 0.15 dB/m (0.05 dB/ft), a duct with a 25-mm (1-in.) lining will yield 3 dB/m (1 dB/ft). Low-frequency attenuation is not as good. Ducts made entirely of fiberglass exhibit

Figure 2.5 These fiberglass boards, when applied to the interior of ventilation ducts, will attenuate noise. (Courtesy of Owens-Corning Fiberglas.)

similar properties, but due to their very low mass they allow sound to escape into the surrounding space.

Fiberglass. Fiberglass, which is available in the form of batts, blankets, and boards, is an excellent sound *absorber*. The manufacturing process ensures consistent porosity at a very fine scale. Applications include a great many sound-absorbing treatments, insulation as in stud walls and ducts, and various applications in industrial noise control. Compressed blocks or sheets of fiberglass are also used to form resilient supports/hangers (see Section 2.6) or as joint fillers where rigid ties are to be avoided.

The absorptivity of fiberglass depends on flow resistance, which, in turn, is affected by the material's thickness, its density, and the diameter of the fibers. In most applications, the thickness of the board or blanket is the most important parameter.

Fibrous Batts and Blankets. Usually made of fiberglass or mineral fiber, fibrous batts and blankets are among the most common forms of acoustical (also thermal) insulation in use today. They serve two distinct acoustical purposes.

If exposed to the room, as a wall finish (behind fabric or an open grillage) or as a ceiling finish (behind perforated pans or spaced slats), they *absorb* sound and thus reduce noise and reverberation in the room. Performance depends on thickness and on the properties of the facing. It can be as high as NRC 0.90.

If used between the two faces of a partition (typically in the stud space, but also above suspended ceilings where the ceiling and the floor above form the partition), batts and blankets improve *attenuation*. They do it by absorbing sound that is in transit through the partition's cavity. If the cavity is bridged by rigid ties (e.g., wood studs), there is little improvement. With light-gauge steel studs, about 6 STC points are gained. Performance again depends on thickness, but the batt or blanket should never completely fill the cavity.

Fibrous Board. Fibrous board works much like batts and blankets but is of higher density—up to approximately 0.32 kg/dm³ (20 lb/ft³), but more usually near 0.1 kg/dm³ (6 lb/ft³). Such rigid or semirigid boards, especially those made of fiberglass, are excellent sound *absorbers*. They are available with a variety of sound-transparent (usually fabric) facings, for use as wall or ceiling panels. Ratings range from approximately NRC 0.75 for 25-mm (1-in.) fiberglass board to NRC 0.90 for 50-mm (2-in.) board. Less porous or thinner boards, such as those made of mineral fiber, are somewhat less absorptive.

Fibrous Plank. Fibrous plank is a rigid (often structural) material, usually made of coarse fibers, such as wood fibers, embedded in a cementitious mix. The structural properties of certain planks allow them to be used as roof decking. The fibrous surface *absorbs* sound. Performance depends on thickness and ranges from approximately NRC 0.40 for 25-mm (1-in.) plank to NRC 0.65 for 75-mm (3-in.) plank. If exposed to the room, fibrous plank reduces noise and reverberation in the room.

Fibrous Spray. Fibrous spray is any of a variety of sprayed-on insulating materials, often specified for fire-proofing reasons. Previously made of asbestos fibers, which are now known to be a health hazard, most contemporary sprays contain cellulose or mineral fibers of various descriptions. Fibrous spray is inherently porous and therefore *absorptive*. However, performance is highly dependent on thickness and application technique. A well-applied coat of 25-mm (1-in.) thickness may achieve or exceed NRC 0.60.

Insulation (Loose). Loose insulation is similar to fibrous batts and blankets, except that it can be blown or dumped in place. It serves much the same purpose as batts within a partition; that is, it improves *attenuation* through the partition.

Laminated Glass. Laminated glass is a sandwich of two or more sheets of glass with viscoelastic interlayers that provide damping as the sandwich is flexed. Certain types of laminated glass offer substantially better sound *attenuation* than an equal thickness of monolithic glass. For example, 13-mm (0.5-in.) plate glass rates in the low STC 30s, whereas 13-mm laminated glass may approach STC 40.

Lead Sheet. In its purest form, lead sheet is sheet metal made of lead or a lead alloy. It is also available in combination with other materials, for example, as leaded vinyl. It is often used to close off the plenum above a room whose partitions extend only to a suspended ceiling.

Lead provides excellent *attenuation* per unit thickness because it is heavy [approximately 11 kg/dm³ (700 lb/ft³)] and limp. Furthermore, lead is easily shaped to conform to irregularities, which helps avoid holes in barriers that must be tightly sealed.

Metal Pans. Perforated metal pans, backed by fibrous batts (see Figure 2.6), are an alternative to acoustical tile ceilings. Similar panels can also be used on walls. The object is to *absorb* sound. The pan itself has little acoustical value, but the size and spacing of its perforations (not just the percent openness) affect performance.

Depending on the perforation pattern and the type and thickness of the batt, absorptivity may range from NRC 0.50 or lower to NRC 0.95. If the batts are encased in plastic, which may be required in some situations, high-frequency absorptivity is impaired. Perforated pans do not attenuate sound unless equipped with a solid backing.

Mineral Fiber. Mineral fibers are a very common family of fibers used in the manufacture of acoustical tile, blankets and boards, fibrous spray, and so on (see the discussion of the respective materials; also see discussion of cellulose fibers).

Sealants. Sealants are nonhardening compounds used to seal joints and cracks in many construction types. They are especially applicable to gypsum board partitions and where services (ducts, pipes) penetrate a partition. The acoustical value of sealants lies in their ability to render partitions airtight. Failing this, *attenuation* may be seriously compromised.

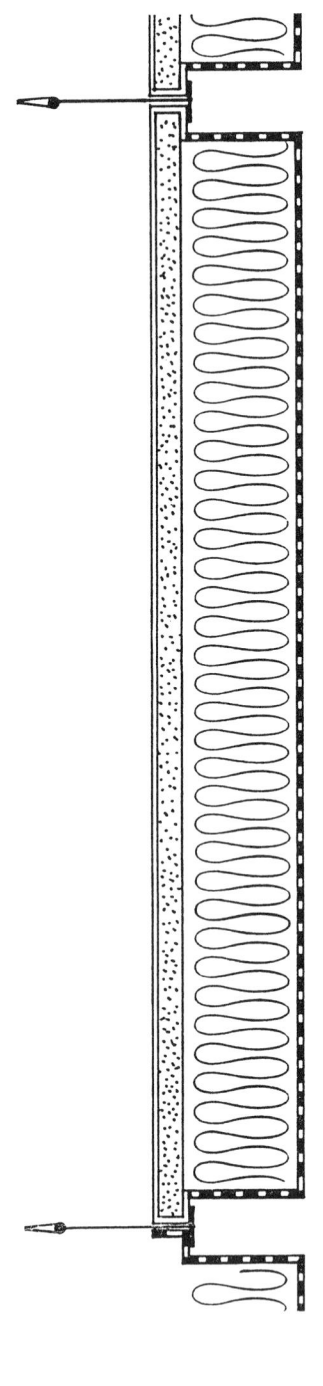

Figure 2.6 Diagram of a perforated metal ceiling panel. The insulation immediately behind the perforated metal absorbs sound. Gypsum board backup is used to stop sound from passing through the panel. (Courtesy Industrial Acoustics Company, Inc.)

Slats and Grilles. Often believed to have acoustical value, slats and grilles (made of wood, metal, etc.) serve only to protect the material behind them. Typically, the material behind is fiberglass, which is *absorptive*. Absorptivity is maintained if the slats or grille members are small and widely spaced. Increasing their size and/or reducing the space between them generally results in some high-frequency reflectivity.

2.6 SPECIAL DEVICES

The products discussed below are manufactured and used for strictly acoustical purposes. A few absorb (or diffuse) sound. Most attenuate sound or vibration. Each is briefly discussed with reference to its properties and its typical applications.

Air Springs. Functionally comparable to steel springs and to those made of elastomers (see below), air springs are probably the most effective *vibration-isolating* devices available today. They are generally custom-designed for critical applications where only extremely low levels of vibration can be tolerated.

In principle, an air spring consists of a trapped volume of air encased in a flexible jacket (see Figure 2.7). There are no mechanical ties between the building structure and whatever is to be isolated. Since air is compressible, it acts as a spring. The stiffness of these springs is controlled by air pressure and the jacket design. Isolation efficiencies well in excess of those practical with steel springs can be engineered.

Duct Silencers. Duct silencers, also called sound traps, are commercially made units designed to fit rectangular or round ventilation ducts of various sizes (see Figure 2.8). They are available for a range of pressure-drop conditions and in several standard lengths. American-made rectangular silencers are 91 cm (3 ft), 152 cm (5 ft), or 213 cm (7 ft) long; round silencers are typically two to 3 times as long as their diameter.

Silencers, typically containing fiberglass-packed baffles, absorb sound and thus *attenuate* duct-borne noise. They perform much like duct lining (see above) but more efficiently. Performance is fair in the low frequencies, best in the middle frequencies, and quite good in the high frequencies. Long, high-pressure-drop silencers generally provide more attenuation than those that are shorter and/or less resistive.

Elastomers. Elastomers, such as neoprene, are a family of elastic, rubberlike materials used in the manufacture of resilient mounts and hangers (see below) and of other devices whose purpose is to avoid rigid contact and thus to *attenuate* noise or vibration. The efficiency of the elastomer is related to its hardness (durometer), geome-

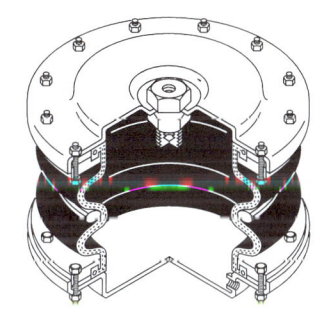

Figure 2.7 Diagram of a twin sphere air spring mount. (Courtesy of Mason Industries.)

72 • *Acoustical Materials and Methods*

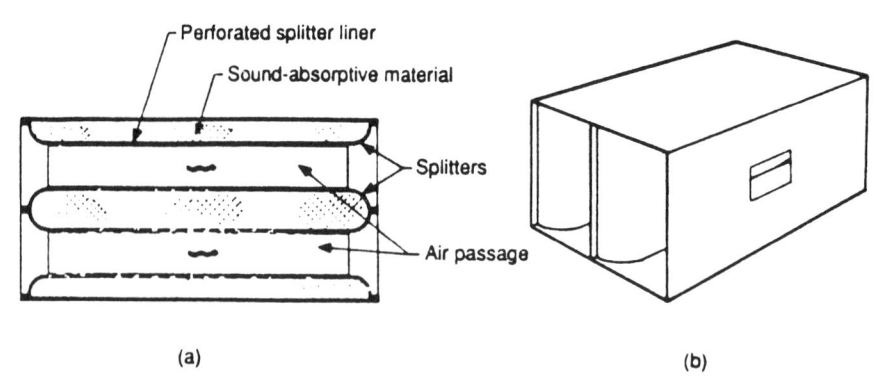

"Round-nosed" rectangular silencer. (*a*) Cross section; (*b*) external view.

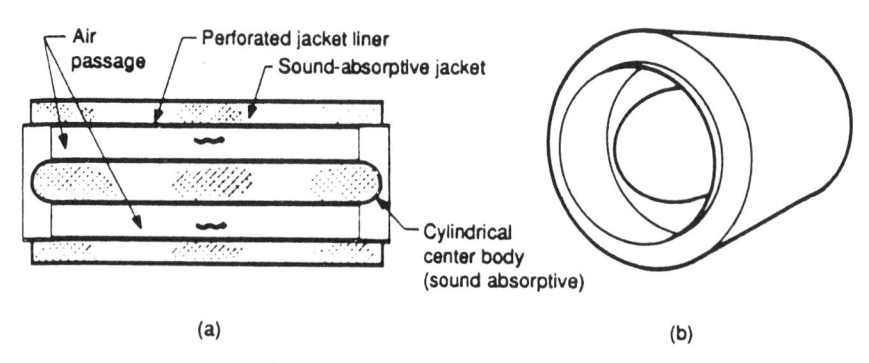

Cylindrical silencer. (*a*) Cross section; (*b*) external view.

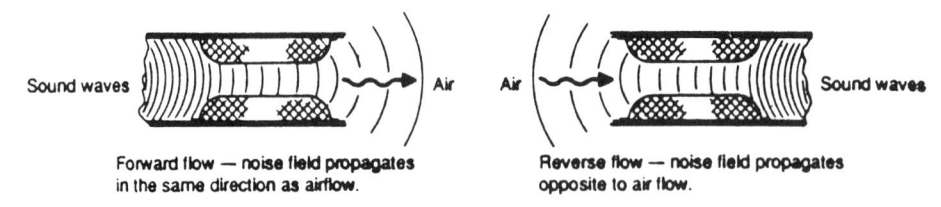

Figure 2.8 Some common duct silencers with descriptions of how they attenuate duct-borne noise. (Courtesy of Industrial Acoustics Company, Inc.)

try, and loading. Durometer typically ranges from 30 (softest) to 70 (hardest). A typical recommended loading for a 40-durometer waffle pad made of neoprene is on the order of 4.2 kg/cm^2 (60 lb/in.^{2}).

Flexible Connections. Flexible connections are flexible inserts made of canvas or leaded vinyl (see discussion of lead sheet in Section 2.5) and typically located between two pieces of metal duct. In a broader sense, flexible connections also include flexible conduit and various types of flexible hose.

The common purpose of all these connectors is to create resilient breaks in ducts and pipes and thus to *attenuate* vibration that is conducted along these otherwise rigid elements. Flexible connections are essential in all duct, pipe, and conduit runs between a piece of vibration-isolated equipment and the building structure.

Functional Absorbers. Made primarily for industrial applications in the form of free-hanging cylinders, functional absorbers combine surface absorptivity with tuned resonances to *absorb* sound. Performance of functional absorbers is usually given in

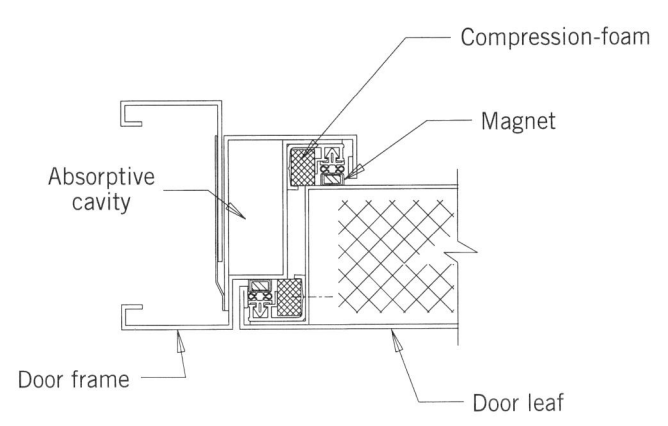

Figure 2.9 Schematic diagram of a double-sealed (gasketed) door panel. (Courtesy of Industrial Acoustics Company, Inc.)

sabins per unit, rather than sabins per square foot of surface area. Like absorptive room finishes, they help reduce noise and reverberation in the room.

Gaskets. Gaskets are airtight seals made of pliable materials such as neoprene or vinyl, especially for acoustical doors and sound-rated partition systems. They are also used for other applications where airtightness cannot otherwise be achieved.

The sole purpose of acoustical gaskets (or sound seals) is to eliminate air leaks, which are also sound leaks, and thus to maximize *attenuation*. A perfect fit is essential if the attenuation capabilities of a door (or other) panel are to be fully realized (see Figure 2.9).

Quadratic-Residue Diffusors. Quadratic-residue diffusors consist of a series of narrow "wells" of unequal depth, separated by even narrower plates, giving them a distinctly ribbed appearance. They have the unique property of *diffusing* sound, that is, of spreading the reflections over a wide arc at right angles to the wells. Applications include broadcast and recording studios, control rooms, and wherever specular reflection, as off a plain surface, is to be avoided.

Quadratic-residue diffusors can be made of any hard material. They can be engineered to work over a wide range of frequencies. Typical depths range from as little as 10 cm (4 in.) to 40 cm (16 in.) and more.

Resilient Clips and Channels. Specially designed mechanical connectors, resilient clips and channels are typically made of light-gauge sheet metal, for use between studs or joists and a finished gypsum board or plaster surface. Resilient channels are very effective in combination with wood studs or joists, which are quite rigid (see Figure 2.10).

Such clips and channels serve to break the rigid connection between a partition's two faces, which appreciably *reduces sound transmission* through the partition. Fibrous batts (see above) are usually recommended for the stud or joist space of partitions, including floor/ceilings, whose faces are resiliently attached.

Resilient Hangers. Resilient hangers are any of a variety of springlike devices designed to support suspended ceilings, suspended pieces of mechanical equipment, or ducts and pipes connected to equipment. The resilient elements may be steel springs, pieces of elastomeric material, or compressed fiberglass.

Resilient ceiling hangers perform much the same task as resilient clips or channels (see above), but more efficiently; they improve *attenuation*. Resilient equipment hangers, on the other hand, are primarily vibration isolators (see Figure 2.11); they are a direct counterpart to the resilient mounts (see below) specified for floor-mounted equipment.

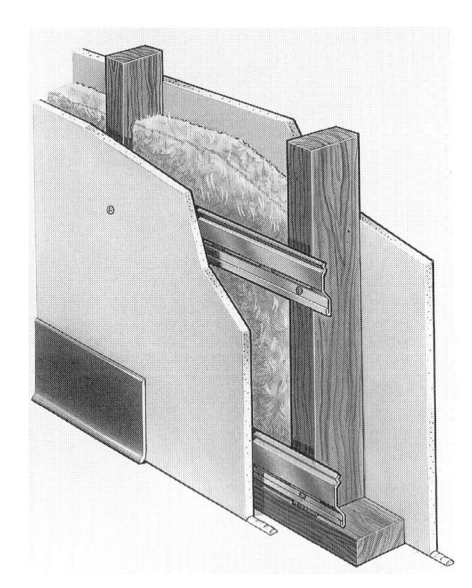

Figure 2.10 This single-layer gypsum board system, with resilient-channel attachment and 3 in. of mineral fiber insulation, delivers STC 50 and a one-hour fire rating. (Illustration courtesy of USG Corporation.)

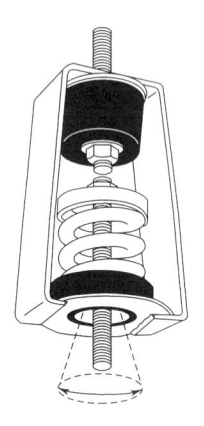

Figure 2.11 Diagram of a high deflection spring hanger. (Courtesy of Mason Industries.)

Resilient Mounts. Resilient mounts are equivalent in purpose to resilient hangers and may also incorporate steel springs (as in Figure 2.12) or elastomer or compressed fiberglass elements (also see discussion of air springs). Aside from their common role as *vibration-isolating* supports for mechanical equipment, there are mounts [usually 50 mm (2 in.) tall, made of solid neoprene or neoprene-covered fiberglass] designed to support so-called floating floors, as shown in Figure 2.13. Such double floors, consisting of a structural slab and a floating slab, offer exceptionally good sound *attenuation*.

Space Units. Space units are blocks of fibrous-porous material made of mineral fibers, foamed glass, or any other substance of comparable porosity. In appearance similar to acoustical tile, but typically about 50 mm (2 in.) thick, these units are intended for spaced application to hard wall and ceiling surfaces. Space units *absorb* sound. In composition not significantly different from acoustical tile, their efficiency is helped by exposure of their sides.

Steel Springs. Generally of the coil type, steel springs form the core of most resilient mounts and hangers. They are sometimes engineered to provide static deflections of up to 130 mm (5 in.), although deflections between 25 and 50 mm (1 and

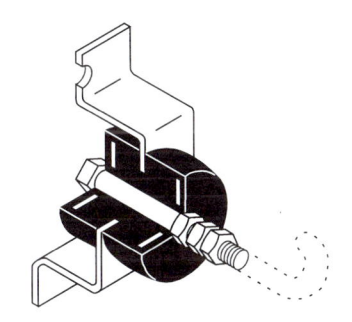

Mason Industries Type **DNSB**
Double Acting Neoprene Sway Brace

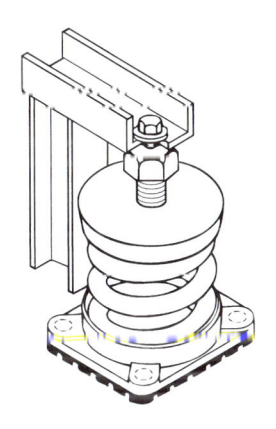

Mason Industries Type **SLF-100**
High Deflection Spring Mount

Figure 2.12 Diagram of a high deflection spring mount. (Courtesy of Mason Industries.)

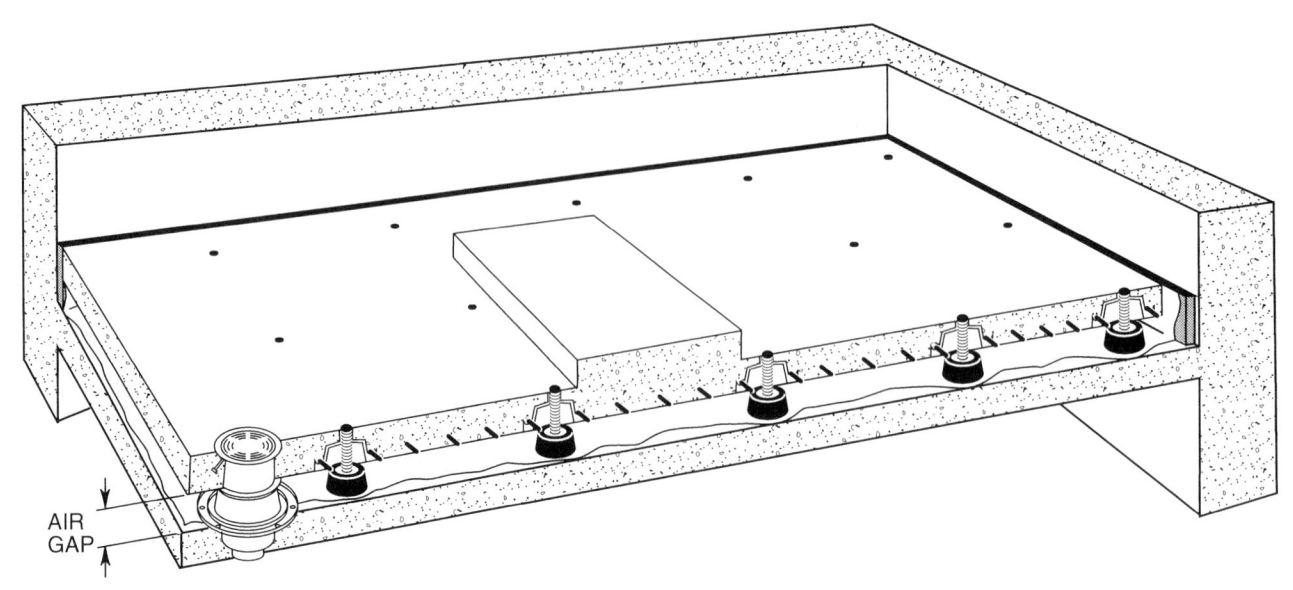

Figure 2.13 Illustration of a floating floor supported by neoprene mounts. (Courtesy of Mason Industries.)

2 in.) are much more common. For comparison, static deflection of elastomer mounts/hangers seldom exceeds 13 mm (0.5 in.). Consequently, steel springs can offer considerably lower natural frequencies, and therefore greater isolation efficiencies, than any other springs except special-purpose air springs.

Steel springs (often in combination with elastomeric inserts, to reduce high-frequency transmission along the coil) are used primarily to *isolate vibrating equipment* such as fans, pumps, compressors, and so on. They are also used in ceiling hangers designed for critical applications. Steel spring mounts for floating floors are available, but their use is confined to unusual conditions.

Sway Braces.   Sway braces include resilient connectors of various designs whose purpose is to provide structural, specifically lateral, support, but without creating any rigid ties (see Figure 2.12). The insulating medium is typically neoprene or fiberglass attached to steel clips or angles.

Sway braces are used to brace free-standing walls in double-wall constructions where rigid ties would seriously harm *attenuation*. Functionally similar angle braces are used to lend stability to masonry walls whose tops must, for sound isolation reasons, be kept free of the slab above.

2.7 PERFORMANCE TABLES

Tables 2.1 through 2.4 provide a general overview of the acoustical performance that can be expected of some typical constructions and materials. Good detailing and execution is assumed in all cases. Attenuation is given in terms of the STC and IIC, and absorptivity is indicated by the sound absorption coefficients and by the NRC values, as discussed earlier in this chapter as well as in Chapters 1 and 3.

TABLE 2.1 Sound Absorption Coefficients of General Building Materials[a]

MATERIALS	125 Hz	250 Hz	500 Hz	1000 Hz	2000 Hz	4000 Hz	NRC
SOUND ABSORPTION COEFFICIENTS (α)							
Brick, unglazed	0.03	0.03	0.03	0.04	0.05	0.07	0.05
Carpet, heavy, on concrete	0.02	0.06	0.14	0.37	0.60	0.65	0.30
Same, on 40-oz hairfelt or foam rubber	0.08	0.24	0.57	0.69	0.71	0.73	0.55
Concrete block, painted	0.10	0.05	0.06	0.07	0.09	0.08	0.05
Fabrics							
Light velour, 10 oz/yd^2, hung straight, in contact with wall	0.03	0.04	0.11	0.17	0.24	0.35	0.15
Medium velour, 14 oz/yd^2, draped to half area	0.07	0.31	0.49	0.75	0.70	0.60	0.55
Heavy velour, 18 oz/yd^2, draped to half area	0.14	0.35	0.55	0.72	0.70	0.65	0.60
Floors							
Concrete or terrazzo	0.01	0.01	0.015	0.02	0.02	0.02	0
Linoleum, asphalt, rubber or cork tile on concrete	0.02	0.03	0.03	0.03	0.03	0.02	0.05
Wood	0.15	0.11	0.10	0.07	0.06	0.07	0.1
Glass, large panes of heavy plate	0.18	0.05	0.04	0.03	0.02	0.02	0.05
Gypsum board, $\frac{1}{2}$ in. nailed to 2 × 4's 16 in. o.c.	0.29	0.10	0.05	0.04	0.07	0.09	0.05
Marble or glazed tile	0.01	0.01	0.01	0.01	0.02	0.02	0
Openings							
Stage, depending on furnishings			0.25—0.75				
Deep balcony, upholstered seats			0.50—1.00				
Grilles, ventilating			0.15—0.50				
Plaster, gypsum or lime, smooth finish on tile or brick	0.13	0.15	0.02	0.03	0.04	0.05	0.05
Plaster, gypsum or lime, rough finish on lath	0.02	0.03	0.04	0.05	0.04	0.03	0.05
Same, with smooth finish	0.02	0.02	0.03	0.04	0.04	0.03	0.05
Plywood paneling, $\frac{7}{8}$ in. thick	0.28	0.22	0.17	0.09	0.10	0.11	0.15
Water surface, as in a swimming pool	0.008	0.008	0.013	0.015	0.020	0.025	0
Air, sabins per 1000 ft^3					2.3	7.2	
ABSORPTION OF SEATS AND AUDIENCE[b]							
Audience, seated in upholstered seats, per ft^2 of floor area	0.60	0.74	0.88	0.96	0.93	0.85	
Unoccupied cloth-covered upholstered seats, per ft^2 of floor area	0.49	0.66	0.80	0.88	0.82	0.70	
Unoccupied leather-covered upholstered seats, per ft^2 of floor area	0.44	0.54	0.60	0.62	0.58	0.50	
Wooden pews, occupied, per ft^2 of floor area	0.57	0.61	0.75	0.86	0.91	0.86	
Chairs, metal or wood seats, each, unoccupied	0.15	0.19	0.22	0.39	0.38	0.30	

Source: W. J. Cavanaugh, "Acoustical Control in Buildings" in *Building Construction: Materials and Types of Construction,* 5th ed., W. C. Huntington and R. E. Mickadeit, Eds., Table 10.1, p. 390, Wiley, New York, 1981.

[a] The performance data in this table have been compiled from a variety of sources including the editor's and chapter author's files. They are intended as a guide, *not* a substitute for laboratory or field data from other reliable sources.

[b] Values given are in sabins per square foot of seating area or per unit.

TABLE 2.2 Sound Absorption Data for Selected Acoustical Materials[a]

DESCRIPTION	THICKNESS, INCHES	MOUNTING TYPE	UNIT SIZE, INCHES	125	250	500	1000	2000	4000	WEIGHT, LB/FT^2	NRC
Board, Blanket, and Batt Products											
Glass fiber blanket, 3 pcf	1	A	Roll	0.06	0.25	0.54	0.75	0.84	0.90	(0.25)	0.60
Glass fiber blanket, 3 pcf	1½	A	48 × 96	0.35	0.46	0.86	0.98	0.99	0.99	(0.38)	0.90
			48 × 120								
Semirigid glass fiber board, 1.7 pcf	3	A	24 × 48	0.43	0.83	0.99	0.95	0.97	0.99	(0.42)	0.95
Glass fiber blanket, 2 pcf	3½	A	Rolls	0.38	0.90	0.99	0.93	0.94	0.99	(0.58)	0.95
Glass fiber blanket, 2 pcf	5	A	Rolls	0.68	0.99	0.99	0.94	0.99	0.97	(0.83)	1.00
Glass fiber blanket, 2 pcf	6	A	Rolls	0.88	0.99	0.99	0.96	0.99	0.99	(1.0)	1.00
Glass fiber blanket, 0.75 pcf	1	A	Roll	0.12	0.18	0.51	0.72	0.78	0.86	(0.75)	0.55
Glass fiber insulation board, 6 pcf	1	A	24 × 48	0.02	0.27	0.63	0.85	0.93	0.95	(0.50)	0.65
Glass fiber insulation board, 6 pcf	2	A	24 × 48	0.16	0.71	0.99	0.99	0.99	0.99	(1.0)	0.95
Glass fiber insulation board, 6 pcf	3	A	24 × 48	0.54	0.99	0.99	0.99	0.99	0.99	(1.5)	1.00
Glass fiber insulation board, 6 pcf	4	A	24 × 48	0.75	0.99	0.99	0.99	0.97	0.98	(2.0)	1.00
Glasswool blanket with perforated foil facing	1	A	Roll	0.55	0.69	0.99	0.99	0.99	0.99	(0.55)	1.00
Glasswool blanket with unperforated foil facing	1	A	Roll	0.56	1.03	0.93	0.41	0.21	0.14	(0.55)	0.65
Ceiling Tiles and Panels											
Fissured mineral fiber ceiling tile	¾	E-400	12 × 12	0.49	0.53	0.53	0.75	0.92	0.99	(1.3)	0.70
Fissured mineral fiber tiles and panels	¾	E-400	24 × 48	0.34	0.36	0.71	0.85	0.68	0.64	(1.3)	0.65
Glass fiber panels	1	E-400	24 × 48	0.55	0.89	0.73	0.99	0.99	0.99	(0.5)	0.90
Mineral fiber panels	⅝	E-400	24 × 48	0.29	0.39	0.54	0.79	0.78	0.60	(1.05)	0.60
Cloth-faced, glass fiber panels	¾	E-400	24 × 48	0.71	0.87	0.62	0.86	0.96	0.99	(0.35)	0.85
Membrane-faced glass fiber panels	1½	E-400	24 × 48	0.57	0.79	0.77	0.90	0.71	0.47	(0.30)	0.80
Prefinished fabric-faced glass fiber panels	1	E-400	24 × 48	0.83	0.88	0.80	0.99	0.99	0.99	(0.25)	0.95
Duct Liners											
Glass fiber duct liner	1	F-25	Roll	0.23	0.47	0.61	0.79	0.88	0.91	(0.25)	0.70
Glass fiber duct liner	1½	F-25	Rolls	0.37	0.63	0.80	0.90	0.89	0.90	(0.37)	0.80
Glass fiber duct liner	2	F-25	Rolls	0.36	0.79	0.96	0.99	0.93	0.97	(0.50)	0.90
Mat-faced duct liner	½	A	Roll	0.09	0.14	0.40	0.60	0.73	0.82	(0.08)	0.45
Mat-faced, duct liner	1	A	Roll	0.08	0.26	0.58	0.84	0.96	0.99	(0.15)	0.65
Mat-faced, duct liner	1½	A	Rolls	0.17	0.53	0.87	0.99	0.99	0.95	(0.19)	0.85
Mat-faced, duct liner	2	A	Rolls	0.22	0.69	0.99	0.99	0.99	0.99	(0.22)	0.95
Foam Materials											
Polyurethane foam panels	1	A	54 × 24 ft	0.17	0.25	0.73	0.99	0.99	0.99	(0.17)	0.75
Polimide foam panels	2	A	96 × 96	0.23	0.51	0.96	0.99	0.93	0.96	(0.25)	0.85
Wedge-shaped polyurethane foam panels	2	A	24 × 48	0.08	0.25	0.61	0.92	0.95	0.92	(0.34)	0.75
			48 × 48								
Wedge-shaped polyurethane foam panels	3	A	24 × 48	0.14	0.43	0.98	0.99	0.99	0.99	(0.51)	0.85
			48 × 48								

Perforated Metal or Fabric-Faced Panels

Material	Thickness	Mounting	Size	125	250	500	1000	2000	4000	NRC
Perforated metal, backed by 1.5 pcf film-faced glass fiber	2	A	24 × 108 (1.5)	0.44	0.62	0.94	0.99	0.79	0.59	0.85
Perforated metal, backed by 1.5 pcf film-faced glass fiber	2	A	24 × 108 (1.5)	0.21	0.66	0.99	0.99	0.99	0.90	0.95
Perforated metal, backed by 1.5 pcf unfaced glass fiber	2	E-400	24 × 48 (1.8)	0.63	0.71	0.98	0.99	0.99	0.99	0.95
Perforated metal, backed by 1.5 pcf unfaced glass fiber	3	E-400	24 × 48 (1.8)	0.56	0.84	0.99	0.99	0.92	0.82	0.95
Structural wood fiber roof deck	$1\frac{1}{2}$	A	48 × 96 (2.4)	0.07	0.22	0.48	0.82	0.64	0.96	0.55
Structural wood fiber roof deck	2	A	48 × 96 (3.5)	0.15	0.26	0.62	0.94	0.64	0.92	0.60
Structural wood fiber roof deck	$2\frac{1}{2}$	A	48 × 96 (4.5)	0.20	0.31	0.72	0.84	0.77	0.90	0.65
Structural wood fiber roof deck	3	A	48 × 96 (5.3)	0.21	0.41	0.99	0.75	0.99	0.97	0.80

Special Masonry Units

Material	Thickness	Mounting	Size	125	250	500	1000	2000	4000	NRC
Slotted concrete blocks, empty	4	A	8 × 16 (23)	0.12	0.85	0.36	0.36	0.42	0.35	0.50
Slotted concrete blocks, empty	6	A	8 × 16 (26)	0.62	0.84	0.36	0.43	0.27	0.30	0.45
Slotted concrete blocks, empty	8	A	8 × 16 (32)	0.97	0.44	0.38	0.39	0.50	0.25	0.40
Slotted concrete blocks, with metal septa in cavities	8	A	8 × 16 (38)	0.99	0.97	0.61	0.37	0.56	0.39	0.70
Slotted concrete blocks, with glass fiber in cavities	12	A	8 × 16 (62)	0.48	0.83	0.86	0.54	0.47	0.44	0.65
Slotted concrete block, with metal septa and glass fiber in cavities	12	A	8 × 16 (70)	0.57	0.76	1.09	0.94	0.54	0.59	0.80
Sound-diffusing concrete block	12	A	8 × 16 (76)	0.76	0.51	0.57	0.34	0.24	0.26	0.40
Hollow perforated ceramic clay tile with glass fiber in cavities	4	A	8 × 16 (25.9)	0.19	0.64	0.73	0.62	0.20	0.10	0.55

Sprayed-on Materials

Material	Thickness	Mounting	125	250	500	1000	2000	4000	NRC
Cellulose fiber, textured finish	$\frac{1}{2}$	A	0.15	0.16	0.46	0.87	0.99	0.99	0.65
Cellulose fiber, textured finish	$\frac{3}{4}$	E-400	0.25	0.36	0.74	0.98	0.99	0.99	0.75
Cellulose fiber, textured finish	1	A	0.12	0.38	0.88	0.99	0.99	0.99	0.80
Acoustical plaster, Portland cement base	1	A	0.18	0.35	0.64	0.73	0.73	0.77	0.60

Unit Panels or Baffles[b]

Material	Thickness	Mounting	Size	125	250	500	1000	2000	4000	NRC
Hanging panels, 1.5 pcf glass fiber in polyethylene casing 0.002 in. thick	$1\frac{1}{2}$	Hung	24 × 48 (0.1)	0.32	0.59	1.49	1.53	1.36	0.59	—
Hanging panels, 2.70 pcf glass fiber in polyethylene casing 0.003 in. thick	$1\frac{1}{2}$	Hung	24 × 48 (0.39)	0.32	0.62	1.27	1.48	0.86	0.46	—
Two-sided wedge-shaped polyurethane foam, panels	3	Hung	31 × 48 (0.8)	0.58	0.47	0.84	1.26	1.61	1.80	—
Two-sided wedge-shaped melamine foam, panels	3	Hung	24 × 48 (0.4)	0.74	0.78	1.42	1.76	1.83	1.84	—

[a] The performance data in this table have been compiled from a variety of sources including the editor's and chapter author's files. They are intended as a guide *not* a substitute for laboratory or field data from other reliable sources.

[b] The sound absorption provided by baffles and panels hung vertically from the structure, depends on how closely they are spaced and other parameters. Therefore, information on such test conditions should be obtained from the manufacturer. There is no standard method for measuring and rating the sound-absorptive characteristics of these units. These tabulated sound absorption coefficients are obtained by dividing the total absorption of each unit by the total area of each unit. NRC values are not appropriate.

TABLE 2.3 Airborne Sound Isolation Performance Data for Selected Partition Systems[a]

CONSTRUCTION DESCRIPTION	SOUND TRANSMISSION CLASS[b] (STC)	SURFACE WEIGHT (PSF)	OVERALL THICKNESS (IN.)	FIRE-RESISTIVE RATING[b] (HR)
Poured Concrete				
3-in. solid concrete	47	39	3	$\frac{1}{2}$ (est.)
6-in. solid concrete, $\frac{1}{2}$ in. plaster both sides	53	80	7	3 (est.)
Masonry Units				
15-in. solid concrete blocks, $\frac{1}{2}$ in. plaster both sides	63	184	16	>4
$4\frac{1}{2}$-in. brick, $\frac{1}{2}$ in. plaster both sides	42	55	$5\frac{1}{2}$	$2\frac{1}{2}$
9-in. brick, $\frac{1}{2}$ in. plaster both sides	52	100	10	>4
12-in brick	56	121	12	>4

Construction				
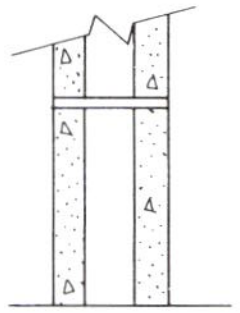 3⅝-in. lightweight hollow cement block painted both sides	44	26.1	3¾	1½ (est.)
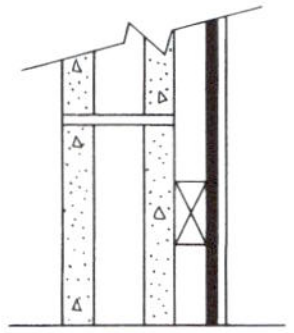 3⅝-in. lightweight hollow cement block, ¼-in. plywood covered lead on 1 × 2 wood furring strips one side	50	31	5	1½ (est.)
3-in. cinder block, ⅝ in. plaster both sides	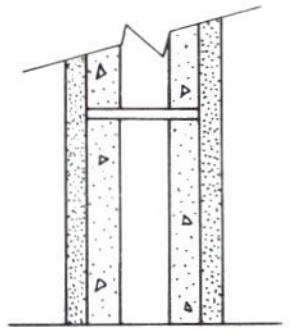 45	32.2	4¼	1½
4-in. cinder block ⅝ in. plaster both sides	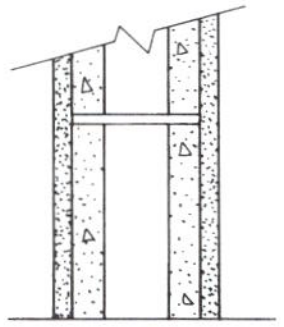 46	35.8	5¼	2
6-in. hollow concrete block	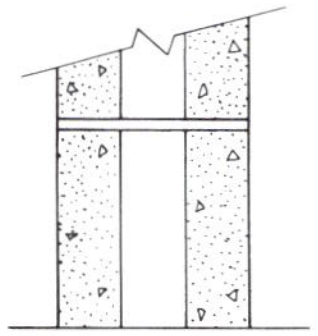 43 (unpainted) 45 (painted)	34	6	1 (est.)
12-in. hollow concrete block	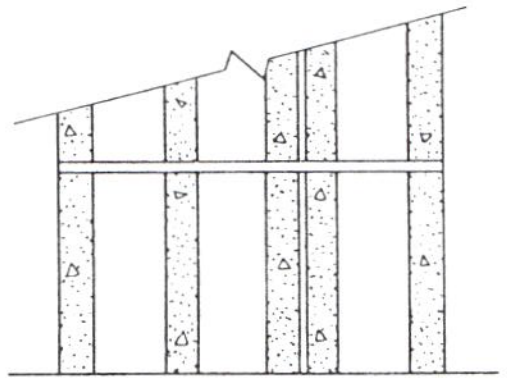 48	79	12	4
3-in. hollow gypsum block, ½-in. plaster both sides	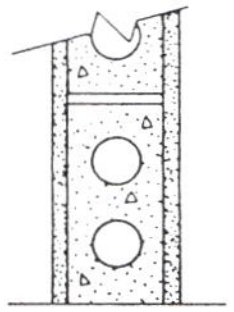 40	21.5	4	3

TABLE 2.3 (*Continued*)

CONSTRUCTION DESCRIPTION	SOUND TRANSMISSION CLASS[b] (STC)	SURFACE WEIGHT (PSF)	OVERALL THICKNESS (IN.)	FIRE-RESISTIVE RATING[b] (HR)
4-in. hollow gypsum block, $\frac{1}{2}$-in. plaster both sides	42	23.4	5	4
4-in. hollow gypsum block, $\frac{1}{2}$-in. plaster one side, $\frac{1}{2}$-in. resilient plaster other side	53	31	6	4
Double $4\frac{1}{2}$-in. brick wall, 2-in. cavity $\frac{1}{2}$-in. plaster both outsides	49 (w/wire ties) 54 (no wire ties)	100	12	>4
Double $4\frac{1}{2}$-in. brick wall, 6-in. cavity (no wire ties), $\frac{1}{2}$-in. plaster on wood fiberboard both outsides	62	120	18	>4
Double $4\frac{1}{2}$-in. hollow clay tile wall, $2\frac{1}{2}$-in. cavity (w/wire ties), $\frac{1}{2}$-in. plaster both outsides	43	50	12	3 (est.)

Description				
Double 4-in. solid cinder block wall, 2-in. cavity (no wire ties), $\frac{1}{2}$-in plaster both outsides	52	70	11	>4
Wood Stud Walls				
2 × 4 wood studs 16 in. o.c. $\frac{1}{2}$ in. gypsum board both sides	32	5.9	$4\frac{5}{8}$	$\frac{3}{4}$
As above w/2-in. cavity insulation	35	6	$4\frac{5}{8}$	$\frac{3}{4}$ (est.)
2 × 4 wood studs 16 in. o.c., $\frac{1}{2}$-in. gypsum board laminated to $\frac{1}{2}$-in. fiber board both sides	37	8.5	$5\frac{1}{4}$	—
2 × 4 wood studs 16 in. o.c., $\frac{1}{2}$-in. plaster on $\frac{3}{8}$-in. gypsum lath both sides	46	13.4	$5\frac{3}{4}$	$\frac{3}{4}$
2 × 4 wood studs 16 in. o.c., two layers $\frac{1}{2}$-in. gypsum board both sides	40	8.2	$5\frac{1}{2}$	1
2 × 3 staggered wood studs 8 in. o.c., $\frac{1}{2}$-in. gypsum board both sides	44	6.2	5	$\frac{1}{2}$ (est.)

TABLE 2.3 (*Continued*)

CONSTRUCTION DESCRIPTION	SOUND TRANSMISSION CLASS[b] (STC)	SURFACE WEIGHT (PSF)	OVERALL THICKNESS (IN.)	FIRE-RESISTIVE RATING[b] (HR)
2 × 3 staggered wood studs in 8 in. o.c., two layers $\frac{1}{2}$-in. gypsum board both sides	44	13.4	$6\frac{1}{2}$	$1\frac{1}{2}$ (est.)
2 × 4 staggered wood studs 8 in. o.c., $\frac{1}{2}$-in. gypsum board both sides. 1-in. cavity insulation	46	13.8	$5\frac{3}{4}$	$\frac{1}{2}$
2 × 4 slotted wood studs 16 in. o.c. $\frac{1}{2}$-in. plaster on $\frac{3}{8}$-in. gypsum lath both sides, 3-in. cavity insulation	45	14.2	$5\frac{5}{8}$	1 (est.)
2 × 4 wood studs 16 in. o.c., $\frac{5}{8}$-in. gypsum board on resilient channels both sides	47	6.7	$6\frac{1}{4}$	1 (est.)
2 × 4 wood studs 16 in. o.c., $\frac{1}{2}$-in. gypsum board laminated to $\frac{5}{8}$-in. gypsum board on resilient channels both sides	48	9	$6\frac{3}{4}$	1 (est.)
Metal Stud Walls				
$3\frac{1}{4}$-in. wire studs 16 in. o.c., $\frac{7}{8}$-in. plaster on metal lath both sides	39	19.6	$5\frac{1}{4}$	$1\frac{1}{4}$ (est.)

Construction	STC	lb/ft²	Thickness (in.)	(in.)
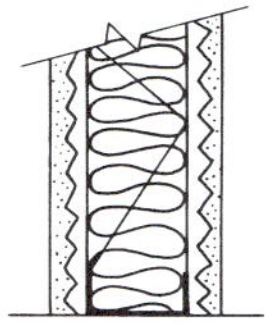 3¼-in. wire studs 16 in. o.c., ⅞-in. plaster on metal lath both sides, 3-in. cavity insulation	39	21.1	5¼	1½ (est.)
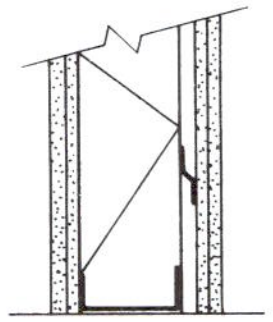 3¼-in. wire studs 16 in. o.c., ½-in. plaster on ⅜-in. gypsum lath both sides (one side on resilient clips)	43	13	5½	¾ (est.)
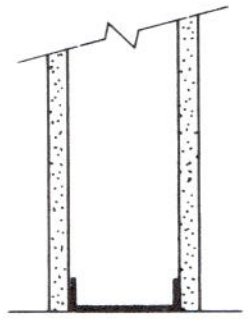 3⅝-in. metal channel studs, 24 in. o.c., ⅝-in. gypsum board both sides	41	6	4⅞	1
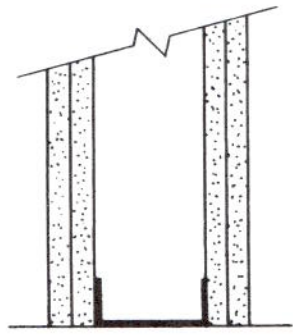 3⅝-in. metal channel studs 24 in. o.c., two-layer ⅝-in. gypsum board both sides	47	11.4	6⅛	2
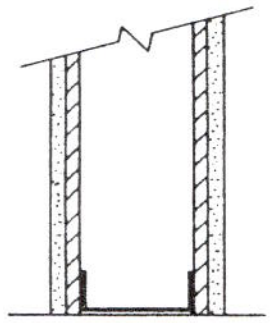 3⅝-in. metal channel studs 24 in. o.c., ½-in. gypsum board laminated to ½-in. fiberboard both sides	50	6.2	5⅝	1½ (est.)
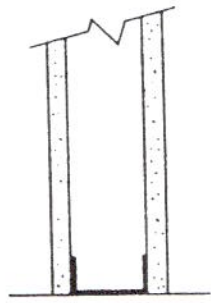 2½-in. metal channel studs 24 in o.c., ⅝-in. gypsum board both sides	36	6	3¾	1 (est.)
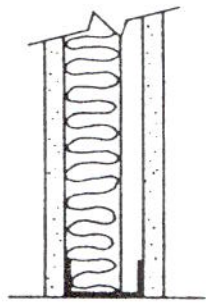 2½-in. metal channel studs 24 in o.c., ⅝-in. gypsum board both sides, 2-in. cavity insulation	41	6.1	3¾	1 (est.)

85

TABLE 2.3 (*Continued*)

CONSTRUCTION DESCRIPTION	SOUND TRANSMISSION CLASS[b] (STC)	SURFACE WEIGHT (PSF)	OVERALL THICKNESS (IN.)	FIRE-RESISTIVE RATING[b] (HR)
$2\frac{5}{8}$-in. metal channel studs 24 in. o.c., two layers $\frac{5}{8}$-in. gypsum board both sides	45	10.2	$4\frac{7}{8}$	$1\frac{1}{2}$ (est.)
$2\frac{5}{8}$-in. metal channel studs 24 in. o.c., two layers $\frac{5}{8}$-in. gypsum board both sides, 2-in. cavity insulation	51	10.2	$4\frac{7}{8}$	2 (est.)
Double $2\frac{5}{8}$-in. metal studs 24 in. o.c., w/$\frac{1}{2}$-in. gap, $\frac{5}{8}$-in. gypsum board both outsides, 2-in. cavity insulation	52	6.2	7	$1\frac{1}{2}$ (est.)
Double $2\frac{5}{8}$-in. metal studs 24 in. o.c. w/$\frac{1}{2}$ in. 2 layers gap. $\frac{5}{8}$-in. gypsum board both outsides, 2-in. cavity insulation	57	10.3	$8\frac{1}{8}$	2 (est.)
Solid Plaster Walls				
2-in. sanded gypsum plaster on metal mesh	36	18.3	2	1
2-in. gypsum perlite plaster on metal mesh	31	8.8	2	$1\frac{1}{4}$ (est.)

Construction	STC	Approx. weight (lb/ft²)	Approx. thickness (in.)	Fire rating (hr)
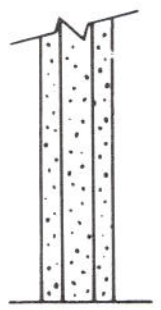 2¼-in. laminated gypsum board wall, ⅝ in. both sides of 1-in. core	36	10.2	2¼	1
Interior Doors				
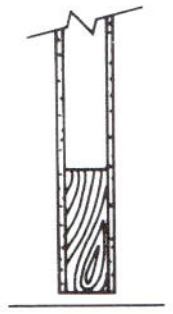 1¾-in. hollow core wood door	17 20 (w/gasketing)	1.5	1¾	—
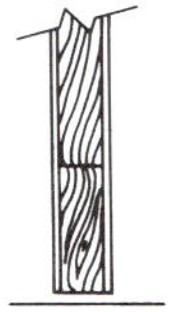 1¾-in. solid core wood door	20 26 (w/gasketing)	4	1¾	—
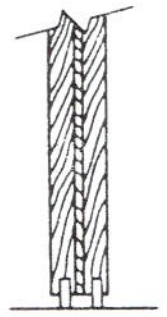 1¾-in. rated acoustical wood door	38	8	1¾	—
4-in. rated acoustical metal door	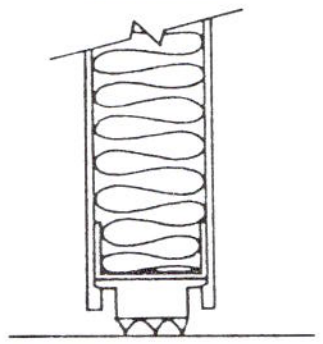 53	23	4	—
Windows				
3/32-in. single glass	 27	2	3/32	—
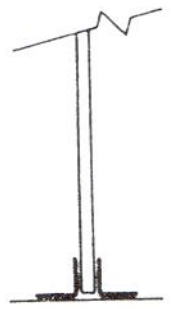 ¼-in. single glass	30	4	¼	—

TABLE 2.3 (*Continued*)

CONSTRUCTION DESCRIPTION	SOUND TRANSMISSION CLASS[b] (STC)	SURFACE WEIGHT (PSF)	OVERALL THICKNESS (IN.)	FIRE-RESISTIVE RATING[b] (HR)
$\frac{1}{4}$-in. single laminated glass	33	4	$\frac{1}{4}$	—
$\frac{3}{8}$-in. single laminated glass	36	5	$\frac{3}{8}$	—
$\frac{1}{2}$-in. thermal insulating glass ($\frac{1}{8}$-in. glass, $\frac{1}{4}$-in. airspace $\frac{1}{8}$-in. glass)	26	4	$\frac{1}{2}$	—
$2\frac{1}{2}$-in. double glass ($\frac{1}{4}$-in. glass, 2-in. airspace, $\frac{1}{4}$-in. glass)	42	8	$2\frac{1}{2}$	—
$4\frac{1}{2}$-in. double glass ($\frac{1}{4}$-in. glass, 4-in. airspace, $\frac{1}{4}$-in. glass)	45	8	$4\frac{1}{2}$	—

[a] The performance data in this table have been compiled from a variety of sources including the editor's and chapter author's files. They are intended as a guide, *not* a substitute for laboratory or field data from other reliable sources.
[b] Sound isolation and fire-resistive ratings are sensitive to variations in details of erection of the partition system from the sample tested in the laboratory. Values given in this table are for general guidance in determining performance ranges. In preparing contract specifications refer to the laboratory test reports for specific constructions.

Source: W. J. Cavanaugh, "Acoustical Control in Buildings," in *Building Construction: Materials and Types of Construction*, 5th ed., W. C. Huntington and R. E. Mickadeit, Eds., Table 10.2, pp 398–407, Wiley, New York, 1981.

TABLE 2.4 Impact and Airborne Sound Isolation Performance Data for Selected Floor/Ceiling Assemblies[a]

CONSTRUCTION DESCRIPTION	SOUND TRANSMISSION CLASS[b] (STC)	IMPACT INSULATION CLASS[b] (IIC)	SURFACE WEIGHT (PSF)	OVERALL THICKNESS (IN.)	FIRE-RESISTIVE RATING[b] (HR)
3-in. reinforced concrete slab, exposed ceiling, vinyl tile floor finish	45	42	35	$3\frac{1}{8}$	$\frac{3}{4}$
Above w/"cushion" foam-backed tile floor finish	45	49	35	$3\frac{3}{16}$	$\frac{3}{4}$
Above w/$\frac{1}{4}$-in. carpet on underpadding	45	70	35	$3\frac{3}{4}$	$\frac{3}{4}$
3-in. reinforced concrete slab with $\frac{5}{8}$-in. gypsum board ceiling on resilient channels, $\frac{1}{2}$-in. fiberglass blanket between channels, vinyl tile floor finish	56	51	40	$4\frac{1}{8}$	1
Above w/$\frac{1}{4}$-in. carpet on underpadding	56	70	40	$4\frac{1}{4}$	1
5-in. reinforced concrete slab, exposed ceiling, vinyl tile floor finish	51	46	55	$5\frac{1}{8}$	2
Above w/"cushion" foam-backed tile floor finish	51	50	55	$3\frac{3}{16}$	2
Above with $\frac{1}{4}$-in. carpet on underpadding	51	70	55	$5\frac{1}{4}$	2

TABLE 2.4 (*Continued*)

CONSTRUCTION DESCRIPTION	SOUND TRANSMISSION CLASS[b] (STC)	IMPACT INSULATION CLASS[b] (IIC)	SURFACE WEIGHT (PSF)	OVERALL THICKNESS (IN.)	FIRE-RESISTIVE RATING[b] (HR)
$2\frac{1}{2}$-in. concrete over 12-in. open-web steel joists, $\frac{5}{8}$-in. gypsum board ceiling, vinyl tile floor finish	57	45	40	$15\frac{1}{8}$	1 (est.)
Above w/$\frac{1}{4}$-in. carpet on underpadding	57	70	40	$15\frac{1}{4}$	1 (est.)
Above w/2 layers $\frac{1}{2}$-in. gypsum board ceiling and 2-in. cavity insulation	65	71	45	$15\frac{3}{4}$	1 (est.)
18-in particle board on $\frac{5}{8}$-in. plywood subfloor on 2 × 8 wood joists 16 in. o.c. (no ceiling finish)	25	23	8	$8\frac{1}{2}$	—

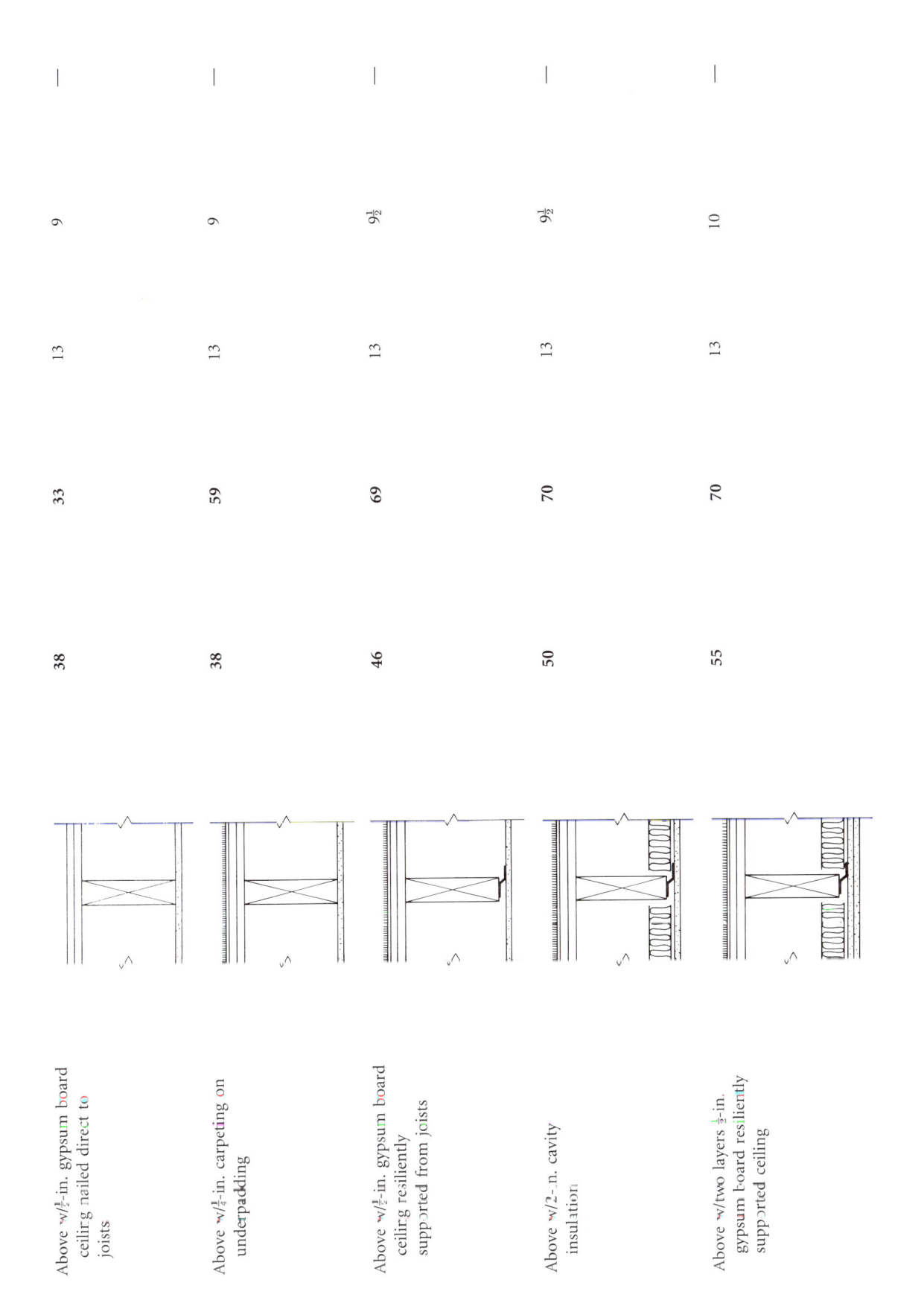

Above w/½-in. gypsum board ceiling nailed direct to joists		38	33	13	9
Above w/¼-in. carpeting on underpadding		38	59	13	9
Above w/½-in. gypsum board ceiling resiliently supported from joists		46	69	13	9½
Above w/2-in. cavity insulation		50	70	13	9½
Above w/two layers ½-in. gypsum board resiliently supported ceiling		55	70	13	10

TABLE 2.4 (*Continued*)

CONSTRUCTION DESCRIPTION	SOUND TRANSMISSION CLASS[b] (STC)	IMPACT INSULATION CLASS[b] (IIC)	SURFACE WEIGHT (PSF)	OVERALL THICKNESS (IN.)	FIRE-RESISTIVE RATING[b] (HR)
$\frac{5}{8}$-in. particle board on $\frac{5}{8}$-in. plywood subfloor on 2 × 8 wood joists, 2 × 6 separated joists support $\frac{1}{2}$-in. gypsum board ceiling, 2-in. cavity insulation, vinyl tile floor finish	50	45	15	$11\frac{1}{4}$	—
Above w/$\frac{1}{4}$-in. carpet on underpadding	50	75	15	$11\frac{1}{2}$	—
2-in. lightweight concrete on $\frac{5}{8}$-in. plywood subfloor on 2 × 8 wood joists, $\frac{1}{2}$-in. gypsum board ceiling nailed direct to joists, vinyl tile floor finish	46	41	33	$10\frac{1}{2}$	—
Above w/$\frac{1}{4}$-in. carpet on underpadding	46	61	33	$10\frac{3}{4}$	—

Construction		STC		IIC	
Above w/resiliently supported $\frac{1}{2}$-in. gypsum board ceiling and 2-in. cavity insulation		51	74	33	$10\frac{3}{4}$	—
6-in. reinforced concrete slab with "floated" wood floor on fiberglass mattress		55	57	65	$8\frac{1}{2}$	3
6-in. reinforced concrete slab with "floated" 3-in. concrete slab on 2-in. fiberglass mattress		65	67	95	11	4
Above w/resiliently suspended $\frac{5}{8}$-in. gypsum board ceiling 8 in. below slab		75	70	100	$19\frac{1}{2}$	74

[a] The performance data in this table have been compiled from a variety of sources including the editor's and chapter author's files. They are intended as a guide, *not* a substitute for laboratory or field data from other reliable sources.

[b] Sound isolation and fire-resistive ratings are sensitive to variations in details of erection of the floor/ceiling assembly from the sample tested in the laboratory. Values given in this table are for general guidance in determining performance ranges. In preparing contract specifications, refer to laboratory test reports for specific constructions.

Source: W. J. Cavanaugh, "Acoustical Control in Buildings" in *Building Construction: Materials and Types of Construction*, 5th ed., W. C. Huntington and R. E. Mickadeit, Eds., Table 10.3, pp. 409–413, Wiley, New York, 1981.

REFERENCES AND FURTHER READING

1. *Classification for Rating Sound Insulation,* ASTM E413-87(1994) (Reapproved 1994).
2. *Test Method for Laboratory Measurement of Airborne Sound Transmission Loss of Building Partitions,* ASTM E90-97.
3. *Test Method for Measurement of Airborne Sound Insulation in Buildings,* ASTM E336-97.
4. *Test Method of Laboratory Measurement of Impact Sound Transmission Through Floor-Ceiling Assemblies Using the Tapping Machine,* ASTM E492-90(1996).
5. *Test Method for Sound Absorption and Sound Absorption Coefficients by the Reverberation Room Method,* ASTM C423-90a.
6. *Standard Test Method for Impedance and Absorption of Acoustical Materials by the Impedance Tube Method,* ASTM C384-95.
7. *Practices for Mounting Test Specimens During Sound Absorption Tests,* ASTM E795-93.
8. *Test Method for Airborne Sound Attenuation Between Rooms Sharing a Common Ceiling Plenum,* ASTM 1414-91a(1996).

(Note: Referenced Standards 1-8 above are available from the American Society of Testing and Materials (ASTM) 100 Barr Harbor Drive, West Conshohocken, PA 19428)

Beranek, L. L., Ed., *Noise and Vibration Control,* McGraw-Hill, New York, 1971.

Beranek, L. L., *Acoustics,* McGraw-Hill, New York, 1954.

Egan, M. D., *Architectural Acoustics,* McGraw-Hill, New York, 1988.

Egan, M. D., *Concepts in Architectural Acoustics,* McGraw-Hill, New York, 1972. U.S. Gypsum Company, *Gypsum Construction Handbook,* United States Gypsum Company, Chicago, 1982.

Harris, C. M., Ed., *Handbook of Noise Control,* McGraw-Hill, New York, 1979.

Jones, R. S., *Noise and Vibration Control in Buildings,* McGraw-Hill, New York, 1984.

Kinsler, K. E., and Fry, A. R., *Fundamentals of Acoustics,* Wiley, New York, 1962.

Knudsen, V. O., and Harris, C. M., *Acoustical Designing in Architecture,* American Institute of Physics, New York, 1978.

Kuttruff, H., *Room Acoustics,* Wiley, New York, 1973.

Lawrence, A., *Architectural Acoustics,* Applied Science, Barkin, Essex, England, 1970.

Lord, P., and Templeton, D., *Detailing for Acoustics,* Architectural Press, London, 1983.

Office of Noise Control, *Catalog of STC and IIC Ratings for Wall and Floor/Ceiling Assemblies,* Office of Noise Control, California Department of Health Services, Berkeley, CA, 1980.

Parkin, P. H., and Humphreys, J. R., *Acoustics, Noise and Buildings,* Faber and Faber, New York, 1979.

Rettinger, M., *Acoustic Design and Noise Control,* Chemical Publishing, New York, 1972.

Rosenberg, C. J., "Sound Control," pp. 42–48 in *Architectural Graphics Standards,* 9th ed. Hoke, J. R., E.I.C. Wiley, New York, 1994.

Sabine, W. C., *Collected Papers on Acoustics,* Dover, Mineola, NY, 1964, reissued by Peninsula Publishing, Los Altos Hills, CA, 1993.

Templeton, D., and Saunders, D., *Acoustic Design,* Von Nostrand Reinhold, New York, 1987.

U.S. Department of Health, Education and Welfare, *Compendium of Materials for Noise Control,* U.S. Department of Health, Education and Welfare, Washington, DC, June, 1975, HEW Publication No. (NIOSH) 75-165.

Yerges, L. F., *Sound, Noise and Vibration Control,* Van Nostrand Reinhold, New York, 1969.

Case Study

DUKE UNIVERSITY CHAPEL: A LESSON ON ACOUSTICAL MATERIALS*

It is a historical fact that many of the great spaces built in the United States between the 1880s and the 1940s would not exist if it were not for Rafael Guastavino, a Spaniard who brought with him a thorough knowledge of Catalan vaulting techniques, which allowed him to build structures as diverse as the vaults of the Metropolitan Museum and supports for the Henry Hudson Parkway. But it was the younger Guastavino, Rafael II, who, in collaboration with Harvard University's Wallace Clement Sabine, produced the material *Akoustolith*—a porous mix of Portland cement and pumice particles that looked like stone, yet absorbed sound, by design. It was a brilliant invention at a time when stone was the material of choice and "good acoustics" meant that there was little reverberance to confuse the listener. Against this backdrop, it is hardly surprising that the interior of more than a few monumental spaces—among them, the Duke University Chapel—was fashioned out of Akoustolith.

The neo-Gothic chapel, designed by Horace Trumbauer of Philadelphia, was built between 1930 and 1932. The building is cruciform in plan (see Figure 2.14) with the following interior dimensions: combined nave and chancel length of 264 ft; widths of 30 ft in the chancel, 54 ft in the nave, and 112 ft at the transepts. The attached Memorial Chapel is approximately 54 ft long and 26 ft wide, and the octagonal narthex is approximately 30 ft wide. The crown of the crossing vault is 75 ft from the floor, the nave vaulting centerline is 73 ft from the floor, and the Memorial Chapel vaulting centerline is 50 ft from the floor. The volume of the space is approximately 1,000,000 ft^3.

Structurally, the building is one of the purer examples of revival-style architecture in that it makes use of steel only as a means of transferring thrust to exterior buttresses. The columnar supports are of limestone, as are the vaulting ribs, arches, and window mullions and quoins. From floor level to a height of about 8 ft, the nave sidewalls are finished in limestone, from which point they are finished in a material composed of one-inch Akoustolith tile bonded to a 1-in. slab of concrete. This material covers the remainder of the walls in the nave, narthex, and Memorial Chapel. It also serves as a soffit in the vaulting throughout the building. There are approximately 25,000 ft^2 (almost half the total surface area) of Akoustolith in the room.

In 1969, Duke University contracted with D. A. Flentrop of Holland to build a new organ at the nave–narthex junction and it agreed to improve the acoustics—that is, increase the reverberance—according to the recommendations of Bolt Beranek and Newman, Inc. (BBN) and to the satisfaction of Flentrop.

A preliminary examination of the chapel revealed that the only reasonable means of achieving the goal was to seal the porous surface of the Akoustolith. At that time, the only known treatment for sealing Akoustolith tile had been a somewhat limited project undertaken at Riverside Church in New York City, in consultation with BBN. Thus, while the nature of the treatment for Duke may have been clear, the prescribed formula was not. Also, the chapel administration was told at the outset that improvements of the acoustics for music would degrade hearing conditions for speech, thus necessitating the replacement of the existing speech reinforcement system with a new system using distributed directional loudspeakers and time delays.

After consulting a member of the Civil Engineering faculty at MIT in October, 1969, it was concluded that an acrylic emulsion (Rhoplex) manufactured by Rohm and Haas would be the most appropriate sealant. After considerable experimentation, impedance tube testing had shown that it took two undiluted coats of the emulsion to seal the surface so its absorption coefficient matched that of the limestone that covered the rest of the chapel interior. Aesthetically, however, this solution left something to be desired in that there was a glossy finish on the treated surfaces. Subsequent investigation by the Kyanize Paint Company resulted in a satisfactory second coat of sealant having a matte finish that matched the limestone. As a practical matter, it was not feasible to extrapolate spreading rates from the 4-in. tile sample used in the impedance tube tests, so the painting contractor was instructed that the treated Akoustolith "must have no remaining pin holes or voids of any sort, and must really be filled."

Before the walls were treated, it was felt that both archival interest and the necessity to measure acoustical change required assessment of the existing chapel acoustics. Accordingly, reverberation time experiments using balloon bursts and organ chords in the chancel and narthex areas were conducted in September, 1970. (See Figure 2.15.)

In March, 1971, a second reverberation time measurement was taken (shown in Figure 2.15). At this point the nave had been treated according to the initial specification with two coats of sealer, but the chancel, narthex, and Memorial Chapel were yet to be painted, and the crossing area was obstructed by the painters' scaffolding. The measurement confirmed Flentrop's observation that the change was far less than he had expected and certainly not satisfactory for the organ he was to build. In fact, the data showed that the absorption coefficient of the Akoustolith at 500 Hz had been reduced from approximately 0.5 to 0.25—a far cry from the limestone's coefficient of 0.02.

---

*This case study is based on a paper titled "Gothic Sound for the Neo-Gothic Chapel of Duke University" by Robert B. Newman and James G. Ferguson, Jr., which was presented at the 98th meeting of the Acoustical Society of America, Salt Lake City, November, 1979.

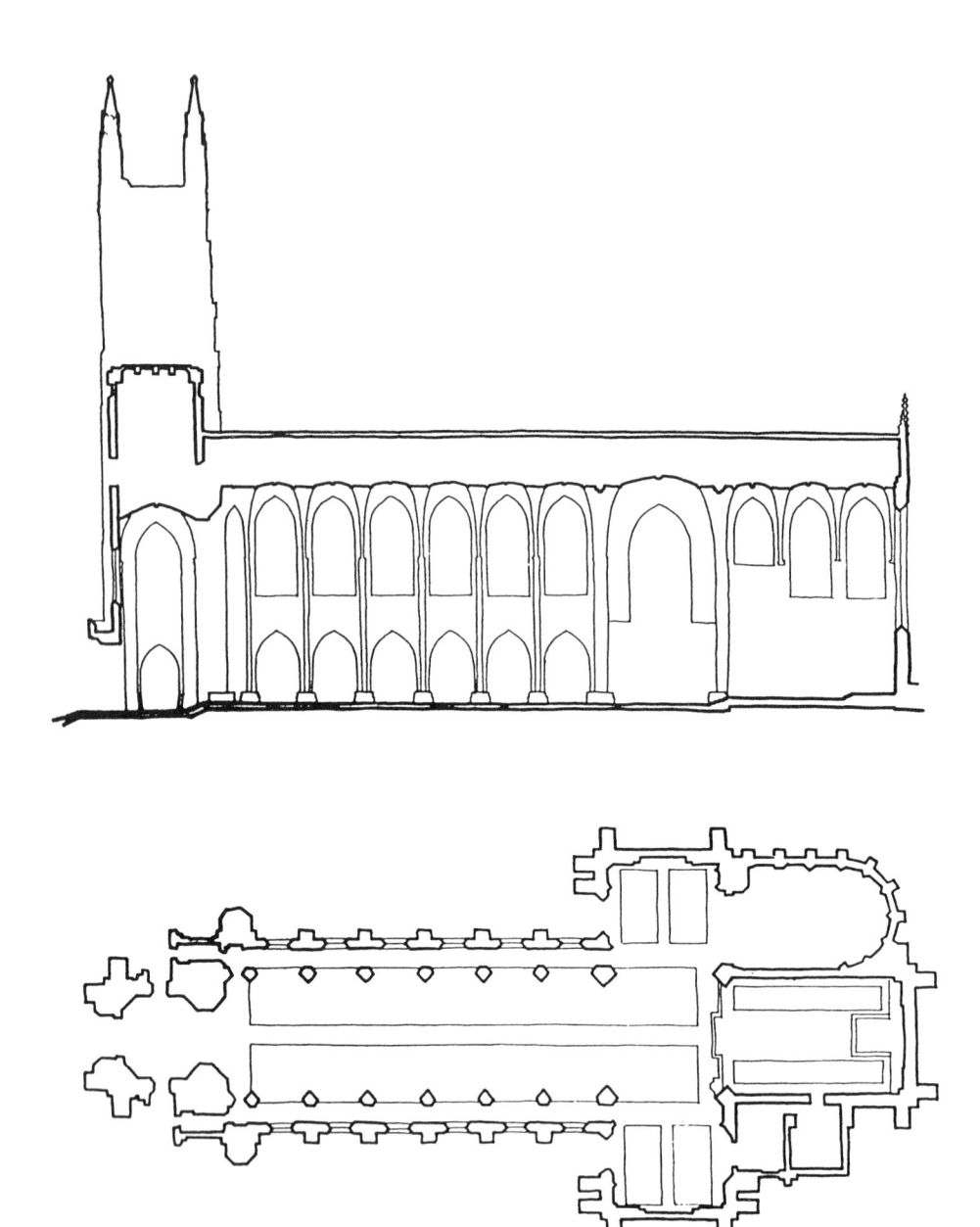

Figure 2.14 Section and plan of the Duke University Chapel.

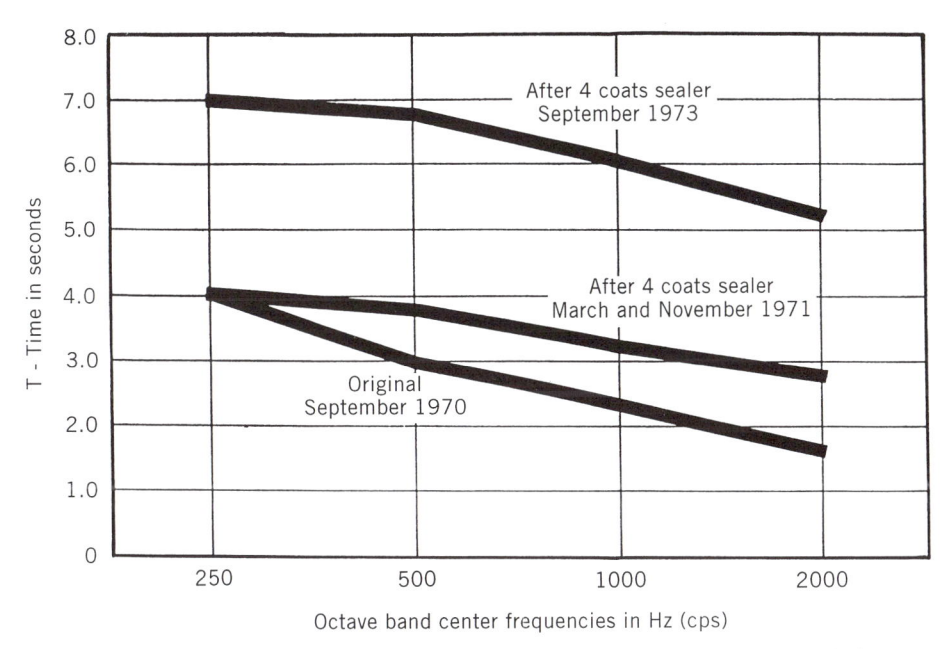

Figure 2.15 This graph shows the difference in measured reverberation times before and after the application of sealer.

It was clear that the all too common difference between laboratory and field procedure was at fault. While the painters had attempted to achieve a complete visual seal, discussion with the foreman revealed that achieving this seal by using two coats of paint was a practical impossibility for painters lying on their backs on a slightly unsteady scaffold 75 ft off the stone floor in 90°F + temperature. In addition to dealing with less than optimal working conditions, the painters were also concerned about the paint "running" down the walls and creating visual problems. All of these considerations led inadvertently to less paint being applied to the in situ Akoustolith than had been applied to the test samples—both the one prepared by BBN and the one prepared by the painting contractor, which was also tested and found satisfactory.

When the university administration became aware of the developing situation, they were understandably reluctant to commit additional funds to a costly project whose outcome seemed unclear. The acoustics had been improved following the recommendations of BBN—if not to the satisfaction of Flentrop. Without the vision of three men, the Duke chapel project might well have ended here. However, former University Chaplain Howard Wilkinson, former Chancellor John Blackburn, and University Architect James Ward resolved the quandary of how to proceed.

Using impeccable experimental technique, Wilkinson devised a number of trial applications of the sealant in order to come up with a practical solution for the apparent impasse. By simulating the interim condition of the painted tile, he was able to demonstrate that two additional coats—a heavy one of base sealer and a light one of the flat sealer—would provide a visual seal. Furthermore, his studies provided a practical directive that the painters could follow as well as a criterion that would enable the university to monitor the quality of the second treatment.

Although Wilkinson's experimental evidence seemed conclusive, the university was cautious and asked that an additional set of acoustical measurements be made on the test patches on the chapel walls. Accordingly, a field impedance tube test was performed in January, 1973, on samples treated with additional coats of sealer. As Figure 2.16 indicates, the experimental data show that the two additional coats of paint resulted in absorption coefficients slightly lower than those of limestone, but the difference is within the range of measurement error and therefore not significant. On the basis of these confirmatory acoustical measurements, a contract was let for the retreatment of the entire Akoustolith interior according to Wilkinson's findings—one heavy coat of base sealant and one light coat of flat sealant.

When the second treatment was concluded, a final reverberation time measurement was taken by BBN in September, 1973 (see Figure 2.15). There was no doubt that the desired

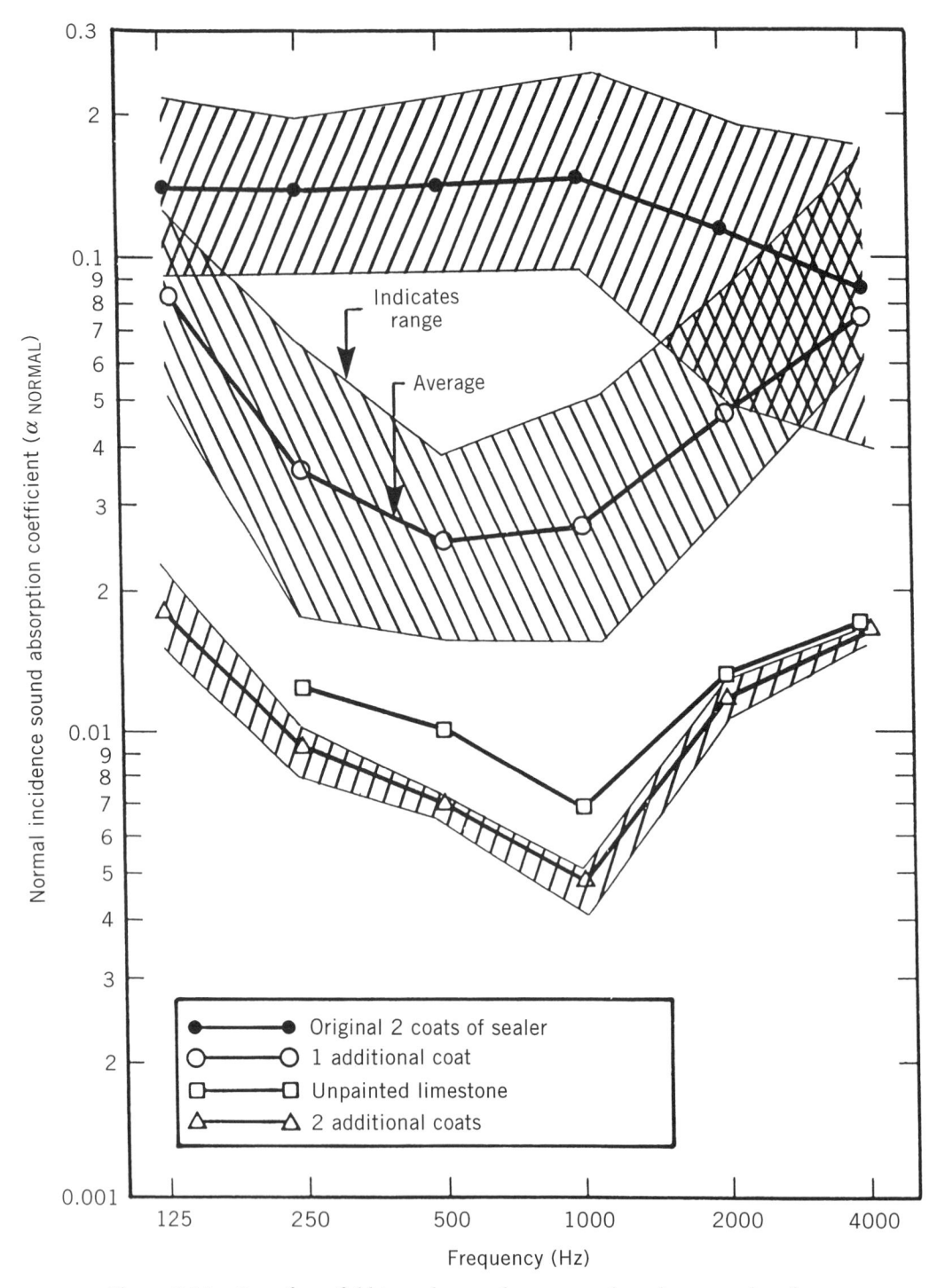

Figure 2.16 Data from field impedance tube tests conducted on samples of Akoustolith *treated with varying coats of sealer.*

dramatic results had been achieved. Reverberation time now ranged from slightly over 5 sec at 2 kHz to 6.75 sec at 500 Hz and 8 sec at 40 Hz. The test data confirmed what the choir members and organists already knew—Duke University Chapel had been transformed into a liturgical space without equal in the United States. Gothic sound had come to a neo-Gothic building.

Since the Duke project, the acoustics of several other spaces whose interior was finished in Akoustolith have been transformed using much the same technique. These include St. Thomas Church in New York City (actually, St. Thomas was finished in Rumford tile, which preceded Akoustolith), the Princeton College Chapel, and The Cathedral of St. Philip in Atlanta.]

3

Building Noise Control Applications

Gregory C. Tocci

3.1 INTRODUCTION

This chapter addresses the most common noise control issues confronted in building design. To this end the chapter presents methods of analysis and discusses the implementation of noise control products and materials discussed in Chapters 1 and 2. The material in this chapter is divided into 3 sections: acoustical analysis, building noise criteria, and noise control methods. Section 3.2 formulates noise problems from the perspective of source, path, and receiver. It continues by discussing the application of this perspective to open- and closed-plan speech privacy. Section 3.3 presents discussions of standards and principal organizations publishing standards, guidelines, and issuing recommendations for the acoustical design of buildings. The chief descriptors for evaluating noise in building spaces—NC, NCB, and RC ratings—are described in detail. And finally, acoustical requirements of the 3 chief building codes used in the United States—BOCA, ICBO, and SBCCI—are presented. Section 3.4 discusses noise barriers, building partition, floor/ceiling, and exterior envelope design. It also includes an extensive discussion of noise control methods for building ventilation systems and piping systems. This section is concluded with a discussion of vibration isolation systems.

The scope of these discussions is intended to provide the reader with an overview of acoustical design practice and how the acoustical design of buildings is addressed. It is not intended to provide analysis methods and algorithms used in acoustical analysis, but it is thoroughly referenced to enable the reader to obtain more detail from publications widely available to the profession.

3.2 ACOUSTICAL ANALYSIS

Source, Path, and Receiver . . . a Systems Approach

The analysis of sound in buildings, whether desired or undesired (noise), can be most expeditiously viewed from the standpoints of source, path, and receiver. This divides the analysis of an acoustical problem into components that can be more easily addressed.

Chief sources of noise in buildings can be divided into 3 categories: (1) sources associated with occupant activity and office equipment, (2) sources associated with the operation of building services, and (3) sources of environmental sound from outside a building. Similarly, these sources of noise can also be sources of vibration that may interfere with occupant activities or comfort. A source of noise in a building may or may not also be a source of vibration. Conversely, a source of vibration in a building may or may not be a source of significant noise.

Sound and vibration produced by occupants includes door slams, footfall, cart roll-bys, conversation, paging, radios, and warning signals. Some of these are almost always undesirable, while others are most often desirable, or at least needed to be heard.

Door slams are an everyday occurrence with which everyone is familiar. When a door slams nearby, the loud impact noise is short in duration and characterized by sound energy over a wide frequency range. A door slam at a more remote location in a building is much less loud and sounds different because it is predominately low-frequency noise. Hence, besides being much less loud, the low-frequency sound of a door slam at a remote location in a building is much less likely to interfere with speech intelligibility. This can be acceptable in many types of buildings, but in others where extraneous sound is unwanted, such as in studios or rooms dedicated to listening, even distant impact sound of a door slam can be unacceptable.

Footfall is a nemesis in multiresidential buildings. Among the most sensitive to footfall, are people moving from single-family homes who may only be familiar with footfall sound produced by other family members. Once in a multiresidential building, they often are surprised at hearing noise produced by persons in other units, even at low levels. Footfall can be described as having two distinctive characteristics. These are the "click" of hard heels impacting hard floor surfaces and the "boom" of bare feet or soft-soled shoes on hard floors or even on carpet. Each type of footfall sound is controlled by different means, as discussed below.

Paths of Sound Transmission

There are two paths of sound transmission between spaces or more specifically between noise sources and spaces. These are airborne and structureborne. An example of an airborne path is one where a machine excites air in the source room. The air in the source room excites the demising wall between the source room and the receiving room, causing bending waves in the demising wall. These bending waves correspondingly excite the air on the receiving room side as they propagate over the surface of the demising wall. A second example is sound transmitted from the same source room, through small openings in a demising wall into the receiving room. In both cases, airborne sound excites a structure that correspondingly excites air on the receiving side or transmits through a wall opening. Structureborne sound most often begins as vibration transmitted from a machine to a building structure. Vibration propagates through the structure to other spaces and reradiates as sound in these other spaces. These other spaces may not be directly adjacent to the space containing the source, thus sometimes making determining the source of structureborne sound difficult.

Often, sound transmission between spaces is by both airborne and structureborne paths. In such cases, it may be necessary to quantify the relative importance of sound transmitted along both paths.

Acoustical Privacy Applications

Closed-Plan Spaces

Studies of acoustical privacy conditions in offices, schools, dormitories, hotel function rooms, and the like have generally confirmed that people are annoyed and feel a sense of loss of privacy when the sound transmitted from an adjacent space reveals even small amounts of intelligible information concerning the adjacent activity. In considering transmitted speech from a source room to a receiver room, speech privacy is the opposite of speech intelligibility. The lower the speech intelligibility the greater the speech privacy. Speech privacy has been formulated as a signal-to-noise problem. The lower the signal-to-noise rating the less the speech intelligibility, the greater the speech privacy. There are 6 factors that influence speech privacy between enclosed rooms:

1. *Background Sound Level in Listener's Space (the receiving room).* The background sound in a receiving room acts to mask unwanted speech sounds and renders them less intelligible.
2. *Strength of Sound Source (vocal effort).* Obviously, the louder the speech signal, the more likely that it will be intelligible in an adjacent room (the receiving room).
3. *Amount of Sound Absorption in Receiving Room.* The less the reverberant buildup of speech sound in the receiving room the lower the speech signal level, and consequently the less the speech intelligibility.
4. *Relative Sizes of Source and Receiving Rooms.* The larger the receiving room relative to the source room, the lower the speech signal level in the receiving room.
5. *Sound Transmission Characteristics of Intervening Construction Separating Two Spaces (i.e., wall between the source and receiving rooms).* Obviously, the higher the sound transmission loss of the wall separating the source and receiving rooms, the lower the speech signal in the receiving room.
6. *Required Speech Privacy.* As discussed below, this is a rating that accounts for the amount of speech privacy required in specific circumstances. The higher this rating, the lower the speech signal must be relative to the background sound level in order to ensure adequate masking of speech signals to minimize speech intelligibility.

Extensive laboratory and field studies in speech privacy have led to the development of speech privacy acceptability criteria. These criteria are the basis for an analysis method that uses the 6 factors described to evaluate the acceptability of speech privacy. One such analysis method, extended to include generally higher levels of vocal effort and other sources in addition to speech, is shown in Figure 3.1.

The speech privacy analysis method allows one to quantify each of the 6 principal factors involved in typical room-to-room situations in buildings. The summation of these factors yields a single privacy rating number. Each factor represents one component of the source–path–receiver model of a speech privacy problem. The privacy rating number is then compared with case history experience to determine the level of satisfaction or dissatisfaction a typical building occupant is likely to express. The format may be used for both the analysis of existing situations and the design of new room-to-room configurations.

Figure 3.2 is an example of the use of the analysis technique applied to a typical office building. If unsatisfactory conditions are found, the given situation can be reevaluated with modified assumptions or improved components [e.g., a partition of

Figure 3.1 Acoustical privacy analysis worksheet for closed-plan offices (room to room). (from W. J. Cavanaugh, in Building Construction: Materials and Types of Construction, 5th ed., W. Huntington and R. Mickadeit, Eds., Chapter 10. Copyright © 1981 John Wiley & Sons. Reprinted by permission of John Wiley & Sons.)

WORKSHEET
CLOSED-PLAN ACOUSTICAL PRIVACY ANALYSIS

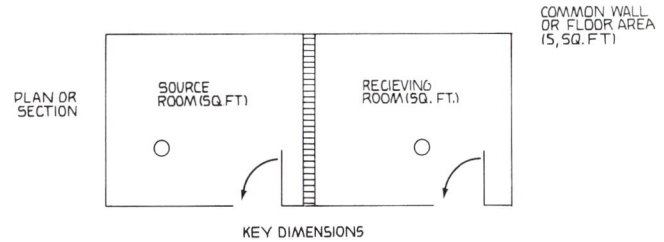

PLAN OR SECTION: SOURCE ROOM (SQ.FT) | RECIEVING ROOM (SQ.FT.)
COMMON WALL OR FLOOR AREA (S, SQ.FT)

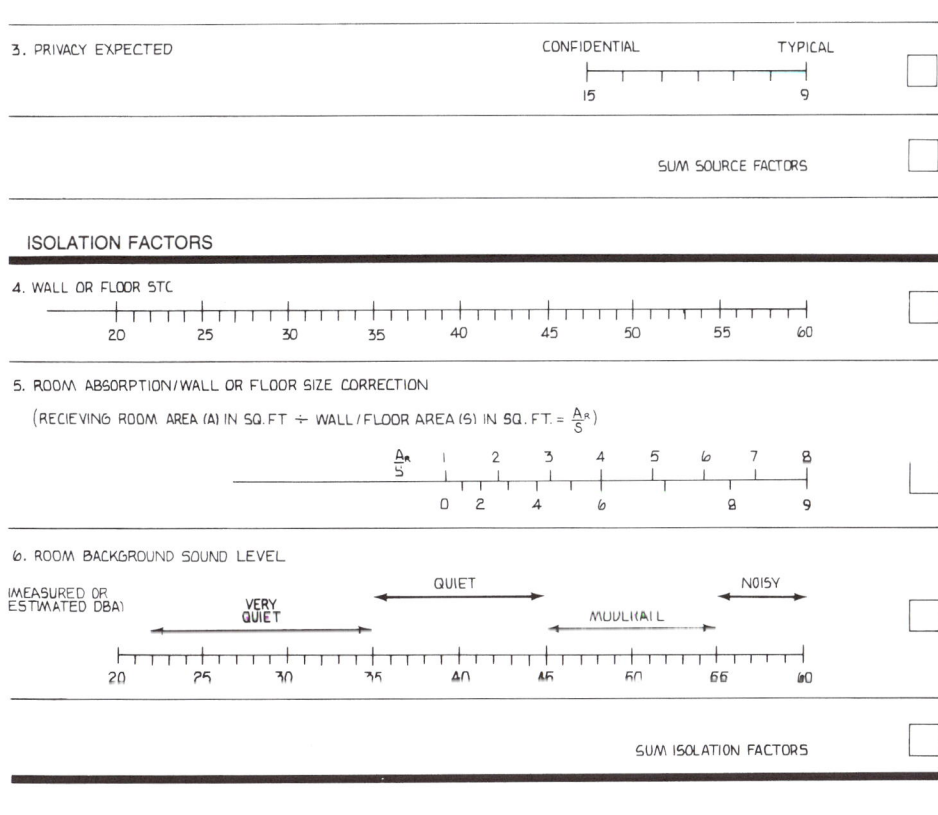

KEY DIMENSIONS

SOURCE FACTORS

1. SOUND SOURCE

ROCK BAND — DISCO 98 — LARGE BAND — SMALL BAND 88 — AMPLIFIED MOVIES / SOFT MUSIC 78 — LOUD VOICE 72 — RAISED VOICE 66 — CONVERSATIONAL VOICE 60 — SOUND LEVEL (DBA)

2. SOURCE ROOM ABSORPTION

FLOOR AREA (SQ. FT) 125 | 250 | 500 | 1000 | 2000 | 4000
7 6 5 4 3 2 1 0 -1 -2 -3 -4 -5 -6 -7 -8

3. PRIVACY EXPECTED

CONFIDENTIAL 15 — TYPICAL 9

SUM SOURCE FACTORS

ISOLATION FACTORS

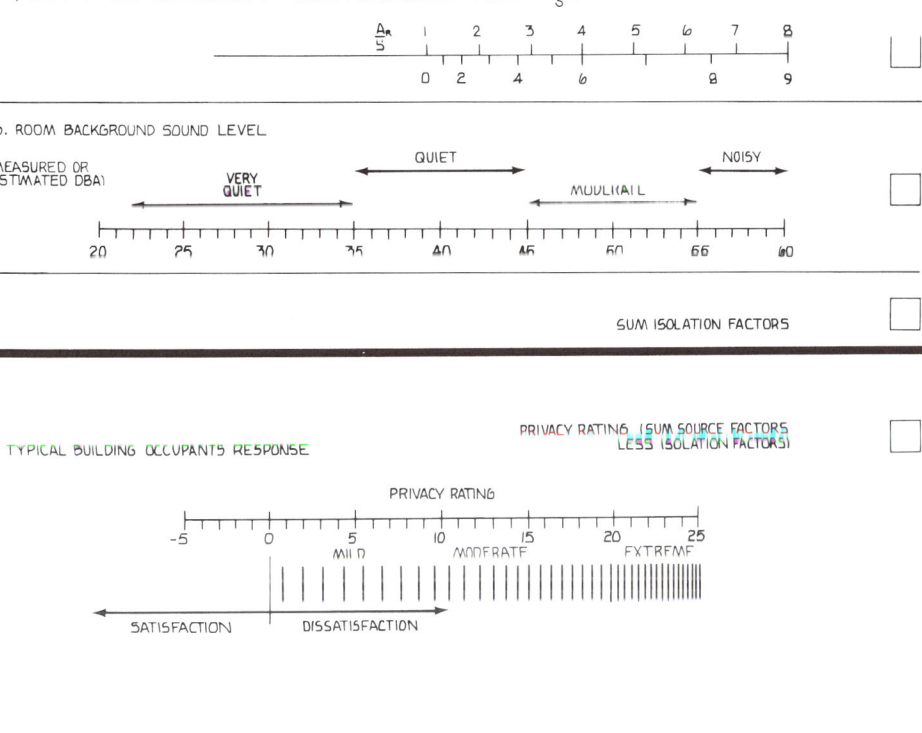

4. WALL OR FLOOR STC

20 25 30 35 40 45 50 55 60

5. ROOM ABSORPTION/WALL OR FLOOR SIZE CORRECTION

(RECIEVING ROOM AREA (A) IN SQ.FT ÷ WALL/FLOOR AREA (S) IN SQ. FT. = $\frac{A_R}{S}$)

$\frac{A_R}{S}$ 1 2 3 4 5 6 7 8
0 2 4 6 8 9

6. ROOM BACKGROUND SOUND LEVEL

(MEASURED OR ESTIMATED DBA) — VERY QUIET — QUIET — MODERATE — NOISY

20 25 30 35 40 45 50 55 60

SUM ISOLATION FACTORS

TYPICAL BUILDING OCCUPANTS RESPONSE

PRIVACY RATING (SUM SOURCE FACTORS LESS ISOLATION FACTORS)

PRIVACY RATING
-5   0   5   10   15   20   25
     MILD   MODERATE   EXTREME

← SATISFACTION | DISSATISFACTION →

EXAMPLE
CLOSED-PLAN ACOUSTICAL PRIVACY ANALYSIS

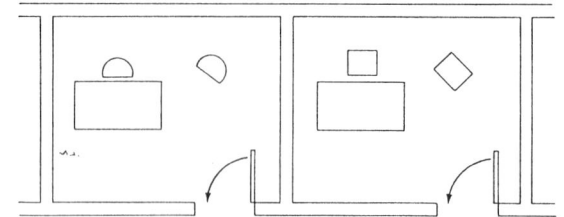

GIVEN:
SOURCE: TYPICAL CONVERSATIONAL SPEECH
SOURCE ROOM: 200 SF
RECIEVING ROOM: 200 SF
WALL AREA: 200 SF
WALL STC: 35
MEASURED
BACKGROUND SOUND: 35 DBA

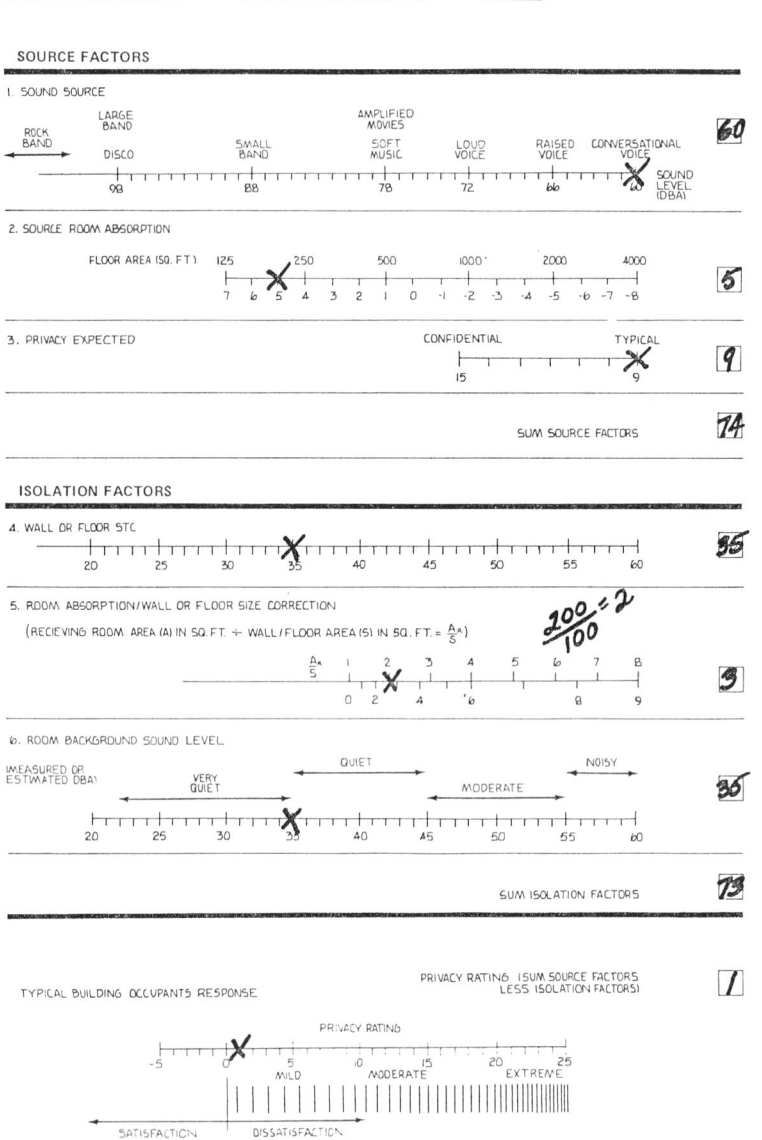

SOURCE FACTORS

1. SOUND SOURCE — 60

ROCK BAND — DISCO (98) — LARGE BAND — SMALL BAND (88) — AMPLIFIED MOVIES — SOFT MUSIC (78) — LOUD VOICE (72) — RAISED VOICE (66) — CONVERSATIONAL VOICE (X) — SOUND LEVEL (DBA)

2. SOURCE ROOM ABSORPTION — 5

FLOOR AREA (SQ. FT): 125 (X=6,5) 250 500 1000 2000 4000
7 6 5 4 3 2 1 0 -1 -2 -3 -4 -5 -6 -7 -8

3. PRIVACY EXPECTED — 9

CONFIDENTIAL (15) — TYPICAL (9) X

SUM SOURCE FACTORS — 74

ISOLATION FACTORS

4. WALL OR FLOOR STC — 35

20 25 30 35(X) 40 45 50 55 60

5. ROOM ABSORPTION/WALL OR FLOOR SIZE CORRECTION — 3

(RECIEVING ROOM AREA (A) IN SQ. FT. ÷ WALL/FLOOR AREA (S) IN SQ. FT. = $\frac{A}{S}$)

$\frac{200}{100} = 2$

$\frac{A}{S}$: 1 2(X) 3 4 5 6 7 8
0 2 4 6 8 9

6. ROOM BACKGROUND SOUND LEVEL — 35

(MEASURED OR ESTIMATED DBA)
VERY QUIET — QUIET — MODERATE — NOISY
20 25 30 35(X) 40 45 50 55 60

SUM ISOLATION FACTORS — 73

TYPICAL BUILDING OCCUPANTS RESPONSE

PRIVACY RATING (SUM SOURCE FACTORS LESS ISOLATION FACTORS) — 1

PRIVACY RATING:
-5 0(X) 5 10 15 20 25
MILD — MODERATE — EXTREME

SATISFACTION | DISSATISFACTION

Figure 3.3 Typical prefabricated music practice rooms. (Courtesy of Wenger Corporation, 1996.)

higher sound transmission class (STC) rating, higher background sound level, etc.] may be selected.

There is an increasing trend toward the use of preengineered and prefabricated sound isolation rooms. These are designed using procedures such as those above and may be used for music practice rooms, laboratory testing, industrial hearing protection booths, and so forth. Figure 3.3 shows one such application.

Open-Plan Spaces

There are obvious architectural and cost advantages for certain kinds of activities, if the individual subarea activities need not be fully enclosed with walls that extend to

Figure 3.2 Speech privacy analysis for a typical closed-plan example. The Xs mark the appropriate selections on each of the analysis scales. (from W. J. Cavanaugh, in Building Construction: Materials and Types of Construction, 5th ed., W. Huntington and R. Mickadeit, Eds., Chapter 10. Copyright © 1981 John Wiley & Sons. Reprinted by permission of John Wiley & Sons.)

the ceiling, as shown in Figure 3.4. The open-office landscape has become an increasingly popular way to provide office space for large corporations and for the major federal government builder of facilities, the General Services Administration (GSA). A significant feature in the degree of success such spaces enjoy is the acoustical environment. The open plan is really not new. Many church basement Sunday school classrooms functioned without walls. The classical bank lobby where loan officers discuss confidential financial information with customers has been around for many years. Why is it, then, that some of these spaces function adequately from the standpoint of acoustics and others do not? Distressed bank loan officers feel cheated when they move from the old monumental bank lobby where they enjoyed complete privacy to a new, low-ceilinged, heavily acoustically treated open-office landscape where they can readily hear conversations from the adjacent desks.

The basic acoustical problem is really no different from the case of adjacent enclosed spaces. Adequate privacy requires sufficient attenuation of the speech signal from the source to the listener location (low signal-to-noise ratio) where its relative annoyance will depend on the masking provided by the continuous background sound present. Figure 3.5 illustrates schematically the similarities and subtle differences between the two situations. Clearly, the containment and buildup of reverberant sound in enclosed rooms is no longer present in open-plan spaces. Sound in the open space continually falls off with distance at a rate dependent on the sound absorption of the floor/ceiling surfaces. The relatively large attenuations of the partitions of an enclosed office give way to lesser attenuation of the open-plan office where barriers just break the line of sight between the source and receiver.

The sound attenuation of noise barriers, including those used in open-plan offices, is discussed in a later section in this chapter. Briefly though, noise barriers reduce sound at the receptor location if they break the direct path between the source and receptor, that is, the line of sight. If the line of sight is not broken, little or no noise reduction results. If the barrier just grazes the line-of-sight path between the source and receiver, a 3- to 5-dB reduction can be obtained. The more the barrier penetrates

Figure 3.4 Typical open-office plan. (Courtesy of Owens-Corning Fiberglas)

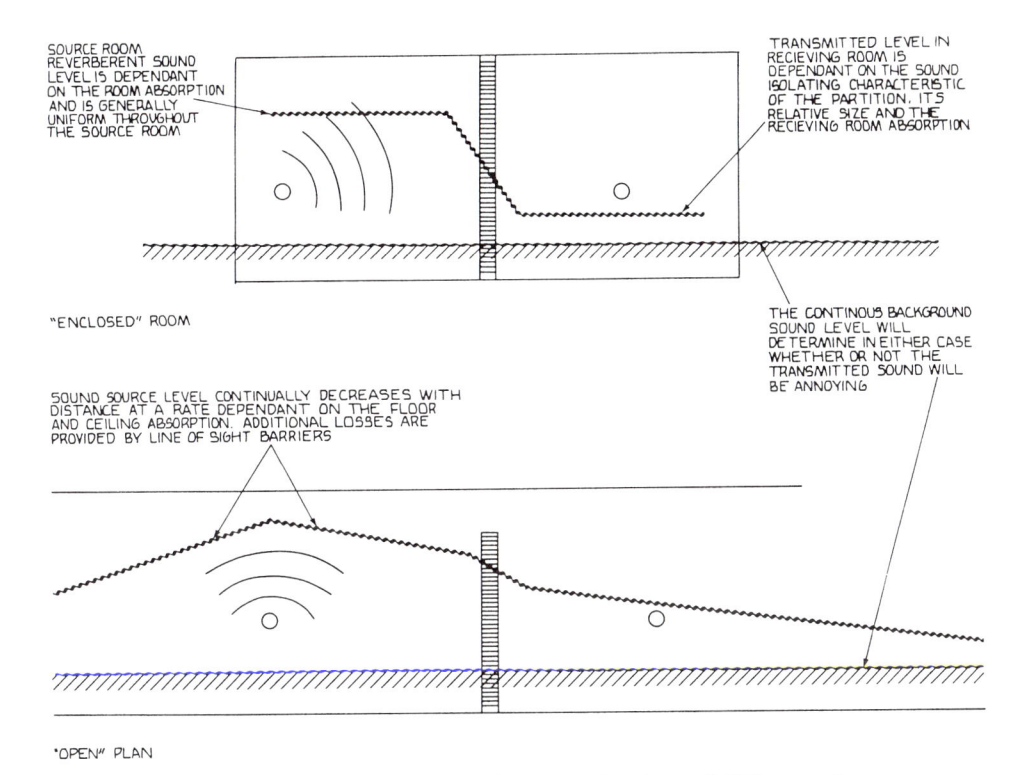

Figure 3.5 Schematic illustration of the relative similarities and differences between enclosed and open-plan arrangements. (from W. J. Cavanaugh, in Building Construction: Materials and Types of Construction, 5th ed., W. Huntington and R. Mickadeit, Eds., Chapter 10. Copyright © 1981 John Wiley & Sons. Reprinted by permission of John Wiley & Sons.)

through the line of sight, the greater the barrier noise reduction. The practical upper limit of barrier noise reduction in an open-office plan is about 10 dBA.

Figure 3.6 is an analysis worksheet for evaluating and quantifying the several factors involved in open-plan office speech privacy. These are similar to the analysis method for closed-plan offices, especially by arriving at a single speech privacy rating number. As with the closed-plan office analysis, the open-plan office privacy rating is compared with the typical response data from building case histories of speech privacy to yield the anticipated reaction of the occupants. The example in Figure 3.7 clearly shows the relative importance of each of the factors and the very important "masking" role that the background sound level plays in the privacy achieved.

Experience indicates that the background sound levels normally produced by the building air distribution system rarely provide the spatial uniformity necessary to achieve acceptable privacy conditions in typical office spaces. In fact, practically all open-plan installations use carefully designed electronic masking systems to assure uniform background sound at the proper level and with good tonal character. Typically, such systems in open-plan installations must be operated at near the upper limits of acceptability for average building occupants [about noise criteria (NC) 45 or 52 dBA]. Figure 3.8 describes criteria for the characteristic frequency spectrum shape for electronic sound-masking systems. Such criteria are an important part of the performance specifications for electronic sound-masking systems.

The Public Building Service (PBS) of the US General Services Administration (GSA) has promulgated criteria and standards for the design, specification, and evaluation of

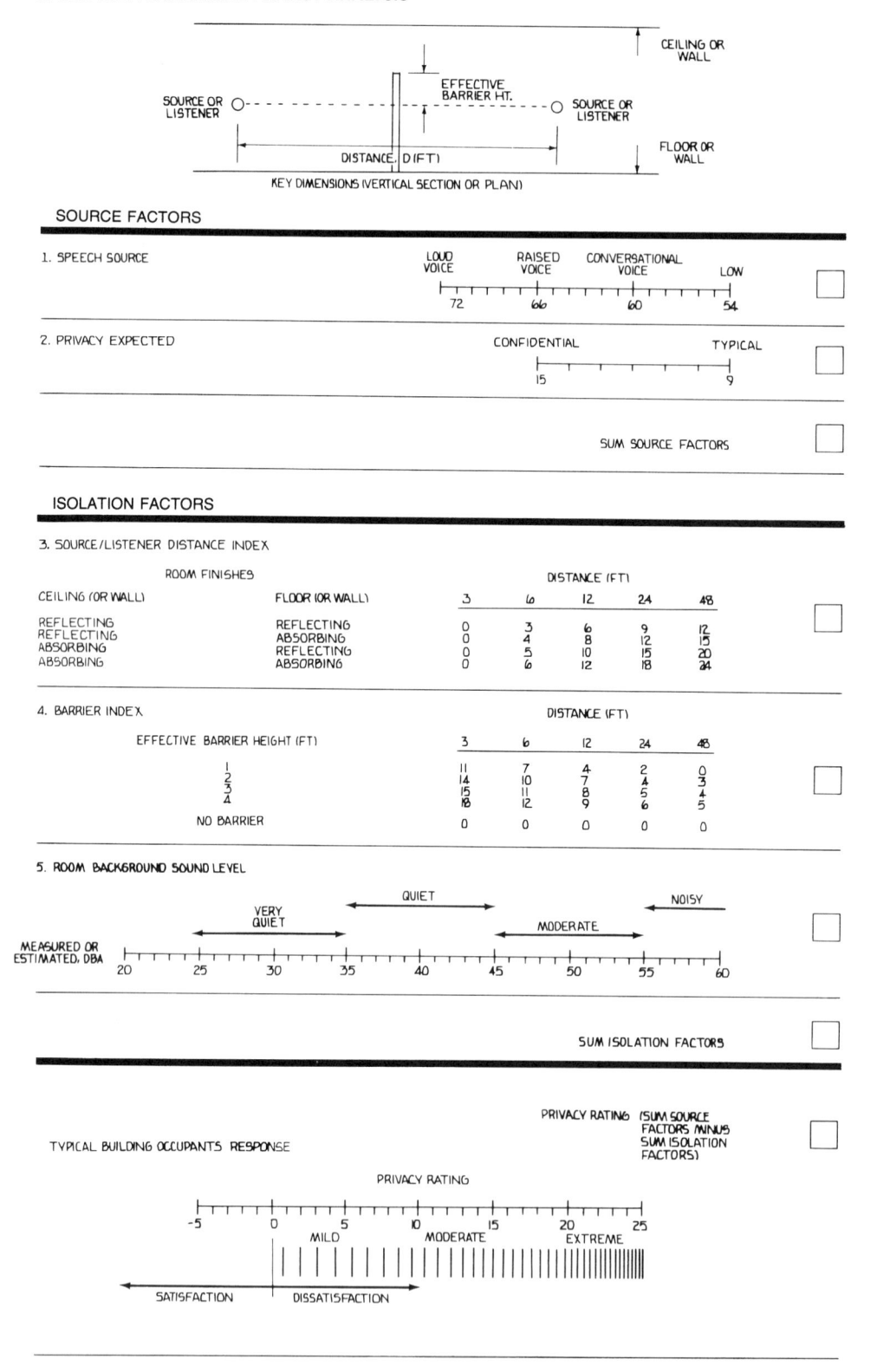

Figure 3.6 *Acoustical privacy analysis worksheet for open-plan offices. (from W. J. Cavanaugh, in* Building Construction: Materials and Types of Construction, *5th ed., W. Huntington and R. Mickadeit, Eds., Chapter 10. Copyright © 1981 John Wiley & Sons. Reprinted by permission of John Wiley & Sons.)*

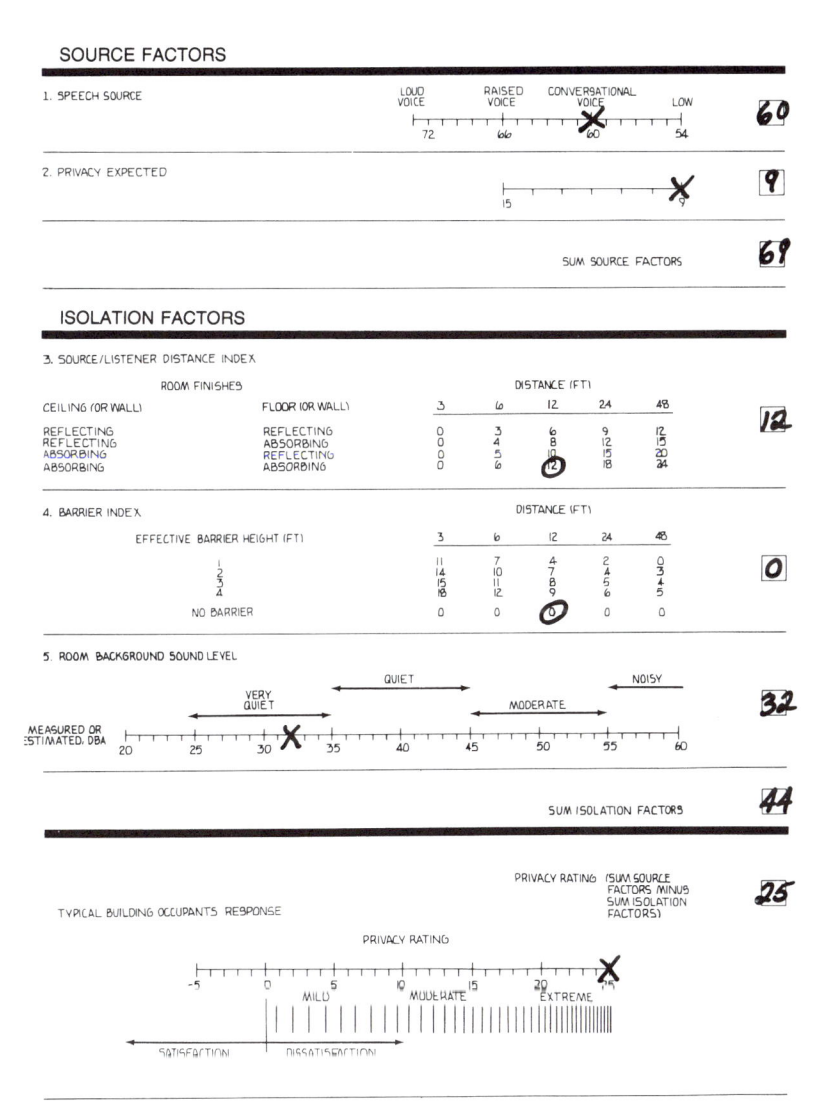

Figure 3.7 Speech privacy analysis for a typical open-plan example. Appropriate selections on each of the analysis scales are marked. (from W. J. Cavanaugh, in Building Construction: Materials and Types of Construction, 5th ed., W. Huntington and R. Mickadeit, Eds., Chapter 10. Copyright © 1981 John Wiley & Sons. Reprinted by permission of John Wiley & Sons.)

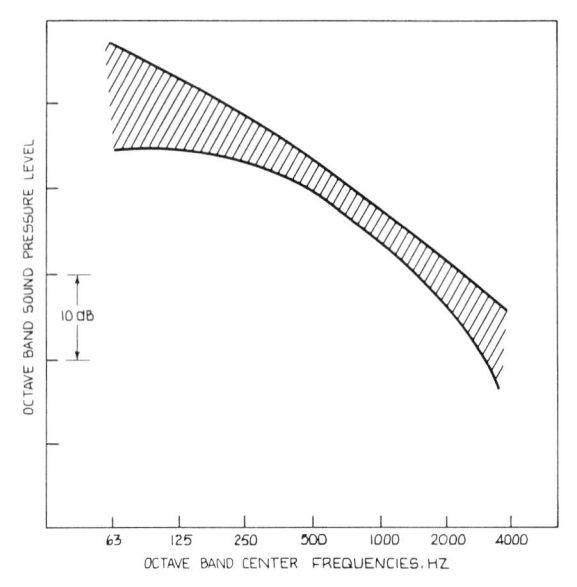

Figure 3.8 Range of recommended frequency spectrum shape for background-masking sound systems in open-plan spaces. (from W. J. Cavanaugh, in Building Construction: Materials and Types of Construction, 5th ed., W. Huntington and R. Mickadeit, Eds., Chapter 10. Copyright © 1981 John Wiley & Sons. Reprinted by permission of John Wiley & Sons.)

systems and components used for open office spaces in federal buildings. In essence, the PBS/GSA criteria of acceptability are expressed as the speech privacy potential (SPP). The SPP is the sum of the background sound level and the attenuation between typical source and listener locations. The background sound level is quantified using an NC rating modified to include only the significant speech privacy frequency bands and is designated NC'. The workstation-to-workstation sound reduction is an average attenuation value taken along the source–receiver path also rated using a modified noise insulation class (NIC') method. Satisfactory acoustical privacy with the PBS method of evaluation is taken to be when the SPP equals 60 or greater (i.e., SPP ≥ 60). For example, an office with an NC-40 background sound level in combination with a source-to-receiver sound attenuation of NIC' 20 would satisfy the PBS criterion.

3.3 BUILDING NOISE CRITERIA

Standards and Organizations

In the United States, standard organizations can be divided into two groups. The first group includes those that foster the development and unification of analysis and the unification of test methods and methods to quantify performance. The second group establishes standards of performance for specific applications. With respect to acoustics in the building industry, the first group, dedicated to analysis and test methods, includes the American Society for Testing and Materials (ASTM), the American National Standards Institute (ANSI), and the American Society of Heating, Refrigerating, and Air-Conditioning Engineers (ASHRAE). The second group of standards organizations, dedicated to establishing standards of performance for buildings, includes Building Officials and Code Administrators International (BOCA), International Conference of Building Officials (ICBO), and the Southern Building Code Congress International

(SBCCI). The first group produces standards and methods of design in buildings. The second establishes minimum building performance standards. Obviously, there are many other organizations that promote standards for the built environment. But these are among those most widely encountered in day-to-day acoustical building applications.

The American Society for Testing and Materials has developed approximately 50 standards and guides that establish methods for analyzing, measuring, and otherwise quantifying a variety of acoustical properties of building materials and systems. For example, ASTM E90 establishes a method for the laboratory measurement of sound transmission loss of materials and building systems. It defines what is to be measured, how the material will be tested, and how test results (in this case sound transmission loss data in $\frac{1}{3}$-octave bands) are to be reported. ASTM E413 is a method for classification of sound transmission loss data to determine a single-number rating called the sound transmission class (STC) rating.

With a slightly different purpose, ANSI defines scientific parameters and criteria used in acoustical analysis. For example, ANSI S12.2 establishes methods for evaluating sound in rooms. It does so by defining two distinct methods, balanced noise criteria (NCB) curves and room criteria (RC) curves. The methods evaluate octave band sound pressure level spectra measured in rooms using NCB and RC ratings. The appendix to the standard provides guidance on acceptable sound levels in a variety of types of building spaces.

On the other hand, ASHRAE is a professional organization whose mission is to disseminate to its membership heating, ventilating, and air-conditioning (HVAC) design methods and criteria considered accepted practice in the building industry. It has long preceded ANSI in defining criteria for sound inside building spaces by recommending sound levels for mechanical systems in a variety of building spaces. These recommended sound levels were first expressed as A-weighted sound levels and NC curve ratings. In the early 1980s, ASHRAE introduced RC curves and will shortly be including NCB curves for evaluating mechanical system sound in building spaces. At its June 1997 meeting ASHRAE Technical Committee (TC) 2.6—Sound and Vibration, voted to drop NC curves from the *ASHRAE Guide* but to include NCB curves as defined in ANSI S12.2 and to include a modified version of the RC method that will include new quantity, a *sound quality index*. Since the NCB and RC curves are defined in the ANSI standard and used in current *ASHRAE Guides,* they are discussed in this book. The *sound quality index* has not as yet been published, as of this writing, it has not been included.

Noise Criteria Curves

Noise criteria curves were first described by Beranek[1] and were developed on the basis of equal loudness contours. Figure 3.9 presents a set of NC curves along with a typical room sound pressure level spectrum. It was originally presumed that octave band spectra of background sound in buildings that generally follow NC curve shapes would be perceived as equally balanced in low-, mid-, and high-frequency sound energy (although this was shown not to be quite the case, leading to the development of other curve sets). NC curves continue to be the most widely used means for specifying criteria and evaluating background sound in buildings. Often, recommended noise criteria are accompanied by an equivalent A-weighted value assuming that the A-weighted sound level will approximate the NC curve spectrum. (Also, see Chapter 1.)

Rating a specific sound pressure level spectrum using NC curves is most commonly achieved using what has become known as the tangency method. By the tangency method, the NC rating of a spectrum is determined to be the value of the highest NC

112 • *Building Noise Control Applications*

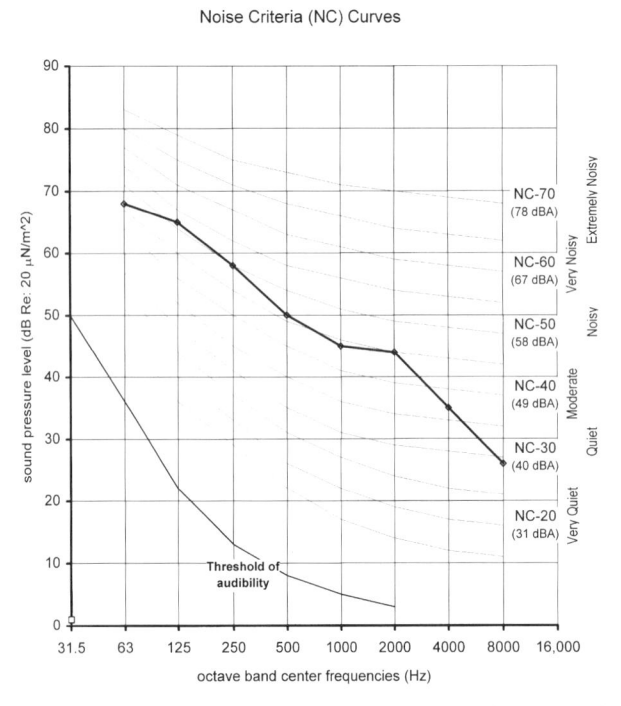

Figure 3.9 NC (noise criteria) curves, corresponding A-weighted sound pressure levels, and an example NC-51 spectrum. The A-weighted sound pressure levels are for octave band spectra that are congruent with the NC curves indicated.

curve reached by any spectrum value. The NC rating of the spectrum shown in Figure 3.9 is approximately NC-51. The tangency method does not attempt to evaluate the tonal or temporal character of an octave band spectrum. Noise criteria curves were originally defined in the old octave bands by Beranek[1]. The NC curves shown in Figure 3.9 are an interpolation of the original curves into the preferred octave bands, as published by Harris[2].

Balanced Noise Criteria Curves

ANSI standard S12.2 provides a table defining balanced NC curves in 1-dB increments. These curves extend from the 16- to the 8000-Hz octave bands. The standard defines the values for each individual curve from NCB-10 to NCB-65. In addition, the NCB-0 curve is defined as the threshold of audibility for continuous sound in a diffuse field and is derived from the ANSI threshold of audibility for pure tones.

The NCB method involves, first, computing the speech interference level (SIL) for the spectrum being evaluated. The SIL is defined as follows:

$$\text{SIL} = \tfrac{1}{4} \left(L_{500} + L_{1000} + L_{2000} + L_{4000} \right) \quad (\text{dB})$$

In the preceding expression for SIL, L_{500}, L_{1000}, L_{2000}, and L_{4000} are the spectrum octave band sound pressures levels in the 500-, 1000-, 2000-, and 4000-Hz octave bands, respectively. The SIL is a measure of sound pressure level in the speech frequency range and is used to evaluate the interference of background sound on speech communication. The NCB rating of a spectrum is equal to the SIL rounded to the nearest decibel. For example, the spectrum shown in Figure 3.10 has an SIL of 44 dB and is therefore an NCB-44 spectrum.

The method then involves determining the perceived balance between low- and high-frequency sound. A spectrum rich in low-frequency sound (16 to 500 Hz) is defined as "rumbly." The rumble criterion is defined as the NCB curve with a value 3 dB higher than the curve determined on the basis of SIL. The rumble criterion curve extends only between 16 and 500 Hz. Figure 3.10 presents the rumble criterion curve corresponding to the NCB-44 spectrum shown. Note that the spectrum exceeds the NCB-47 rumble criterion, therefore the spectrum shown would be characterized as "rumbly."

A spectrum rich in high-frequency sound (1000 to 8000 Hz) is defined as "hissy." The hiss criterion is the arithmetic average of the 3 NCB curve values intersecting the spectrum at 125, 250, and 500 Hz, in this case NCB-49. Note that the spectrum does not fall above the NCB-49 hiss criterion curve, therefore, the spectrum is not "hissy" according to ANSI S12.2.

The ANSI S12.2 standard also presents two criteria for determining the likelihood of low-frequency sound producing vibration that produces audible rattling in lightweight building elements such as suspended ceilings, light fixtures, doors, windows, ductwork, and the like. These are identified in Figure 3.10 as spectra labeled "moderately noticeable vibration" and "clearly noticeable vibration." These extend only over the frequency range 16 to 63 Hz. Sound level spectra with values in these frequency bands that exceed

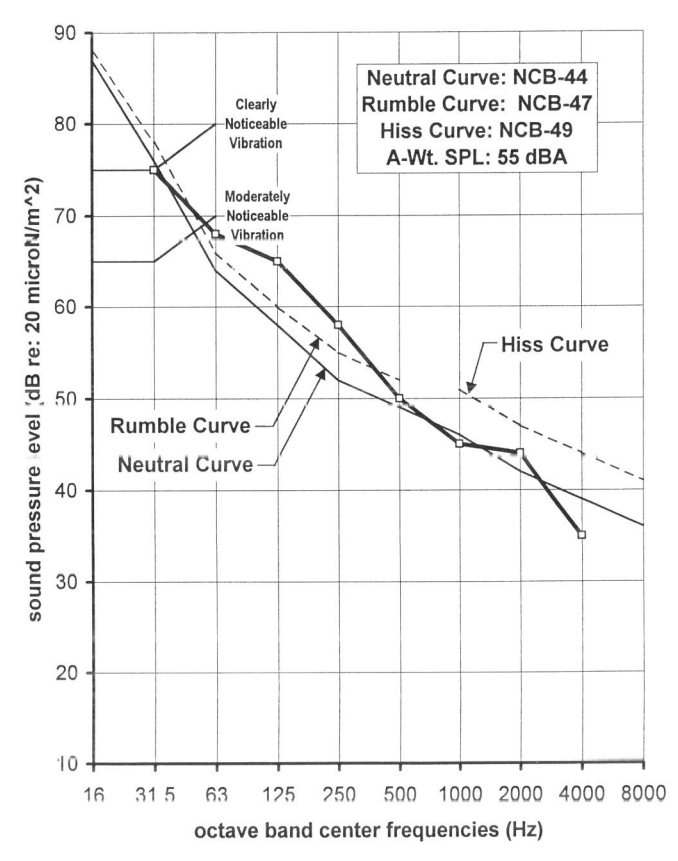

Figure 3.10 Typical sound pressure level spectrum with corresponding NCB (balance noise criteria) neutral, rumble, and hiss criteria curves.

these values are expected, according to ANSI 12.2, to produce moderately or clearly noticeable vibration in lightweight structures, windows, and furnishings.

The NCB rating, for example NC-44 in the example of Figure 3.10, is more completely stated by appending to the *neutral* rating one or more of the following designators: neutral (N), rumbly (R), exceeding the moderately noticeable vibration criterion (RV), and hissy (H). The balanced noise criteria rating of the spectrum in Figure 3.10 is NCB-44(R, RV).

Room Criteria Curves

The room criteria curve rating method, first proposed by Blazier[3] is similar to that of the NCB method. The ANSI S12.2 standard defines the RC curves extending from RC-25 to RC-50 (typical range of background sound in buildings) over the frequency range 16 to 4000 Hz. The RC curves are parallel lines of constant −5 dB per octave slopes. They have been devised to define optimally neutral characteristics of background mechanical system sound in building spaces and have been developed on the basis of perceived background spectrum "quality" rather than on equal loudness contours.

Rating a sound pressure level spectrum using the RC method involves determining the mid-frequency average level (L_{MF}), which is similar to the SIL and defined as follows:

$$L_{MF} = \tfrac{1}{3}(L_{500} + L_{1000} + L_{2000}) \quad \text{(db)}$$

as before, the L_{500}, L_{1000}, and L_{2000} are the spectrum octave band sound pressures levels at 500, 1000, and 2000 Hz, respectively. The RC rating of a spectrum is equal to the mid-frequency average level: LMF. For the spectrum shown in Figure 3.11, the neutral room criteria curve is RC-46. The rumble criterion is defined as the RC curve that is 5 dB higher and extends from 16 to 500 Hz. If low-frequency sound levels exceed the rumble criterion curve, the spectrum is judged to be rumbly. The noticeable vibration criteria curves in Figure 3.11 for the RC method are the same as those shown in Figure 3.10 for the NCB method. The hiss criterion is the RC curve that is 3 dB higher than the neutral criterion and extends from 1000 to 4000 Hz. As with the NCB method, the designators (N), (R), (RV), (H) are used to identify the spectrum balance characteristics. The room criteria rating of the spectrum in Figure 3.11 is RC-46(R, RV).

Building Codes

Noise ordinances, regulations, criteria, and standards are promulgated by organizations to serve the needs of their membership or the public. Cities and towns publish noise ordinances that appear as part of their town regulations or zoning bylaws. These usually control environmental sound by focusing on the community, or portions of the community, as a receptor. These are sometimes referred to as *imission* limits, that is, they limit the amount of sound received by a receptor. Many ordinances set explicit imission limits, such as, a maximum allowed sound level at a property line. Others set implicit imission limits, for example, controlling sound in adjacent residential units by requiring minimum sound isolation between residential spaces, rather than by directly specifying maximum sound limits in either space.

Emission-type limits set constraints on sound produced by a source. Emission limits are more prevalent among agencies setting standards for specific industries. Often, these are expressed using descriptors other than sound pressure level; for example, sound power level is sometimes used.

Building codes have been in use since before the 1900s and set standards of quality and uniformity for all aspects of building construction. Since their development of

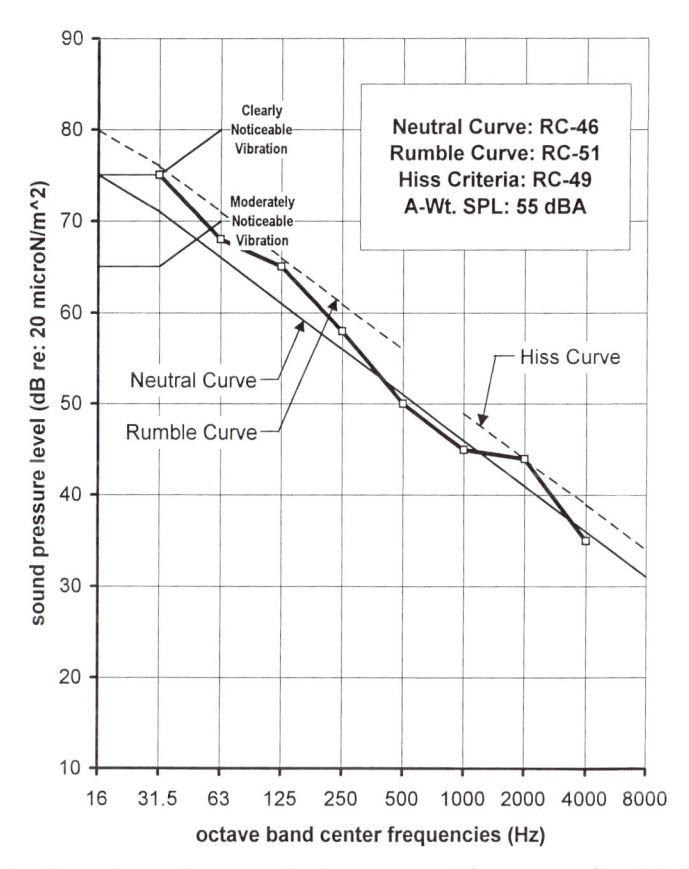

Figure 3.11 Typical sound pressure level spectrum with corresponding RC (room criteria) neutral, rumble, and hiss criteria curves.

comprehensive technical codes soon exceeded the abilities of most cities, even large cities, to administer them, several organizations were formed between 1920 and 1940 to assist local building regulators. A large part of the work of these organizations was to develop model building codes that could be adopted by localities and states. Over the years, most of these organizations have combined into the three chief building code organizations that presently represent most U.S. jurisdictions. As already mentioned, these are the Building Officials and Code Administrators International, International Conference of Building Officials, and the Southern Building Code Congress International.

All three model code groups are private organizations whose memberships comprise local and state building officials responsible for enforcing building codes in their states and localities. The federal government does not require states or localities to have building codes. Federal agencies use their own building codes or adopt one of the model codes for their own projects.

The three major building code organizations serve their memberships by producing bodies of code that are adopted and administered by states and localities. Hence, a large part of the activity of these model code organizations is to educate their members in code practices and enable them to make judgments regarding the conformance of local building construction practice to code requirements.

Code usage in the United States tends to be regional. States and localities in the Northeast and northern part of the Midwest are most often members of BOCA. Those in the western part of the United States most often are members of ICBO and those in the southern states, SBCCI. There are many jurisdictions in the United States that do not have building codes. Others have their own building code, for example, New York City and Chicago. Each state determines whether it will adopt the building code on a statewide basis or leave it to the localities to make the decision whether or not to adopt a building code.

Although building codes are intended to be comprehensive, it is only recently that model codes have included limits on noise. Noise limits and other kinds of limits in these codes share several characteristics. Briefly, every standard:

- Indicates its *status,* that is, whether it is voluntary or obligatory, and, if it is obligatory, under what circumstance individuals or organizations must conform to the standard.
- Identifies its *object,* that is, whether it limits sound imissions at a receiver or emissions by a source.
- Indicates *scope,* that is, the specific type of noise exposure and conditions under which the ordinance is applicable.
- Expresses its *limits* either as numerical values for one or more noise descriptors or in some other specific way such as establishing hours of operation, or in some subjective manner such as not allowing noise that constitutes an "annoyance."
- Establishes a means to *evaluate* conformance with limits.

The following details the significant noise control provisions of each model code. Further information, or the full code texts, can be obtained from each organization at the addresses indicated:

BOCA
Building Officials and Code Administrators International
4051 W. Flossmoor Road
Country Club Hills, IL 60477–5795

ICBO (UBC)
International Conference of Building Officials
5360 South Workman Mill Road
Whittier, CA 90601

SBCCI
Southern Building Code Congress International
900 Montclair Road
Birmingham, AL 35213–1206

The BOCA standard for sound in its entirety appears as Section 714.0—Sound Transmission Control in Residential Buildings. It is part of the body of the BOCA code, therefore its *status* is obligatory, that is, a jurisdiction that adopts the BOCA building code must include this section on building noise control or pass specific legislation to remove or modify it.

The *object* of the BOCA Section 714.0 is the receiver, that is, it seeks to limit noise at a receptor as opposed to controlling sound emitted by a source. The *scope* of the code is limited to the sound isolation performance of "all common interior walls, partitions, and floor/ceiling assemblies between adjacent dwelling units or between a dwelling unit and adjacent public areas." This section of the BOCA code deals only with sound transmission between dwelling units even though the overall BOCA code covers all types of building construction.

The *limits* of Section 714.0 are in terms of sound transmission class (STC) and impact insulation class (IIC) ratings of constructions separating dwelling units. The required wall and floor/ceiling STC rating is 45. The required floor/ceiling IIC rating is 45. A building official responsible for *evaluating* conformance of residential building partitions and floor/ceiling constructions with BOCA noise provisions can do so by comparing partition and floor constructions with lists of test assemblies provided in the code. Hence, the code does not explicitly require the involvement of an engineer or some other professional and does not require conformance testing. However, the building official may, at his or her discretion, require testing of wall or floor/ceiling constructions not listed in the code, or may require testing out of a concern for construction quality.

The International Conference of Building Officials has produced the Uniform Building Code (UBC). Acoustical requirements for buildings appear in UBC Appendix Chapter 35—Sound Transmission Control. The appendix of the UBC contains only the recommendations of ICBO and is not part of the code; thus its *status* is voluntary. For any provisions of the appendix to become obligatory, a state or locality must adopt all or portions of the appendix into its code. Presently, the State of California has adopted most of the UBC Appendix, including Chapter 35, into its code body. Hence, in California provisions of Appendix Chapter 35—Sound Transmission Control are obligatory.

The *object* of the UBC Appendix Chapter 35 provisions for sound transmission is the receiver. The *scope* of Appendix Chapter 35 is to "establish uniform minimum noise insulation performance standards to protect persons within new hotels, motels, dormitories, long-term care facilities, apartment houses and dwellings other than detached single-family dwellings."

The *limits* of the Chapter 35 fall into three categories: sound isolation, impact sound isolation, and allowable interior sound levels. The airborne sound isolation limit is expressed as an STC rating of not less than 50 or a field sound transmission class (FSTC) rating of not less than 45. FSTC is similar to the STC rating except it is applied to noise reduction data collected in the field instead of in a laboratory. Because of limitations associated with building conditions, the field performance of a demising construction is usually 0 to 5 dB less than the corresponding laboratory performance. The standard recognizes this and permits the FSTC to be up to 5 points less than the corresponding laboratory-measured STC rating.

Similarly, the IIC rating must not be less than 45. The standard allows the measurement of FIIC, that is, the field impact insulation class rating, but indicates that it must also not be less than 45. The aforementioned discrepancy between laboratory- and field-measured STC ratings does not occur in the impact insulation test. Hence, the laboratory- and field-measured IIC ratings should be in close agreement. The chief difference between the IIC and the FIIC ratings relates to absorption in the receiving room, the number of measurements conducted, and the placement of the tapping machine on the floor tested. Note that STC, FSTC, and IIC ratings are described in ASTM standards; however, FIIC is not. Moreover, Appendix Chapter 35 only indicates that FIIC is the measurement of impact insulation class in the field.

The UBC limit on interior sound levels is expressed as a *day–night average sound level* (DNL) not to exceed 45 dB (also, see Chapter 1). The standard permits the alternative use of the *community noise exposure level* (CNEL) and indicates that the interior CNEL sound level must not exceed 45 dB. The day–night average sound level is the 24-hr equivalent sound level with a 10-dB penalty added to sound levels occurring between 10 PM and 7 AM. The community noise exposure level is similar in that it is a 24-hr equivalent sound level, but with a 5-dB penalty applied to sound levels occurring between 6 PM and 10 PM, and a 10-dB penalty applied to sound levels occurring between 10 PM and 7 AM. The UBC standard encourages the use of day–night average sound level in lieu of the community noise exposure level but indicates that a project should be evaluated by either one or the other, but not both.

A building official responsible for *evaluating* the conformance of a building with UBC Appendix Chapter 35 sound limits may do so by comparing wall and floor/ceiling constructions with those assemblies listed in UBC Standard No. 35–1, 35–2, or 35–3. Or the official can request certification through field testing by an acoustician. The acoustician can judge conformance or nonconformance of a building construction with the UBC Appendix Chapter 35 standard through the use of one or more of the following noise descriptors:

- Noise isolation class (NIC)
- Normalized noise isolation class (NNIC)
- Field impact insulation class (FIIC)
- Normalized A-weighted sound level difference (DNL)

The NIC and the NNIC are defined in ASTM standards. The FIIC and Dn are described in UBC Appendix Chapter 35.

The Appendix Chapter 35 evaluation of interior sound levels requires an acoustical analysis to be conducted before construction if the DNL or the CNEL exceeds 60 dB. The standard also alternatively requires an evaluation report by an acoustician after construction is completed.

The Southern Building Code Congress International promulgates a separate document as its standard for noise control. It is composed of five chapters. Chapters 1 and 2 are the introduction and a description of terms. Chapter 3 is entitled Noise Disturbances and sets limits on source noise emissions. Chapter 4, entitled Airport Noise, sets limits on aircraft noise at the receiver (immissions). Chapter 5, entitled Sound Isolation in Multi-family Dwellings, sets limits on partition sound isolation performance. The entire standard is voluntary and is only a recommendation of SBCCI for those communities wishing to implement a noise regulation.

The *scope* of Chapter 3 (Noise Disturbances) indicates that it "shall apply to any sound producing source referenced in this standard or any sound producing source that can effect the health, safety, and comfort of those persons within audible range of the sound." A variety of *limits* are set in Chapter 3. They include sound pressure level limits for certain types of sources, for example, air compressors, motor vehicles, and so forth. For animals and motor boats, subjective limits are set. For construction, materials handling, and the like, time limits are set. This chapter also defines geographical limits, that is, noise-sensitive zones or zoning districts. The *evaluation* process for Chapter 3 includes permits for certain types of activities. It also establishes a "Board of Sound Control," which is composed of five members including one architect or professional engineer, one attorney, one general contractor, two members-at-large, and a nonvoting "sound control official" who is responsible for evaluating acoustical conditions and implementing decisions of the Board of Sound Control. This requirement to establish a board of sound control is unique among model codes.

The scope of the SBCCI Chapter 4 standard for airport noise defines "an Airport Noise Impact Overlay District for the purpose of controlling conflicts between land uses [and] noise generated by aircraft, . . . and . . . establishes sound isolation requirements for exterior walls and roofs of buildings located in the Airport Noise Impact Overlay District." The limits used in this standard are given in Table 3.1 (Table 403.1 of the standard)—SBCCI Noise Level Standards. The table indicates minimum required day–night average sound level noise reductions of exterior building facades. Minimum required noise reductions depend upon exterior sound level (expressed as ranges of DNL) and depend upon building use. This table is very much like the table used by the Federal Aviation Administration (FAA) in federal regulation, 14 CFR Part 150, for airport land use compatibility studies. The *evaluation* procedure involves certification by a registered professional architect or engineer that the exterior walls, roofs, and windows conform to the minimum STC ratings specified in Table 3.2 (Table 403.2 of the standard).

TABLE 3.1 SBCCI Noise Level Standards (Table 403.1 of the Standard)[a]

OCCUPANCY	\>75	70–75	65–70	<65
Group A				
Public assembly halls, restaurants, motion picture theaters, churches, museums, heaters for stage production, auditoria	NP	30	25	NSR
Other	NP	NSR	NSR	NSR
Group B				
Libraries	NP	30	25	NSR
Other	30	25	NSR	NSR
Group E & I	NP	30	25	NSR
Group M	30	25	NSR	NSR
Group R				
Hotels and motels	35	30	25	NSR
Other	NP	30	25	NSR
Other	NSR	NSR	NSR	NSR

Columns show EXTERIOR DAY–NIGHT AVERAGE SOUND LEVEL (DNL) IN dB.

[a] NP = not permitted; NSR = no special requirement.

To use Table 3.2 for a specific building:

1. Determine the appropriate range of opening percentage (percentage of the total wall/ceiling occupied by fenestration) in the first column.
2. Locate the corresponding required minimum noise reduction in the second column.
3. For this method to be valid, the exterior walls and roofs must have the minimum STC ratings specified in the third column.
4. The required fenestration (exterior doors, windows, and sloped glazing) should have the minimum STC rating specified in the fourth column.

If, in the unlikely event, the exterior wall of a building has a STC rating less than specified in the third column of Table 3.2, then a higher STC rating than shown in the fourth column may be needed. It should be noted, however, that the STC ratings given in the third column are lower than the STC ratings for most standard exterior wall constructions, so that the fenestration STC ratings required in the fourth column are conservatively high. In lieu of using Table 3.2, some economy in the selection of

TABLE 3.2 SBCCI Minimum Sound Transmission of Assemblies (Table 403.2 of the Standard)

PERCENT OPENINGS	MIN. REQ'D NOISE REDUCTION (DNL) IN dB	MIN. TL OF EXTERIOR WALLS AND ROOFS	MIN. TL OF EXTERIOR DOORS, WINDOWS AND SLOPED GLAZING
1–25	35	50	42
	30	45	37
	25	39	28
26–70	35	55	45
	30	50	41
	25	45	37

windows can be obtained by an acoustical analysis during building design. This can often result in fenestration STC ratings that are less than those shown in the fourth column of Table 3.2 while still achieving the required noise reduction of Table 3.1. The purpose of Table 3.2 is to assure that building components with acceptable STC ratings are used in the event that a more sophisticated analysis by an acoustical consultant is not available. The risk to a building owner using this table is that it may require a window sound isolation performance that is greater than actually needed to achieve the noise reductions of Table 3.1.

For its *scope,* SBCCI Chapter 5 indicates that "this chapter shall apply to the minimum allowable sound isolation for partitions that separate adjacent units in multi-family dwellings, and similar partitions that separate a dwelling unit from public areas, service areas, or commercial facilities." This chapter *limits* partition sound isolation between adjacent dwelling units, or between dwelling units and public areas such as hallways, service areas, or commercial facilities, to a minimum STC rating of 45. Unlike other chapters of the SBCCI standard for sound control, the evaluation process to be undertaken by a building official is not clearly established. It does imply, however, that several publications may be used for reference. These present STC ratings for a variety of partition constructions. Those having a minimum STC rating of 45 can be judged acceptable under SBCCI Chapter 5.

Comparison of Model Building Codes

Table 3.3 summarizes the provisions of the 3 model codes by briefly indicating status, object, scope, limits, and evaluation procedures for each model code.

3.4 NOISE CONTROL METHODS

Noise Barriers

A noise barrier is a wall, screen, building, or other impervious structure that breaks the line of sight between a source and receiver. Figure 3.12 contains a chart that can be used to estimate the insertion loss of a noise barrier in octave band frequencies ranging between 63 and 8000 Hz. Insertion loss is the amount in decibels that sound is reduced when a barrier, or other noise reduction method, is added, or inserted, into a noise source–receiver path. The horizontal axis in the chart of Figure 3.12 is octave band frequency in Hertz. The vertical axis indicates the noise barrier insertion loss (IL) in decibels. The chart presents a family of lines with corresponding quantities ranging between 0.5 and 32. These quantities are the ratio of the square of the effective barrier height divided by the distance between the source and the barrier. This chart only applies to a point source and conditions where the receiver is located much further from the barrier than the source. It should be noted that the chart can also be used for the inverse situation where the receiver is close to the barrier and the source is much further away. The chart cannot be used where the barrier is located midway between the source and receiver. For this more general situation, a more complex mathematical relationship is needed and can be found in Beranek and Ver[4] or Bell and Bell.[5]

The IL relationship of Figure 3.12 is limited to outdoor or "outdoorlike" situations where the receiver is a point source. Indoors, sound reflections from ceilings reduces barrier IL from what would have otherwise been obtained outdoors. Some perspective on this can be obtained by considering the noise barrier reductions obtained using partial height partitions in the open-plan analysis described earlier in this chapter (see Figure 3.7). In the case of a line source, the estimated noise reduction using Figure

TABLE 3.3 Summary of Model Noise Codes

	BOCA	ICBO	SBCCI
Title	Sec 714.0 Sound Transmission Control in Residential Buildings	Appendix Ch. 35 Sound Transmission Control	The Standard for Sound Control Ch. 3—Noise Disturbances Ch. 4—Airport Noise Ch. 5—Sound Isolation in Multifamily Dwellings
Status	Obligatory	Voluntary (obligatory in California)	Voluntary
Object	Receiver	Receiver	Ch. 3—Source Chs. 4 & 5—Receiver
Scope	Interior walls and floors between dwelling units	Interior walls and floors between dwelling units, exterior. Also interior sound levels produced by exterior noise sources.	Ch. 3—Exterior noise sources Ch. 4—Exterior wall, windows, doors of buildings near airports Ch. 5—Interior walls, and floors between dwelling units
Limits	STC ≥ 45 IIC ≥ 45	STC ≥ 50 (FSTC ≥ 45) IIC ≥ 45 (FIIC ≥ 45) DNL or CNEL ≥ 45 dB (App. to CA Amend.)	Ch. 3—Various Ch. 4—Exterior wall/window STC depending on exterior DNL Ch. 5—STC ≥ 45
Evaluation	Comparison with listed assemblies	Comparison with listed assemblies and report by acoustical consultant (testing if needed)	Ch. 3—"Board of Sound Control" Ch. 4—Certification by P. E. or Registered Architect, conformance with specified values Ch. 5—Comparison with listed assemblies

3.12 would be about 3 dB lower than determined for a point source. Examples of a line source are continuous traffic on a road or a row of exhaust fans along the roof of a building.

Building Partition Design

Building interior partitions are generally of three types. These are (1) gypsum wallboard (GWB) on a wood or metal stud frame, (2) masonry block, and (3) some combination of types 1 and 2. Figure 3.13 presents a chart for estimating the STC rating of various GWB construction combinations. The chart covers wood and metal stud systems, systems using resilient channels, staggered studs, and multiple layers of gypsum wallboard. Figure 3.14 presents a chart relating masonry wall surface weight and STC rating.

When drywall and masonry unit walls are used in combination, varying results can be obtained. The STC rating of a construction can be best determined by first determining the STC rating of the masonry wall alone. If gypsum wallboard is directly attached to the masonry wall, either mechanical fastening by means of strapping or using an adhesive to directly attach gypsum wallboard to the masonry wall, an improvement of 2 to 5 dB in STC rating can be obtained depending on the masonry block surface weight and porosity. The lighter the wall the greater the improvement; conversely, the heavier the wall the lower the improvement. If the gypsum wallboard is installed on

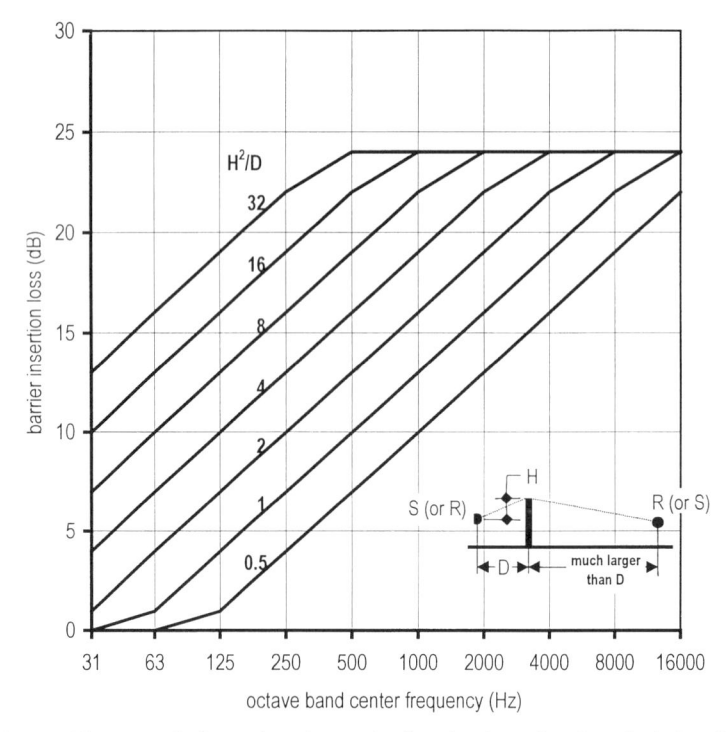

Figure 3.12 *Nomograph for estimating noise barrier insertion loss (IL) for the barrier located close to the source and far from the receiver, or conversely, close to the receiver and far from the source.*

a stud frame *not having contact* with the masonry wall, the improvement in STC rating will be between 5 and 10 dB, depending on the clearance between the GWB and the masonry wall, and on the number of layers of GWB used. If glass fiber insulation is added to the cavity formed by the GWB and masonry walls, an additional 5 STC rating points can be obtained. (Also see tables of sound isolation performance in Chapter 2.)

Building Floor/Ceiling Design

With building floor and ceiling constructions, there are two aspects that need to be considered: (1) airborne sound isolation using the STC rating and (2) impact sound isolation using the IIC rating. The airborne sound transmission loss is determined in a laboratory in a manner identical to that of partitions. The impact isolation performance of a floor/ceiling system is determined by testing a construction using a standard tapping machine. The two ratings can be quite different for a given floor/ceiling construction. For example, a floor/ceiling construction composed of heavy materials may have a high STC rating and therefore provide good airborne sound isolation between spaces but provide relatively poor impact isolation, particularly if the floor surface is hard. Conversely, a lightweight floor/ceiling construction may have a relative low STC rating and high IIC rating if the floor is carpeted. Hence, it is not possible to develop simple, reliable guidelines for estimating STC and IIC ratings for floor/ceiling construction similar to that provided above for the STC ratings of walls. For this purpose, the reader is directed to sound isolation performance tables of Chapter 2 for STC and IIC ratings of a variety of floor/ceiling constructions.

There are a variety of features that influence the sound isolation performance of floor/ceiling constructions. Among these is the weight of the floor system, the weight of the ceiling system suspended beneath, presence of sound absorption in the cavity

STC Rating: 39

½" Fire Rated GWB

3-5/8" Light Gauge Metal Channel Stud, 24" O. C.

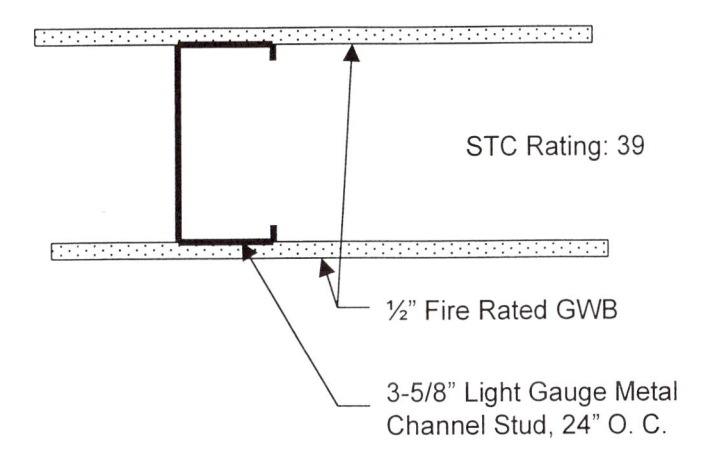

From (or Add)	To	Change in STC Rating
½" GWB	5/8" GWB	0
Studs 24" o. c.	Studs 16" o. c.	-1
Add glass fiber batt		+5
Metal studs	Wood studs	-5
Metal studs	Wood studs w/ resilient channels	+1
Light gauge metal studs	Add resilient channels	0
Heavy gauge metal studs	Add resilient channels	+2
Add one layer of GWB		+2
Add two layers of GWB		+4
Stagger studs		+5
3-5/8" studs	6" studs	+3

Note:
Light gauge: 24 to 28 ga.
Heavy gauge: 16 to 20 ga.
Increasing GWB from ½" to 5/8" thickness produces little or no improvement in STC rating but does improve low frequency sound transmission loss, particularly below the STC contour range.

Figure 3.13 Table for estimating the STC (sound transmission class) rating of typical gypsum wallboard (GWB) partitions.

between the floor and ceiling, and whether or not the floor surface is carpeted. Another significant factor is the type of structural system used in a building. Floor/ceiling constructions in buildings with wood structural systems and wood stud bearing walls tend to have lower IIC ratings than in post-and-slab-type steel buildings. This is because vibration produced by footfall in the floor above transmits into the bearing walls of the unit below. Reradiation of sound from these walls produces sound in the unit that is not controlled by any type of ceiling treatment. In this case, footfall noise can only be treated using carpet or other resilient floor finishes at the point of impact or by special detailing to "decouple" the part of the structure directly impacted from the radiating surfaces in the space below.

There are a number of special treatments that are used, particularly in residential buildings for improving the sound isolation performance of floor/ceiling spaces. The following is a list of guidelines for building design that minimize the perception of unwanted sound from adjacent spaces. These are mostly pertinent to multiresidential buildings, but have some relevance in all buildings, and address a wide variety of miscellaneous structure-borne noise issues:

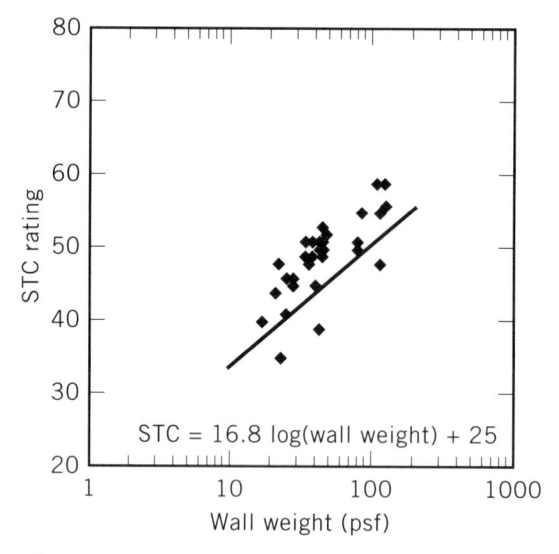

Figure 3.14 *Sound transmission class ratings as a function of wall surface weight for a population of masonry walls and corresponding regression curve. The regression curve has been devised to be conservative (i.e., results in a lower STC rating than likely to occur), with 90% of the data population falling above the regression line.*

- Install sound absorption in the floor/ceiling cavity. This is normally 3 to 6 in. thick, but should not be packed into the cavity. It should lie either on the ceiling or be attached to the underside of the floor. Blown-in insulation can serve this purpose if even coverage can be assured and if packing of the insulation into the cavity can be avoided.
- Replace the existing ceiling or augment it with two or more layers of GWB on a metal channel frame suspended from vibration isolation hangers (neoprene or precompressed glass fiber). Isolators should be sized to achieve a minimum 0.25-in. static deflection.
- Install a double-layer GWB ceiling on resilient channels or clips. These are drywall hardware items that incorporate a spring action to provide a measure of resilience between the GWB finish and the floor structure above. Here too, glass fiber insulation should also be installed in the floor/ceiling cavity.
- Increase the weight of the structural floor. This is not often feasible and would require a significant increase in mass to accrue a noticeable benefit.
- Install a resilient layer between the structural floor and a hard-finish floor treatment. There are a number of resilient products on the market that act to break the vibration coupling between a structural floor and a hard-finish floor such as marble, ceramic tile, or wood flooring. These resilient systems are often installed beneath a lightweight gypsum concrete or other lightweight leveling material, and then the finish floor is installed on the lightweight treatment.
- Institute a "carpet rule" in living units above other occupied living units. In multiresidential buildings where units are outfitted with hard floors, condominium covenants often require that carpeting be installed in hallways and foyers, and over 50 to 70% of other living areas. This has been found to remove most of the objection to footfall in residential units above other units and is most especially useful where floor structures are exposed to the space below and in buildings with wood bearing-wall constructions.
- Obviously, where possible, install carpeting.

- Stack kitchens and baths. Avoid, almost at all costs, locating kitchens and bathrooms above living rooms or bedrooms.
- Where possible, design mechanical systems with continuous fan operation to provide masking of sound from adjacent units in multiresidential buildings.
- Require that chairs and other moveable furnishings be outfitted with felt sliders to minimize sliding noise.
- Avoid the use of telephones with mechanical ringers. Even with an electronic ringer, wall-hung telephones can produce sufficient vibration to produce audible structure-borne sound in adjacent living.
- Mount sound system speakers on resilient padding. This is needed to avoid, or at least, minimize complaints of low-frequency noise in adjacent living units.
- Install closers on all cabinetry. This is to minimize impact vibration that may transmit to adjacent units and reradiate as sound.

Building Exterior Envelope Design

Building exterior facade design generally focuses on the sound isolation performance of windows. This is because building exterior walls generally provide greater noise reduction than windows, so that most of the sound transmitted into a room is transmitted through the windows. Transportation noise sources—motor vehicle traffic, aircraft, and trains—are the predominant sources of exterior noise. The procedure for evaluating the required STC rating for windows of a building exposed to exterior noise involves these steps:

1. Evaluate the exterior noise level at the building façade. This can be by means of sound measurements or through some evaluation procedure using basic information about the noise source and its distance from the building.
2. Determine the applicable noise criteria for spaces exposed to exterior sound. Sources of criteria include federal transportation agencies and the building criteria discussed in ANSI and other standards for interior noise in buildings. The difference between the exterior sound level and the acceptable interior sound level determined on the basis of some applicable criterion is the minimum required composite noise reduction (NR_c) that must be provided by the exterior wall/window construction. This is called the composite noise reduction because it is the noise reduction associated with both the wall and window together.
3. Determine the exterior wall noise reduction (NR). This can be done by referring to a number of references, among them the *Monsanto Acoustical Glazing Design Guide*. In addition, the outdoor–indoor transmission class (OITC) rating for a wall construction can be used. The OITC is defined in ASTM standards[6] and is a measure of the ability of a wall, window, or building facade element to resist the transmission of transportation sound.
4. Using the exterior wall NR and the composite wall/window NR, determine the required window NR. The window NR can be determined by using the graphical relationship also given in the *Monsanto Acoustical Glazing Design Guide*.
5. Convert the window NR determined in the previous step into a minimum required window STC rating. This relationship is dependent on the type of transportation noise exposure.

Standard windows, selected without special consideration of acoustics in buildings, most often have STC ratings falling between 28 and 32. The addition of laminated glass to common glass configurations increases the STC rating to a range of 34 to 42, depending on the glass thickness, airspace thickness in insulating glass, and number of laminated layers. Using special double glazed window constructions, STC ratings as high as 50 can be obtained through the use of laminated glass, airspace thicknesses of 4 in. or greater, and sound absorptive reveals (sound absorptive lining of the perimeter edges of the space between two lights' insulating glass).

Heating, Ventilating, and Air-Conditioning (HVAC) System Noise Control

When background sound levels in buildings exceed expectations and criteria, it is most often because of excessive noise produced by building heating, ventilating, and air-conditioning systems. Sources of noise and paths of sound transmission are varied. The following discusses the most common sources of sound in building HVAC systems and most frequently encountered paths of sound transmission to occupied spaces. Since ventilation system noise control is generally of most importance in buildings, the following includes a discussion of methods for estimating ventilation system sound levels in buildings as published in the American Society of Heating, Refrigerating, and Air-Conditioning Engineers (ASHRAE) *Handbook*.[7] This publication is intended as a compilation of recognized practice in the field of mechanical engineering. The methods presented for estimating sound in rooms produced by ventilation systems have been widely used for many years, some methods dating back to the 1960s.

Ventilation

Fans

Fans used in building mechanical systems can be divided into two general types: propeller and centrifugal. Generally speaking, among propeller fans no distinction is made between types; however, differences in propeller fan service do dictate features such as blade shape, length, and tip speed.

The differences among centrifugal fans are clearer than the differences among propeller fans. Centrifugal fans are of two types: scrolled and unscrolled. However, all centrifugal fans share a common feature; that is, they all have a fan blade arrangement where air enters the fan at the center of the blade. The spinning blade moves the air in a circular direction. The centrifugal force on the spinning air pressurizes it. The scroll acts to collect the pressurized air and guide it to the fan discharge. The scroll enables the fan to be easily incorporated into a duct system. Its main disadvantage is that strong eddy currents shed by centrifugal fan blades interact with the scroll cutoff plate to produce a discrete pulse each time a blade passes. The continuous train of pulses produces a tonal sound that makes fan noise, particularly for radial blade fans, distinctive, and in the extreme irritating to building occupants. In the case of unscrolled fans, there is no cutoff plate interaction. Air is pressurized in the box or unit cabinet containing the unscrolled fan. Basic types of centrifugal fans are illustrated in Figure 3.15 and are as follows:

Scrolled

Radial Blade.   Centrifugal fans with straight blades. These are generally used in industrial application where the airstream might contain entrained solids or particulates. Radial blade fans are the nosiest type of centrifugal fan and are not generally found in buildings, except for building incineration systems and in industrial buildings. Radial blade fans, besides being noisy, produce distinctive tonal sound that make their use problematic inside buildings and in the community.

Backward Inclined.   These can be viewed as radial blade fans with the blades bent backward, in the opposite direction of fan wheel rotation. This type of centrifugal fan

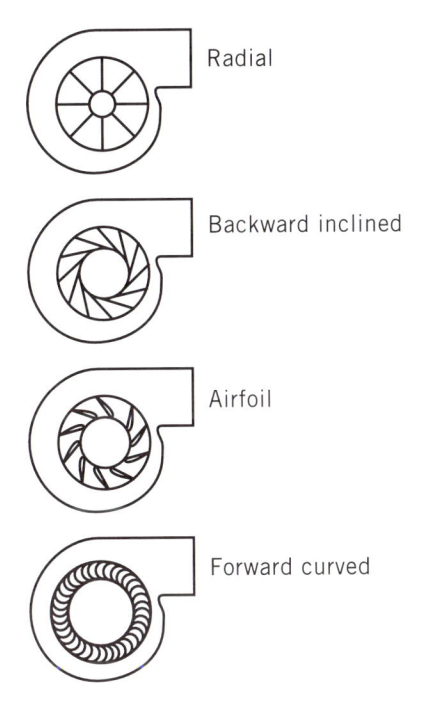

Figure 3.15 Schematic diagram illustrating the 3 basic types of centrifugal fan wheels. (from Harris, C. M., Handbook of Acoustical Measurements and Noise Control, 3rd Ed., McGraw-Hill, New York, N.Y., 1991, pp 41.8–41.9.)

offers the quietest operation of fan types in the vicinity of its maximum efficiency operating point. However, at pressures and airflow rates away from this point, fan operation becomes unstable and markedly more noisy, particularly in midfrequencies. Hence they are best suited for applications where operating conditions are relatively unchanging. Although backward inclined fans can produce significant amounts of midfrequency range sound, sound in this range is usually easily controlled using duct silencers.

Airfoil Blade. These are the same as backward inclined fans except that instead of thin, flat fan blades, airfoil blades are teardrop shaped and contoured to maximize lift, resulting in greater fan efficiency, but at a cost of a somewhat diminished operating range.

Forward Curved. These are the same as a radial blade fans except the fan blades are bent forward in the direction of fan wheel rotation. This type of centrifugal fan is not quite as quiet as backward inclined or airfoil fans; however, forward curved centrifugal fans offer reasonably quiet operation over a much wider range of pressure and airflow volume than do either the backward inclined or airfoil blade fans. As compared with backward and airfoil blade fans, forward curved fans produce more low-frequency noise, noise in the 31.5- and 63-Hz octave bands.

Unscrolled

Plug Fan. Generally a radial blade type wheel. Since it is unscrolled, there is no interaction between a scroll cutoff plate and fan blades, hence, the tonal problem common with radial blade fans is nearly entirely absent with plug fans.

In-line Centrifugal Fans. Essentially the same as a plug fan except, generally, the fan wheel is centered in a fairly closely fitting, round housing.

Boxed Centrifugal Fans. Generally the same as in-line centrifugal fans.

For many years, ASHRAE included a formula relating airflow rate [cubic feet per minute (cfm)], static pressure loss (inches water gauge), adjustment for percent operation away from optimum fan operating point, and fan type. Generally, this relationship was accurate to only ±10 dB. As fan design technology has advanced, and other types of fans have made their way into the building HVAC marketplace, the spread has become even larger. This uncertainty at mid and high frequencies is a contingency that can easily be controlled with duct silencers. However, at low frequencies, this level of uncertainty is unacceptable because of the cost and difficulty in the implementation of low-frequency noise control. Dropping this relationship from the *ASHRAE Handbook*[6] comes at a time when most manufacturers can provide reasonably accurate estimates of fan sound power level on the basis of laboratory test data. Hence, from a practical standpoint, the original fan sound power relationship has been replaced by more accurate fan data.

Silencers

There are a number of types of silencers that are used on a wide variety of building, process, and industrial equipment. These types of silencers include lined ducts and duct silencers; engine exhaust silencers; fluidic silencers used in oil-, water-, and steam-filled systems; and air blow-off silencers (mostly used in industrial processes and equipment). With the exception of air blow-off silencers, all of these attenuate sound traveling down a pipe or duct, allowing the air or fluid to pass through while removing most or part of the sound. Blow-off silencers differ in that they prevent the origination of sound at the discharge of a pipe produced when high-velocity air exits from a pipe. Most types of blow-off silencers do little for controlling sound produced upstream by a fan or compressor. Since most silencers used in buildings are either duct lining or duct silencers, these are discussed below at length.

Dynamic insertion loss (DIL) is the reduction in sound level produced by a silencer at a specific point in a room or inside a duct. The word "dynamic" is used because silencers usually permit the passage of air or gas while impeding the transmission of sound. The air or gas flow velocity, direction, and to some extent temperature affect silencer insertion loss; hence, the term dynamic insertion loss. For most silencers, DIL is quantified in octave bands and expressed in decibels. Typical ranges of DIL are discussed below for different types of silencers.

Depending on airflow velocity, all duct silencers also produce noise. The higher the airflow velocity through a silencer, the more noise that is produced. This airflow noise is called "regenerated noise." When calculating sound produced by and transmitted through an air distribution system, regenerated noise has to be added back into the noise down stream of a duct silencer as illustrated in the duct system noise calculation discussed below.

The simplest type of duct silencer is an internally lined duct, that is, a standard HVAC air duct lined on the *inside* with sound-absorptive materials. Note that external lining of ductwork produces little or no insertion loss. Internal acoustical duct linings are typically 1 in. thick but can range between ½ and 2 in. Most often they are composed of glass fiber impregnated with a rubber or neoprene compound to avoid fiber erosion. Duct lining is generally cemented on the wall of the duct and also secured mechanically in some manner, particularly at the edges to ensure that the airflow does not lift the lining material away from the duct surface. Duct lining is usually only applied in rectangular duct. Its use in circular or oval duct generally requires the installation of an internal perforated screen to retain the duct liner to the outside of the duct. However, there are proprietary products now available that eliminate the cost and complexity of a perforated liner in circular and oval ducts.

Sound absorption, fire resistance, resistance to erosion, and workability are features particular to internal sound-absorptive duct lining that make it unique. Other types of sound-absorptive materials, not specifically designated by the manufacturer and tested for acceptable use as internal acoustical duct liner, should not be used as duct lining.

As of this writing, there are conflicting philosophies regarding the use and safety of internal acoustical duct liners, particularly with respect to potentially friable particulates. The reader is directed to a variety of sources of information on this subject.[8-11] Among these is an information packet collected and published by Schuller International, Inc. This packet includes publications by ASHRAE, The North American Insulation Manufacturers Association (NAIMA), as well as Schuller International, Inc. Most of the concerns that have come about in the past have been a result of misuse of internal acoustical duct lining. Here are a few precautions:

- Do not use duct liner in duct sections that might have air-entrained liquid, such as after humidifiers and cooling coils.
- Do not use duct liner in ductwork carrying contaminated flow such as laboratory hood exhaust ductwork or ductwork in certain types of industrial or laboratory environments.
- Never install duct lining in ducts where flow velocities may exceed the maximum recommended by the manufacturer. Note that the air velocity at the surface of duct lining can significantly exceed the average cross-sectional flow velocity by as much as a factor of 2 to 3 particularly at fan discharges or poorly contoured fittings. Even at duct elbows and transitions, air velocities at duct lining surfaces can exceed the average duct face velocity by a substantial margin.
- Do not use in certain bacterially sensitive health care areas, such as a burn unit if air is not filtered for such contaminants before entering a room.

The longer the length of lined duct, and the thicker the duct lining, the greater the noise reduction obtained. The *ASHRAE Handbook*[7] provides tables of acoustical duct lining insertion losses for straight rectangular and circular duct lined with 1-in.-thick and 2-in.-thick glass fiber duct lining. The *Handbook* also provides the insertion loss of lined elbows as well. For computational purposes, the tables of insertion losses have been modeled as mathematical algorithms. These have been compiled by Reynolds and Bledsoe[12] and are used in most HVAC noise computer software programs.

Rectangular duct silencers are short lengths of duct that are divided by one or more baffles as illustrated in Figure 3.16. These baffles are usually constructed from perforated sheet metal. In dissipative duct silencers, these baffles are filled with glass fiber material. The baffles are generally contoured so that air flows around them with minimum static pressure loss and regenerated noise. Baffle thickness typically ranges between 3 and 12 in., with thicker baffles used in silencers designed to provide greater low-frequency DIL. The airflow passages between duct silencers typically range between 3 and 12 in. depending on the required DIL and restrictions on static pressure losses. Duct silencers are commercially available in lengths from 18 in. to 10 ft, and cross sections from 1 ft^2 and up. Very large silencer cross sections are generally assemblies of two or more silencers with smaller cross-sectional areas. For noncorrosive environments found in most buildings, duct silencer casings are constructed from galvanized steel, and the perforated material is postgalvanized after perforation. The next most robust configuration is galvanized casing with perforated stainless steel baffles. Next is both stainless steel casing and perforated baffles. Duct silencers are also offered in aluminum

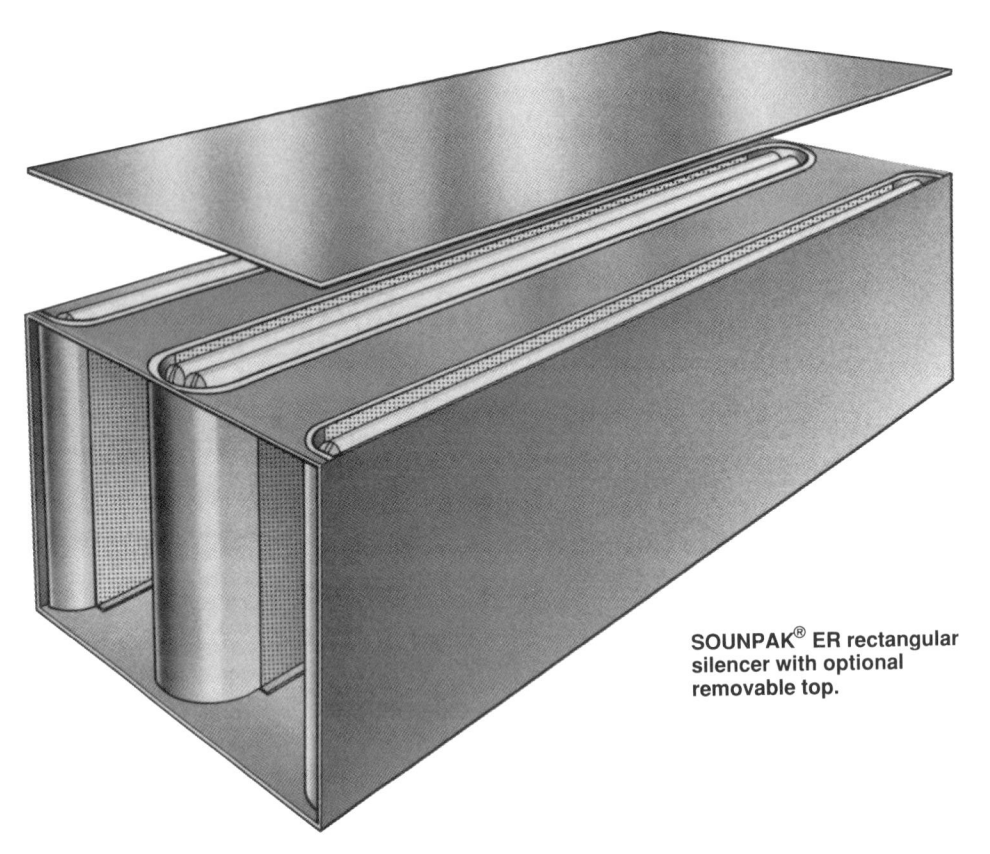

Figure 3.16 Rectangular silencer. (Courtesy of United McGill Corporation.)

and with protective coatings as appropriate. The use of these various materials has little or no influence on silencer DIL, but only on its longevity in corrosive environments.

For clean or hospital environments, most manufacturers can wrap the baffle glass fiber fill in a thin Mylar or Tedlar film to protect the fill and eliminate erosion of fill fibers by the airstream. These *hospital-type duct silencers* generally have a somewhat lower DIL than the same configuration silencer without the wrapped fill. This is because of the finite TL that even thin flexible films have at high frequencies. Most manufacturers now recognize the importance of maintaining a space between the perforated metal and the glass fiber wrap. If the silencer baffle is packed solid, so that the film is pressed hard to the interior surface of the perforated baffle, a considerable loss in DIL can result across the entire frequency range.

In other applications, where baffle fill cannot be used or where there is a need for more low-frequency DIL than is generally provided by dissipative duct silencers, reactive duct silencers are used. Reactive duct silencers have baffles that are specially configured to incorporate Helmholtz resonators. In its most simple form, a Helmholtz resonator can be thought of as a bottle. The bottle has a volume and a connecting neck open to the outside. The air in the bottle acts as a spring while the air in the neck acts as a mass. This spring-and-mass arrangement will have a natural frequency associated with the size of the neck mass and the bottle volume. At the natural frequency, the Helmholtz resonator will efficiently absorb sound. In reactive silencers, the volume of the baffles and the holes in the perforated steel form a neck-and-volume arrangement similar to that of the bottle. By carefully configuring the proportions of the baffle volume and the perforations, the natural frequency of the resonator can be controlled, thus controlling the frequency response of the silencer DIL.

Acoustical louvers, as illustrated in Figure 3.17, are in reality just short duct silencers with baffles that are the louver blades. The topside of the louver blades are unperforated in order to shed water and the lower sides are perforated, exposing glass fiber fill inside the blade. Acoustical louvers range in lengths (or more appropriately thicknesses) between 4 and 12 in. In order to increase the DIL, the louver blades are generally spaced close together resulting in free areas of 40 to 50% or less. Hence, acoustical louvers generally produce higher static pressure losses than standard louvers.

Most manufacturers of duct silencers now offer custom designs to fit the tight confines sometimes entailed in ductwork design. Silencers can be designed as elbows, offsets, or transitions from one cross section to another. Duct silencers that incorporate elbows generally have very high DILs.

Most silencer manufacturers have models that provide convenient stepping with respect to DIL between types: dissipative, reactive, fiber-wrapped, and the like, lengths, and static pressure loss. Figure 3.18 exhibits the typical variation in DIL with increasing length for a model of duct silencer advertised as one providing low static pressure. Note that each of the four silencer lengths has its lowest DIL at low frequencies,

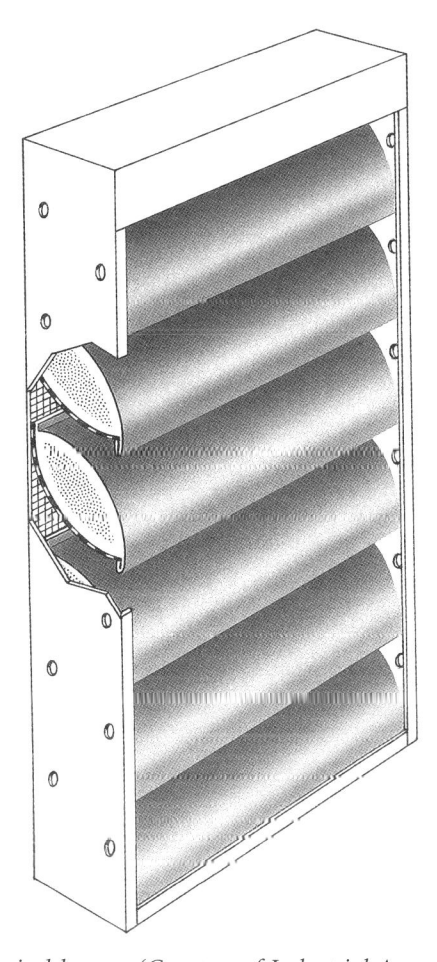

Figure 3.17 Acoustical louver. *(Courtesy of Industrial Acoustics Company, Inc.)*

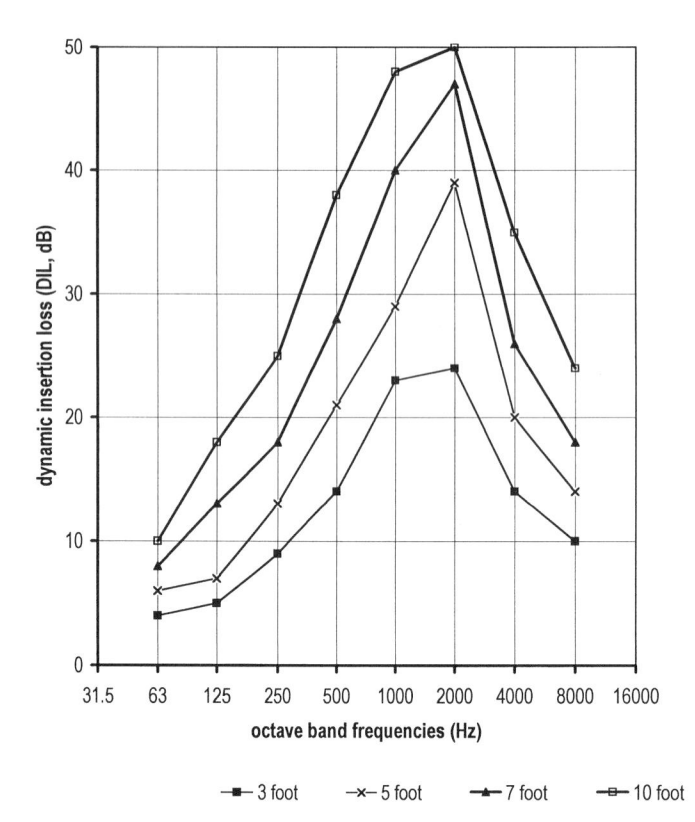

Figure 3.18 Octave band dynamic insertion loss (DIL) for a typical duct silencer design of various lengths. (Industrial Acoustics Company, Inc. type L duct silencer. Courtesy Industrial Acoustics Company.)

increasing to highest DIL in mid-frequencies, and down to lower DILs at very high frequencies.

Figure 3.19 presents DILs for two duct silencers of equal cross-sectional area and length (3 ft long). The higher of the two exerts eight times the static pressure of the lower. This is important to keep in mind when selecting silencers. If a lower static pressure is required, a larger cross-sectional area must be used. If the cross-sectional area of the silencer with the higher static pressure loss and higher DIL in Figure 3.19 is increased by a factor of 8, the resulting static pressure loss will be reduced, down to that of the silencer with the lower DIL. It is generally the experience that silencers in building fan systems need to have larger cross-sectional areas than the duct in which they are installed. This is much more the case if silencers with high DILs, and consequently high static pressure losses, are to be used.

Figure 3.20 presents dynamic insertion losses for dissipative and reactive silencers. As noted above, dissipative silencers have fibrous fill in their baffles to absorb sound passing through the silencer. Reactive silencers have arrangements of Helmholtz resonators in their baffles to achieve the same end. As illustrated by the data in Figure 3.20, reactive silencers typically provide higher DILs at low frequencies than do dissipative silencers. However, the maximum DIL values reached by dissipative silencers are generally higher than those of reactive silencers.

Figures 3.21 and 3.22 present regenerated sound power levels produced by airflow varying in velocity from −4000 fpm to +4000 fpm through a typical 3-ft long, low

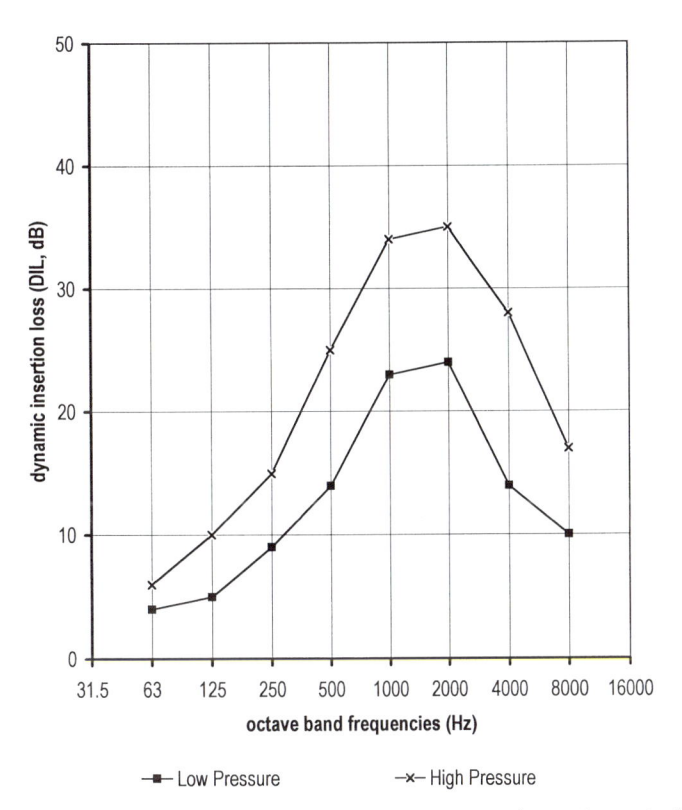

Figure 3.19 Comparison of octave band dynamic insertion losses for typical low- and high-pressure loss silencers. [Industrial Acoustics Company, Inc. types L (low pressure loss) and S (high pressure loss) 3-foot long duct silencers.]

static pressure loss duct silencer. The first of these figures is for flow in the positive direction, that is, sound transmitted downstream of the silencer in the same direction as the airflow. Conversely, the second figure is for flow in the negative direction, that is, sound transmitted upstream of the silencer in the opposite direction of the airflow. Although regenerated sound is somewhat higher at low frequencies for positive direction airflow than for negative, higher frequency regenerated sound in the speech bands is generally more problematic for silencers with negative flow. To provide some perspective on the significance of regenerated sound power levels in Figures 3.21 and 3.22, consider the following hypothetical example. Suppose a silencer is connected to a duct on one end and emptying into a room on the other. Were one to stand 3 ft from the open end of the silencer, regenerated sound levels would vary from 30 to 70 dBA for sound in the positive direction depending on flow velocity, that is, airflow out the silencer into the room. Were airflow to be in the negative direction, regenerated sound would vary from 50 to 70 dBA, depending on flow velocity. In buildings, silencers are generally not used with velocities exceeding about 1500 fpm. This, together with other system attenuation, guards against excessive regenerated noise in many situations when silencers are used with positive airflow. However, with negative airflow, other precautions may be needed. These include the use of lining upstream of the silencer (between the occupied space and the silencer) and/or oversizing the silencer cross-sectional area in order to minimize airflow velocity.

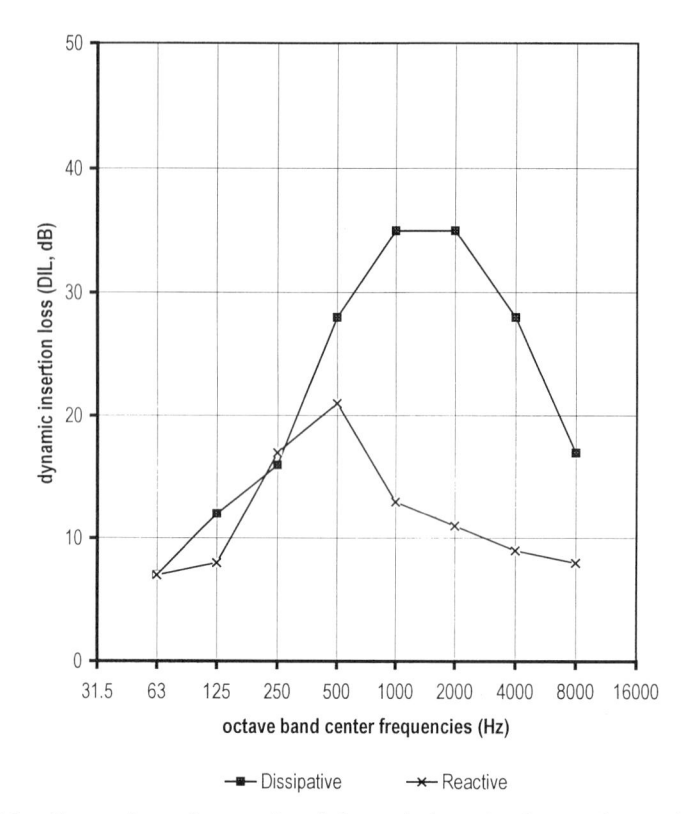

Figure 3.20 *Comparison of octave band dynamic insertion losses of typical 3-ft-long dissipative and reactive duct silencers. (Industrial Acoustics Company, Inc. type S dissipative duct silencer and type XL reactive duct silencer.) These have been selected for comparison since they have about the same static pressure loss. Although having a lower DIL than dissipative silencers, reactive silencers offer the advantage of providing silencing without the use of fibrous baffle fill. (Courtesy Industrial Acoustics Company.)*

Circular duct silencers are generally available in three configurations: (1) essentially a round, internally lined duct, (2) a round, internally lined duct with a bullet in the middle, and (3) a round unlined duct with a bullet in the middle. These are illustrated in Figure 3.23. The first configuration is found in a wide variety of duct silencing products. These include standard steel duct silencers, flexible duct silencers where the casing is a thin aluminum accordion foil, pulse boiler silencers where the casing is PVC pipe, and flexible duct where the casing is paper, impregnated fabric, or a thin foil. The second configuration is an internally lined steel casing with a perforated metal cylinder suspended in the middle of the duct and filled with glass fiber. The third configuration is simply the center bullet suspended in an unlined round duct.

The variation of dynamic insertion loss with frequency for circular silencers is similar to that of rectangular silencers, except that circular silencers generally require a greater length to achieve a given DIL than do rectangular silencers. Elbow silencers, usually rectangular silencers configured with a 90' bend, have generally the same DIL at low frequencies as that of straight silencers of roughly equal length. At mid and high frequencies, however, elbow silencers often have a considerably higher DIL than those of straight duct silencers of equivalent length.

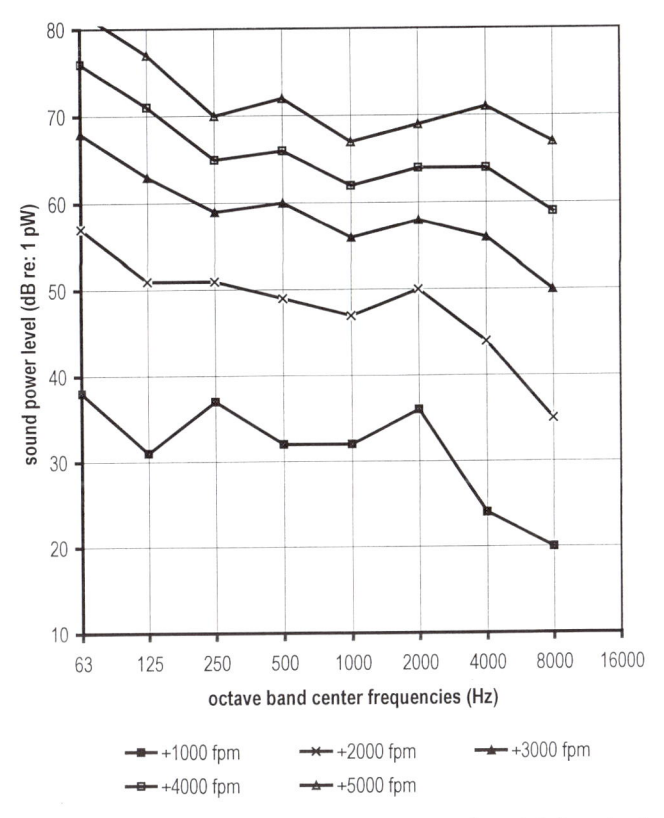

Figure 3.21 Regenerated noise produced by positive airflow (airflow in the same direction as sound transmission, that is, sound transmitted downstream of the silencer) at indicated velocities through a typical duct silencer. (Industrial Acoustics Company, Inc. type L 3-ft-long duct silencer, 4-ft^2 face area. Courtesy Industrial Acoustics Company.)

Airflow

Noise produced by airflow is often as problematic as noise produced by fans, especially airflow noise generated at diffusers and grilles. Often, when mechanical system noise in a space exceeds criteria or expectations, it is because of noise produced downstream of fans, at duct fittings, terminal units, and diffusers and grilles. There are two general approaches to controlling airflow noise: (1) to reduce flow velocities in the vicinity of duct fittings and discontinuities and (2) to install lining or silencers in ductwork between points where flow noise is generated and occupied spaces. Table 3.4 presents maximum recommended airflow velocities in feet per minute (fpm) in lengths of duct preceding supply-air diffusers or return-air grilles[13]. These are intended to help control mid- and high-frequency noise normally perceived as an innocuous broadband sound that, although audible, is not distinctive. In order to take advantage of the benefit that low airflow velocity has on noise, it is necessary that control dampers be eliminated from supply diffusers and return grilles, and that air balancing be achieved by dampers as far upstream of these as possible. In any event, for very quiet spaces, it is important that the air distribution ductwork not be designed to result in a wide pressure variation across ductwork serving a room. This is to avoid situations where some dampers must be closed down significantly while others are left wide open. Dampers that are choked-down to maintain a desired airflow may produce significant amounts of flow noise.

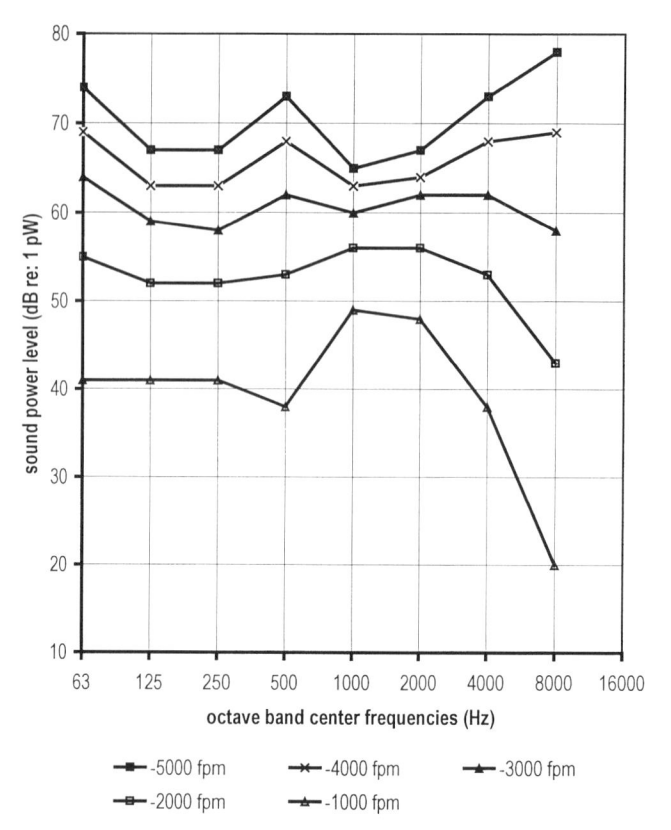

Figure 3.22 *Regenerated noise produced by negative airflow (airflow in the opposite direction as sound transmission, that is, sound transmitted upstream of the silencer) at indicated velocities through a typical duct silencer. (Industrial Acoustics Company, Inc. type L 3-ft-long duct silencer, 4-ft^2 face area. Courtesy Industrial Acoustics Company.)*

Acoustical Modeling of Sound Produced by HVAC Systems

Since the 1970s, the *ASHRAE Handbook*[7] has contained an example of a simple ventilation system and provides a set of computations illustrating how the methods of the *Handbook* are implemented.[14] The example problem is depicted in Figures 3.24, 3.25, and 3.26. Figure 3.24 is a section through a rooftop unit showing the location of a supply air fan, and supply and return air ductwork and the 3 principal paths of sound transmission to the subject room. The isometric sketch of Figure 3.25 provides a little more detail by showing a supply air branch to a variable air volume (VAV) terminal box and a room supply air diffuser. Figure 3.26 provides details of sound transmission through the return-air intake into a ceiling plenum, and through an acoustical ceiling into the room. Each element of the duct system either attenuates sound or adds sound or both. Parenthesized numbers in Figures 3.24, 3.25, and 3.26 identify each element, and correspond to the line numbers of the calculations in Table 3.5. Table 3.5 shows only the calculation of sound pressure level in the room that is transmitted along path 1. This calculation is given without detailed explanation of the specific methods for determining sound produced or attenuated by each element. For this, the reader is

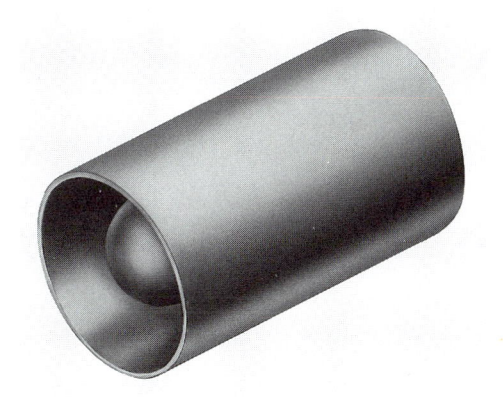

Figure 3.23 Typical circular silencer models showing center bullet in an unlined casing and a center bullet in an outer lined casing providing higher DIL. (Industrial Acoustics Company, Inc. types NS/NL and FCS. Courtesy Industrial Acoustics Company.)

directed to the *ASHRAE Handbook*.[7] The calculation table is given to provide some familiarity with how ventilation sound in rooms is estimated.

There are a few features of the calculations in Table 3.5 that are worth mentioning. These are as follows:

- The fan sound power indicated is hypothetical but is representative of a common size of fan used in rooftop units.

TABLE 3.4 Criteria for Air Distribution Systems Serving Critical Spaces[13]

NC CRITERION FOR OCCUPIED SPACE SERVED	SLOT AIRFLOW VELOCITY AT DIFFUSER/GRILLE	AIRFLOW VELOCITY (FMP) IN LENGTHS OF DUCT (FT) FROM SUPPLY DIFFUSER/ RETURN GRILLE		
		0–10 FT	10–20 FT	20–30 FT
NC-15	250/300	300/350	350/350	425/500
NC-20	300/350	350/425	425/500	550/650
NC-25	350/425	425/500	550/650	700/800
NC-30	425/500	500/600	700/800	850/950
NC-35	500/600	600/700	800/900	1000/1150

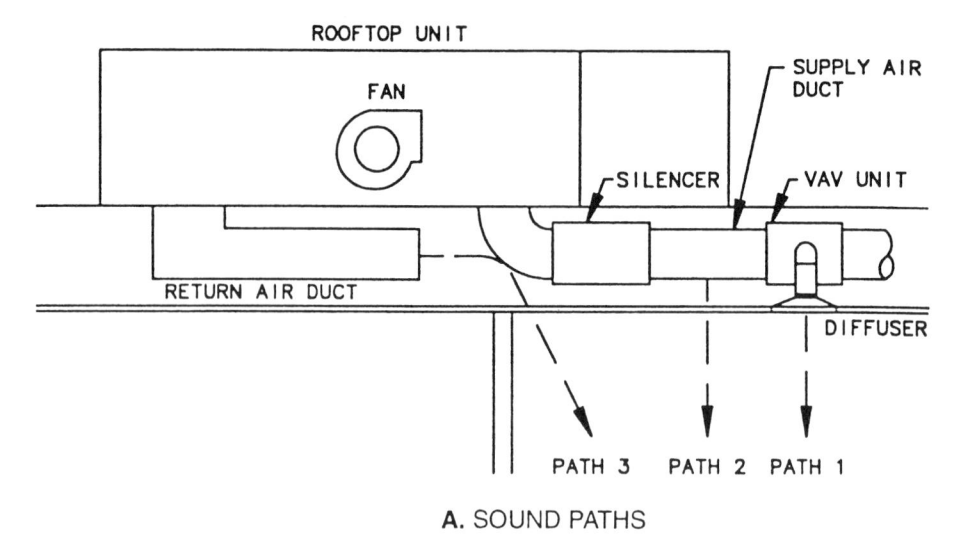

```
                    ROOFTOP UNIT
                                          ┌─ SUPPLY AIR
                          FAN             │   DUCT
                                    ┌─ SILENCER   ┌─ VAV UNIT
                                                
     RETURN AIR DUCT
                                                    DIFFUSER
                           PATH 3   PATH 2   PATH 1

                       A. SOUND PATHS
```

Figure 3.24 Section through rooftop air-handling unit example used in ASHRAE Handbook.[6] (From the 1995 ASHRAE Handbook, Figure 33A. Reprinted with permission.)

- Several line elements are repeated in the second column. In these cases, duct elements both attenuate (DIL or IL) and regenerate noise. Examples of these are the duct silencer and duct elbows.
- The subtotals are indicated as either arithmetic or logarithmic. Insertion losses are arithmetically added to the previous subtotals, regenerated sound powers are added logarithmically to the previous subtotals.
- Under comments, dimensions and features of each element needed to estimate insertion loss or regenerated sound power levels using ASHRAE methods are indicated. Even unlined duct is identified here as it provides some attenuation, though negligible.
- The end reflection loss represents the transmission loss that occurs at the interface between the end of a duct and open space. End reflection loss is most significant at low frequencies.
- The power-to-pressure relationship assumes that the receptor is 5 ft from the diffuser in a 20-by-20-by-9 ft room. ASHRAE discusses 3 basic power-to-pressure relationships, each applicable to different circumstances.

Table 3.5 is the estimated sound level transmitted along path 1. Estimated levels for sound propagation along all 3 paths are shown in Table 3.6. The total shown in the table is a logarithmic sum of the room sound pressure level contributions determined for each path. Figures 3.27 and 3.28 show plots of the estimated octave band sound pressure level with NC curves and RC curves, respectively.

Pumps and Piping. There are generally two types of pumps used in building mechanical systems. The more common of these are end-suction pumps. The less common, and generally more quiet, are split case pumps. End-suction pumps are for general use; split case pumps are generally used where large water flows are needed. If pumps are properly sized and vibration isolated as discussed below, noise is generally not a problem even if located in mechanical equipment rooms above, below, or adjacent to residential units. Location of pumps in or close to special rooms requiring very low noise levels often is not acceptable. Such rooms are, for example, performance

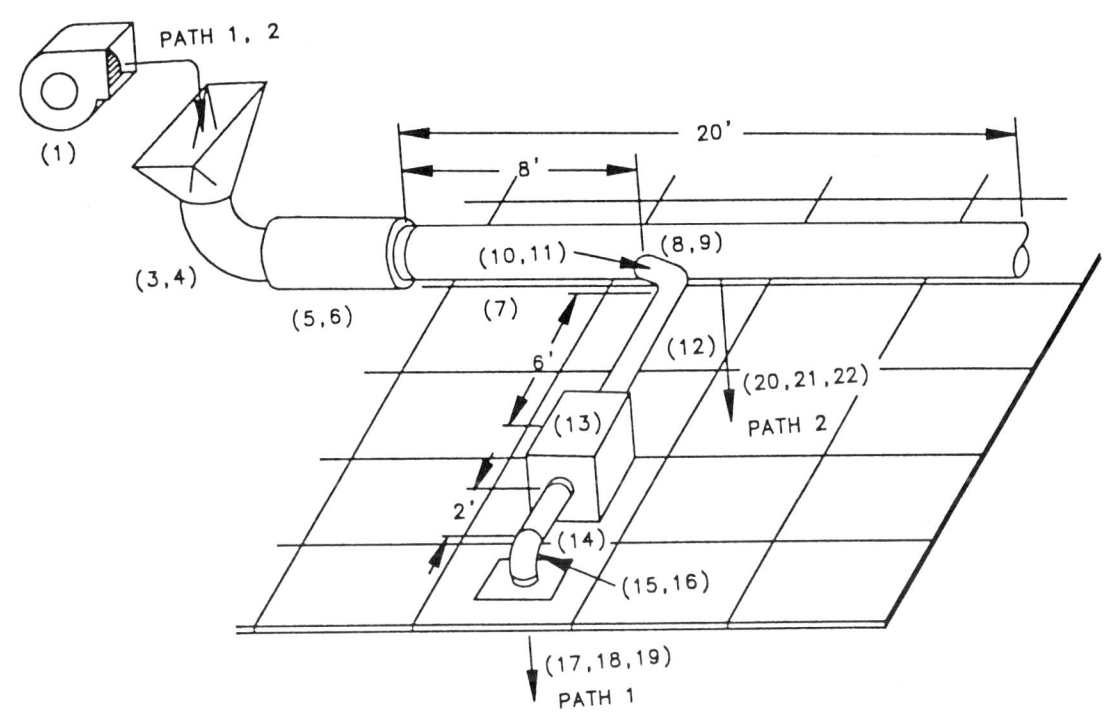

Figure 3.25 Schematic of air distribution ductwork of the example used in ASHRAE Handbook.[6] *(From the 1995 ASHRAE Handbook, Figure 33B. Reprinted with permission.)*

and large lecture spaces, and other special facilities requiring low background sound levels.

Vibration Isolation

Vibration isolation is important from the standpoints of both controlling noise and minimizing perceptible vibration. Vibration isolation is normally accomplished by incorporating resilient elements in connections to and mounting points of machinery that produce vibration. Vibration isolation mounts include steel springs, neoprene mounts and padding, and air mounts. Steel springs are available in freestanding, housed, or restrained. Freestanding springs are used on machines that do not undergo changes in weight, such as fans. Machines that are subject to wind sway or undergo changes in weight, such as water-filled cooling towers, or machines that might undergo significant static displacement at start-up such as reciprocating compressors, are generally mounted on restrained steel spring isolators. These incorporate motion stops that limit displacement of supported loads in vertical and horizontal directions. Housed springs accomplish the same end, but with interlocking top and base pieces that encapsulate the spring. However, if these are subject to a horizontal static load, common in machine mounting, they often bind, reducing their vibration isolation performance. Hence, it is the practice of this author to encourage the use of restrained, unhoused isolators instead of housed isolators. There are a wide variety of air isolators on the market. Although effective, they are generally not needed in building mechanical systems and consequently seldom used, except for special purposes. These isolators and others are generally described in Chapter 2.

TABLE 3.5 Computation of Sound Transmitted along Path 1 of Ventilation System Shown in Figure 3.25

NO. LINE ELEMENT	63	125	250	500	1k	2k	4k	COMMENTS
1 Fan	92	86	80	78	78	74	71	L_w = Rooftop supply air fan, 7000 cfm @ 2.5 in. W.G.
3 Supply duct elbow	0	−1	−2	−3	−3	−3	−3	IL = Unlined radius elbow, 22 in. dia., 12 in. radius
Subtotal (arith.)	92	85	78	75	75	71	68	
4 Supply duct elbow	56	54	51	47	42	37	29	L_w = Regenerated sound from elbow.
Subtotal (log.)	92	85	78	75	75	71	68	
5 Duct silencer	−4	−7	−19	−31	−38	−38	−27	DIL = Duct silencer, 22 in. dia., 44 in. long, high s. p. loss
Subtotal (arith.)	88	78	59	44	37	33	41	
5 Duct silencer	68	79	69	60	59	59	55	L_w = Silencer regenerated noise
Subtotal (log.)	88	81	69	60	59	59	55	
7 Straight duct	0	0	0	0	−1	−1	−1	Il = Unlined circular duct, 22 in. dia., 8 ft long
10 Junction	−8	−8	−8	−8	−8	−8	−8	IL = Junction, unlined circular duct branch, 10 in. dia., 800 cfm (unlined circular main duct, 22 in. dia., 6200 cfm)
Subtotal (arith.)	80	73	61	52	50	50	46	
11 Junction	57	54	51	48	44	38	32	L_w = Regenerated sound from junction.
Subtotal (log.)	80	73	61	53	50	50	46	
12 Straight duct	0	0	0	0	0	0	0	IL = Unlined circular duct, 10 in. dia., 6 ft. long
13 VAV terminal	0	−5	−10	−15	−15	−15	−15	IL = Terminal box
14 Straight duct	0	0	0	0	0	0	0	IL = Unlined circular duct 10 in. dia., 2 ft. long
15 Duct elbow	0	0	−1	−2	−2	−3	−3	IL = 90° unlined circular elbow, 2 in. dia. radius
Subtotal (arith.)	80	68	50	36	32	32	28	
16 Duct elbow	49	45	41	37	31	24	16	L_w = Regenerated sound from elbow.
Subtotal (log.)	80	68	50	39	34	32	28	
17 End reflection	−16	−10	−6	−2	−1	0	0	Il = End reflection loss, 10 in. dia., flush w/ ac. clg.
18 Diffuser	31	36	40	41	39	36	30	L_w = Diffuser, 15 × 15 in., 0.0115 in. W.G. s. p., 800 cfm, generic
Subtotal (arith.)	64	58	45	42	39	37	32	
19 Power to pressure	−5	−6	−7	−8	−9	−10	−11	
Subtotal (arith.)	59	52	38	34	30	27	21	L_p = Sound pressure level in room (40 dBA)

Isolation hangers are used to vibration isolate equipment and piping suspended from overhead. These are available in varieties using steel springs, neoprene elements, and combinations of steel springs and neoprene. The combination steel spring and neoprene isolators are used when there is a concern that higher frequency vibration, coincident with the spring surge frequency, might transmit through an isolator. Spring surge frequency is a frequency where the spring coil resonates, resulting in significantly reduced isolation efficiency at that frequency. Surge frequencies depend on spring size but are usually above 200 Hz, well above the frequency range of concern for vibration produced by most rotational motion, but in a frequency range influenced by

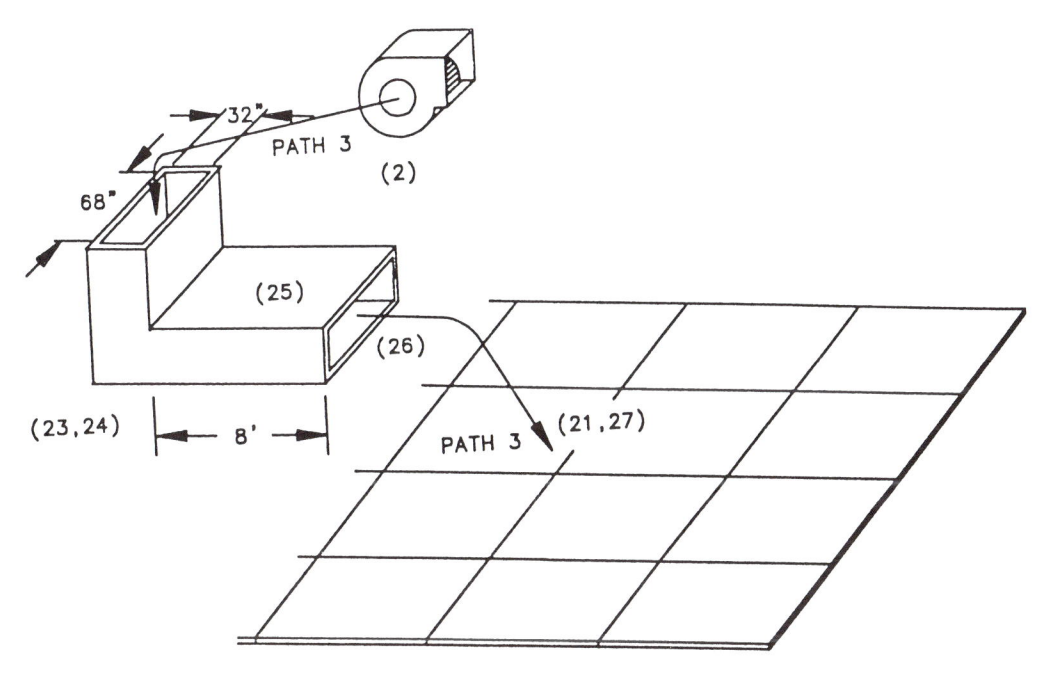

Figure 3.26 Schematic of return-air sound transmission to the room of the example used in ASHRAE Handbook.[6] *(From the 1995* ASHRAE Handbook, *Figure 33C. Reprinted with permission.)*

water flow noise, impacts, or other sources of noise. Pipe suspension is an example of where combination steel spring and neoprene isolation hangers are often used. Many spring mounts also include ribbed neoprene pads for the same purpose, to filter vibration that might transmit unattenuated through an isolator at a spring surge frequency.

Figure 3.29 illustrates a simple dynamic model of a machine, that is, a mass, mounted on a spring. When the machine (the mass) is loaded onto the spring, the spring undergoes a *static deflection*, that is, a downward displacement measured in inches. If the machine, when freely sitting on the spring, is displaced by a force exerted on it, and if the force is quickly removed, the machine will begin to bounce at a fixed rate of oscillation. This bouncing will eventually decay to zero. The rate at which the machine bounces is called the *natural frequency*. Operating the machine at a speed

TABLE 3.6 Total Room Sound Level in Example of Figures 3.24, 3.25, and 3.26

PATH	63	125	250	500	1k	2k	4k	COMMENTS
1. Supply air diffuser	59	52	38	34	30	27	21	
2. Supply air duct break-out	41	29	30	28	24	25	23	
3. Return air duct	62	51	45	26	27	25	19	
Total (Log.)	64	55	46	35	32	31	26	L_p = total sound pressure level in room (44 dBA)

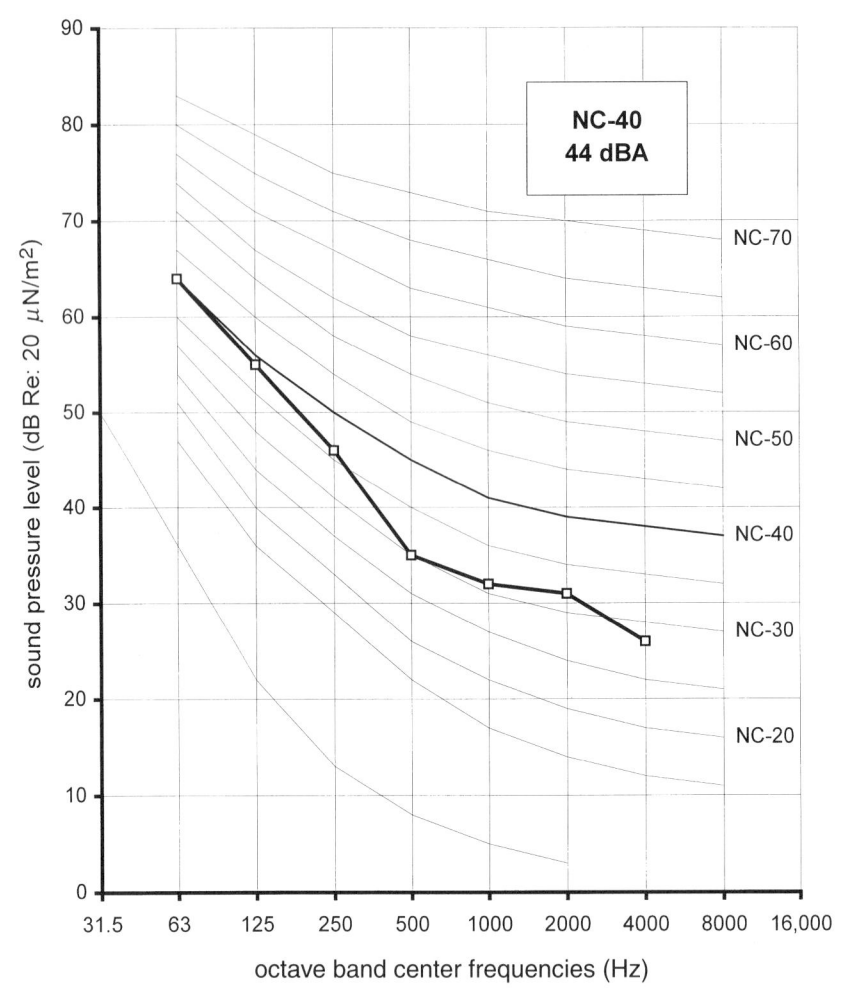

Figure 3.27 Air distribution total sound pressure level in room served by the system depicted in Figures 3.25, 3.26, and 3.27 and NC curves.

equal to the natural frequency may cause the machine to bounce with a large excursion. At machine rotational speeds below the natural frequency, the machine excursion becomes less, but the dynamic force (vibrating force) exerted at the top of the spring by the machine is transmitted directly through the spring to the base, without attenuation, resulting in no isolation. Above the natural frequency, the dynamic force exerted by the machine at the top of the spring is filtered by the spring, resulting in a lower dynamic force exerted at the base of the spring. The amount by which the vibration force at the base of the spring differs from the vibration force at the top of the spring (at the machine base) is defined as the *vibration isolation efficiency*. Generally, when machines require static deflections of less than 0.25 in., one or more layers of ribbed neoprene can be used to vibration isolate equipment. If more than one layer is used to obtain the required static deflection, layers are separated by 16-gauge steel shim plates. Machines requiring 0.25 to 0.75 in. of static deflection are mounted on neoprene isolators. Machines requiring 0.75 to 4.0 in. of static deflection are mounted on steel spring isolators. Springs with static deflections greater than 4.0 in. are not recommended.

Vibration isolation efficiency is related to spring static deflection. The larger the static deflection of the spring, the greater the vibration isolation efficiency; however,

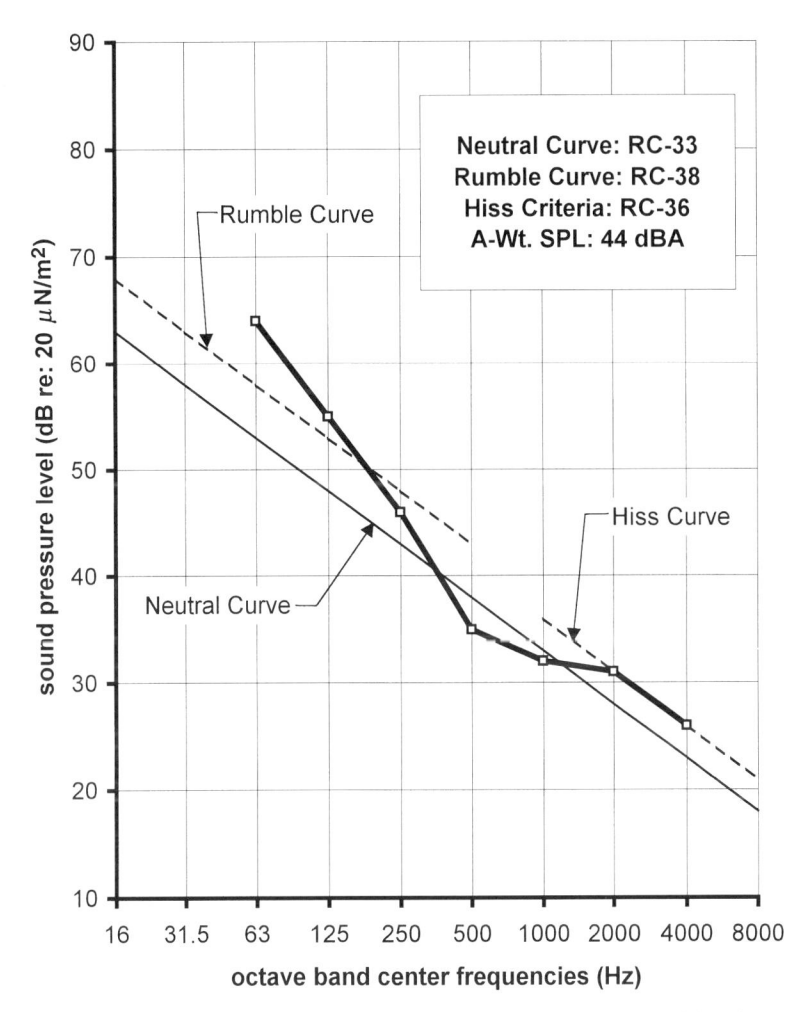

Figure 3.28   Air distribution total sound pressure level in room served by the system depicted in Figures 3.25, 3.26, and 3.27 and RC curves including RC rumble and hiss criteria.

the greater the static deflection needed for an application, the larger, and consequently, the more costly the spring. Hence, it is worthwhile to investigate the range of static deflection that is needed for each type of machinery used in a building project. For most applications, the static deflection of a spring must satisfy the following two guidelines:

1. The spring resonant frequency must be no more than one-eighth the lowest rotational speed of a supported machine. For example, if a fan operates at 1200 rpm and is belt driven by a motor operating at 1750 rpm, the natural frequency of the isolator should be one-eighth the fan's speed, that is, 2.5 Hz (1200 rpm/60 seconds per minute/8). The static deflection can be computed from the spring natural frequency using the following relationship:

$$d = (3.13/f)^2$$

where:

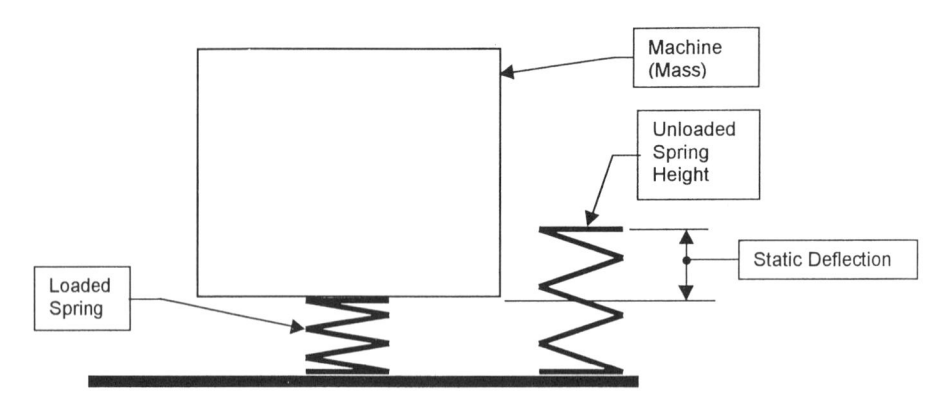

Figure 3.29 Dynamic model for a machine on a spring isolator.

d = static deflection of the spring in inches
f = natural frequency of the spring/mass system in hertz

In the above example, the static deflection d equals 1.6 in. A spring of the proper size is selected from a manufacturer's catalog that lists spring models, maximum design loads, and corresponding static deflections.

2. The static deflection of the building supporting structure must not exceed one-tenth the isolation spring static deflection. In the case at hand, the machine supported on springs must not cause the building floor to deflect by more than 0.16 in. If this is the case, then the static deflection of the spring selected for installation must be resized to be 10 times the building floor static deflection.

In certain cases, machines must first be directly mounted on a concrete inertia base, which is then mounted on spring isolators. An inertia base is usually a steel pan filled with reinforced concrete. The weight and stiffness of the inertia base serve either one or both of two purposes: (1) A concrete inertia base can stabilize the motion of equipment on start-up. This is often needed for the vibration isolation of reciprocating compressors, especially those with compressors mounted at the top of a receiving tank. (2) A concrete inertia base can also be used to obtain a flat, rigid base for equipment items such as base-mounted pumps that need this to maintain shaft alignment. When a pump is mounted on a concrete inertia base, and where one or both pipe risers at

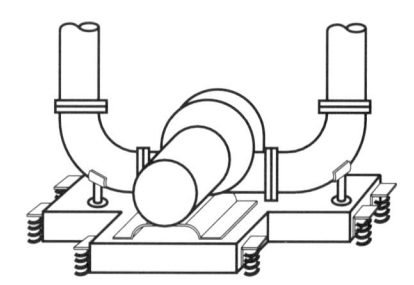

Figure 3.30 Example of a pump with riser stantion supports on a T-shaped inertia base.

TABLE 3.7 Guidelines for Selecting Isolator Static Deflections for Various Types of Building Equipment

EQUIPMENT TYPE	SIZE	RPM	SLAB ON GRADE	SPAN DIMENSION UP TO 20 FT	20 TO 30 FT	30 TO 40 FT
Refrigeration machines						
Reciprocating	All	All	0.25	0.075	1.75	2.5
Centrifugal	All	All	0.25	0.075	1.75	1.75
Absorption	All	All	0.25	0.075	1.75	1.75
Air compressors and vacuum pumps	All	All	0.75	0.75	1.75	1.75
Pumps						
Close coupled	>7.5 HP	All	0.25	0.75	0.75	0.75
	>7.5 HP	All	0.75	0.75	1.75	1.75
Large inline	5–25 HP	All	0.75	1.75	1.75	1.75
	> 30 HP	All	1.75	1.75	1.75	2.50
	>40 HP	All	2.50	3.50	3.50	3.50
End suction & split case	50–125 HP	All	1.75	1.75	2.50	2.50
	>150 HP	All	0.75	1.75	1.75	2.50
Cooling towers		>300	0.25	3.50	3.50	3.50
	All	300–500	0.25	2.50	2.50	2.50
		>500	0.25	0.75	0.75	1.75
Boilers	All	All	0.25	0.75	1.75	2.50
Axial fans						
<22 in. dia.	All	All	0.25	0.75	0.75	0.75
		<300	2.50	3.50	3.50	3.50
>22 in. dia.	All	300–500	0.75	1.75	2.50	2.50
		>500	0.75	1.75	1.75	1.75
Centrifugal fans						
<22 in. dia.	All		0.25	0.75	0.75	1.75
		<300	2.50	3.50	3.50	3.50
>22 in. dia.	<40 HP	300–500	1.75	1.75	2.50	2.50
		>500	0.75	0.75	0.75	1.75
		<300	2.50	3.50	3.50	3.50
>22 in. dia.	>40 HP	300–500	1.75	1.75	2.50	2.50
		>500	1.00	1.75	2.50	2.50
Propeller fans						
Wall mounted	All	All	0.25	0.25	0.25	0.25
Roof mounted	All	All	0.25	0.25	1.75	1.75
Heat pumps	All	All	0.75	0.75	0.75	1.75
Condensing units	All	All	0.25	0.25	1.75	1.75
Packaged units						
	<10 HP	All	0.75	0.75	0.75	0.75
		<300	0.75	3.50	3.50	3.50
	>10 HP and <4 in. WG	300–500	0.75	2.50	2.50	2.50
		>500	0.75	1.75	1.75	1.75
		<300	0.75	3.50	3.50	3.50
	>10 HP and >4 in. WG	300–500	0.75	1.75	2.50	2.50
		>500	0.75	1.75	1.75	2.50
Packaged rooftop units[a]	All	All	N/A	1.75	2.50	3.50

TABLE 3.7 (*Continued*)

EQUIPMENT TYPE	SIZE	RPM	SLAB ON GRADE	SPAN DIMENSION UP TO 20 FT	20 TO 30 FT	30 TO 40 FT
Ducted rotating equipment						
	<600 cfm	All	0.50	0.50	0.50	0.50
	>600 cfm	All	0.75	0.75	0.75	0.75
Internal-combustion-engine-driven electric generators	All	All	0.75	1.75	2.50	3.50
Piping First hanger[b]						
	<2 in. dia.		N/A	N/A	N/A	N/A
All other piping in MER[c]	2–6 in. dia.		0.75	0.75	0.75	0.75
	>6 in. dia.		1.75	1.75	1.75	1.75
	<2 in. dia.		N/A	N/A	N/A	N/A
All piping outside MER[d]	2–6 in. dia.		0.75	0.75	0.75	0.75
	>6 in. dia.		1.75	1.75	1.75	1.75
Electric Transformer	All	All	0.25	0.25	0.25	0.25

[a] The larger of the value indicated or 10 times the deflection of building roof produced by dead load of the unit supported.
[b] Static deflection equal to static deflection of equipment supported.
[c] MER = mechanical equipment room.
[d] For piping within 100 pipe diameters of machinery.

pump connections require vertical support, it is necessary that the inertia base be configured to carry these instead of mounting them on the floor. This condition is as illustrated in Figure 3.30. Note that concrete inertia bases do not improve vibration isolation efficiency. Vibration isolation efficiency is only related to spring static deflection and not to the weight supported.

For structural reasons, most equipment items are mounted on concrete housekeeping pads. The main function of these is to distribute point loads more evenly over the building floor. However, as the name suggests, an important purpose of housekeeping pads is to minimize the chance that floor debris can become jammed under the machine, short-circuiting vibration from the machine, past the isolators, to the building floor.

Table 3.7 is based on recommendations of the *ASHRAE Handbook*.[7] It provides recommended minimum spring static deflections for vibration isolation systems for a variety of equipment commonly used in building mechanical systems. The recommendations are divided by building column span dimension and slab type (structural or grade).

In order to obtain the benefit of vibration-isolation, all connections to vibration isolated equipment must also be resilient. This includes pipe, controls, and electrical connections. Resilient pipe connections are usually either metal braided hose or spherical rubber. The spherical rubber flexible connectors are more effective but can only be used in applications where the fluid temperature is below 200°F; this limit varies slightly from manufacturer to manufacturer. Electrical and control connections can often be attached to vibration isolated machines by means of a "pigtail-loop" connection. Resilient power connections are particularly important in large machines where the conduit and electrical cable used can become quite rigid.

There are a variety of techniques used to control vibration produced by plumbing. Plumbing problems generally breakdown into two groups: domestic water supply and wastewater piping. To control flow noise transmission to adjacent spaces, the following can be used:

- At domestic water pipe mounting locations, first wrap the pipe in $\frac{1}{4}$ in. felt or foam neoprene before securing to a stud frame using a bracket.
- In back-to-back bathrooms and kitchens of adjacent living units or hotel rooms, serve bathrooms with separate water risers for each side. This requires running two sets of hot and cold water risers, so that living unit to unit connections to the riser occur at least one floor apart.
- Use cast-iron soil pipe to minimize wastewater flow noise.
- Do not locate bathroom and kitchen water services on walls common with a living or bedroom space in an adjacent unit.

The application methods discussed in the chapter, and to a large extent throughout this book, require experienced judgment and, when necessary, the assistance of an acoustical consultant. The methods of analyses and information are intended to be guides for use throughout the building design-construction process and to supplement those of other chapters.

REFERENCES

1. Beranek, L. L., "Revised Criteria for Noise in Buildings," *Noise Control,* **3**(1), 19–27 (1957).
2. Harris, C. M., Ed. *Handbook of Acoustical Measurements and Noise Control,* Third Edition, McGraw-Hill, New York, 1984, p. 43.4.
3. Blazier, Jr., W. E., "Revised Noise Criteria for Application in the Acoustical Design and Rating of HVAC Systems," *Noise Control Eng. J.,* March–April 64–73 (1981).
4. Beranek, L. L., and Ver, I., *Noise and Vibration Control Engineering—Principles and Applications,* Wiley, New York, 1992, pp. 128 ff.
5. Bell, L. H., and Bell, D. H.; *Industrial Noise Control—Fundamentals and Applications,* Marcel Dekker, New York, 1993 pp. 116–122.
6. American Society for Testing and Materials, *Standard Classification for Determination of Outdoor Indoor Transmission Class,* ASTM E 1332 90 (Reapproved 1994), West Conshohocken, PA, 1997.
7. ASHRAE, *1995 ASHRAE Applications Handbook,* American Society of Heating, Refrigerating, and Air-Conditioning Engineers, New York, 1995, p. 43.16ff.
8. Schuller International, *Fiber Glass Health & Safety Information Packet,* Schuller International, Inc., Denver, CO (not dated).
9. Guenther, F., "IAQ and Noise Control—Working Together," *Heating/Piping/Air-Conditioning,* January 1996.
10. Hirschorn, M., "Fiberglass & Noise Control—Is it a Safe Combination," in *Sound and Vibration,* Acoustical Publications, Bay Village, OH, 1994, pp. 6–10.
11. Olson, C., "Duct Insulation and Indoor Air Quality—Good Design and Maintenance Keep Duct Insulation Upstream of Air Quality Problems," *Building Design & Construction,* August 1993, pp. 50–54.
12. Reynolds, D. D., and Bledsoe, J. M., *Algorithms for HVAC Acoustics,* American Society of Heating, Refrigerating, and Air-Conditioning Engineers, New York, 1989.
13. Klepper, D. L., Cavanaugh, W. J., and Marshall, L. G., "Noise Control in Music Teaching Facilities," *Noise Control Eng. J.,* **5,**(2) 71–79 (1978).
14. ASHRAE, *1995 ASHRAE Applications Handbook,* American Society of Heating, Refrigerating, and Air-Conditioning Engineers, New York, 1995, pp. 43.30–43.32.

FURTHER READING

AMCA, *Fans and Systems,* Air Movement and Control Association, Arlington Heights, IL, Publication 201.

Beranek, L. L., *Noise and Vibration Control,* McGraw-Hill, New York, 1971.

Fry, A., *Noise Control in Building Services,* Sound Research Laboratories, Pergamon Press, Oxford, England, 1988.

Harris, C. M., *Handbook of Acoustical Measurements and Noise Control,* Institute of Noise Control Engineering, Poughkeepsie, NY, 1998.

Harris, C. M., *Noise Control in Buildings—A Practical Guide for Architects and Engineers,* McGraw-Hill, New York, 1994.

Huntington, W. C., and Mickadeit, R. E., *Building Construction—Materials and Types of Construction,* 5th ed., Wiley, New York, 1980.

Jones, R. S., *Noise & Vibration Control in Buildings,* McGraw-Hill, New York, 1984.

Honsato, *Monsanto Acoustical Glazing Guide,* Monsanto, St. Louis, MO, 1995.

Case Study

MECHANICS HALL, WORCESTER, MASS. COOLING TOWER SOUND ISOLATION

Figure 3.31 shows a photograph of Mechanics Hall in Worcester, Massachusetts. The view is from the rear of the hall and shows a door at the balcony. Figure 3.32 is a part plan sketch of the front stage-right corner of the hall showing the balcony door and stairway accessing the hall. On the roof of the stairway there was located an unisolated cooling tower serving areas of the building. The operation of this cooling tower produced unacceptable levels of sound in the hall. Figure 3.33 shows a section through the balcony and stairway at the balcony door and indicates the location of the cooling tower above the stairway. This section also indicates four airborne and structureborne sound transmission paths. These are as follows:

1. Airborne sound from the cooling tower, transmitted through the stairway roof, into the stairway, and then through the door into the concert hall.
2. Structureborne sound from the cooling tower, transmitted to the stairway roof, and then reradiation of sound from the stairway ceiling into the stairway, and then from the stairway through the door into the hall.
3. Airborne sound from the cooling tower, transmitted through the heavy masonry concert hall wall into the hall.
4. Structureborne sound from the cooling tower, transmitted to the concert hall wall and reradiation into the hall.

There were a number of sound measurements that were conducted. The most revealing of these was that cooling tower sound in the hall increased by only 8 dBA by closing the balcony door. Since the noise reduction of this door is approximately 25 dBA, the more significant path of sound transmission had to be directly through the concert hall wall by means of either an airborne or structureborne path (paths 3 or 4), and not through the stairway (i.e., not via paths 1 or 2). It was separately determined that the airborne sound isolation of the concert hall wall was high enough to ensure that airborne sound directly transmitted into the hall (path 3) is sufficiently low. Then by process of elimination, the path of sound transmission into the hall must be by means of structureborne sound directly through the concert hall wall (path 4). Hence, the proper treatment is to mount the cooling tower on vibration isolators.

Figure 3.31 Interior view of Mechanics Hall, Worcester, Massachusetts. © Steve Rosenthal Architect: Eldridge Boyden, 1857; Architect: Anderson Notter Finegold, 1977 Restoration; Architect: Lamoureux, Pagano Associates, 1990 Restoration.

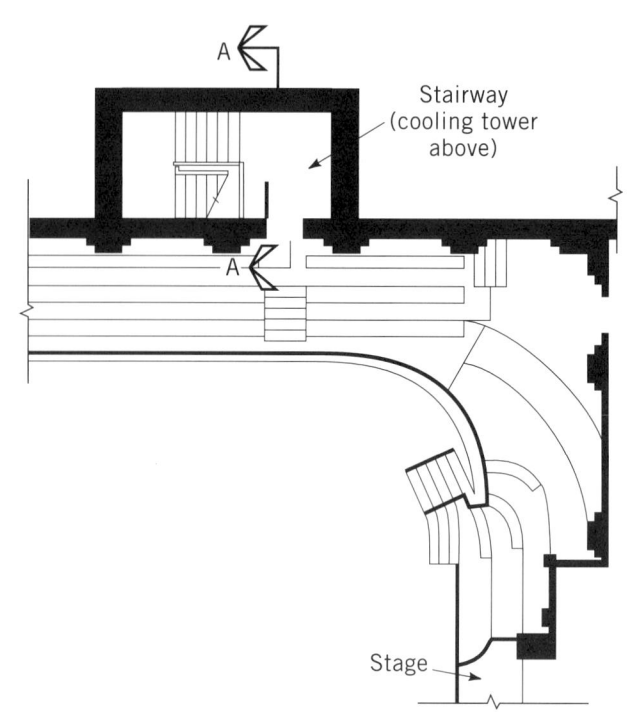

Figure 3.32 Part plan sketch of the front stage-right corner of Mechanics Hall, showing the balcony door and stairway.

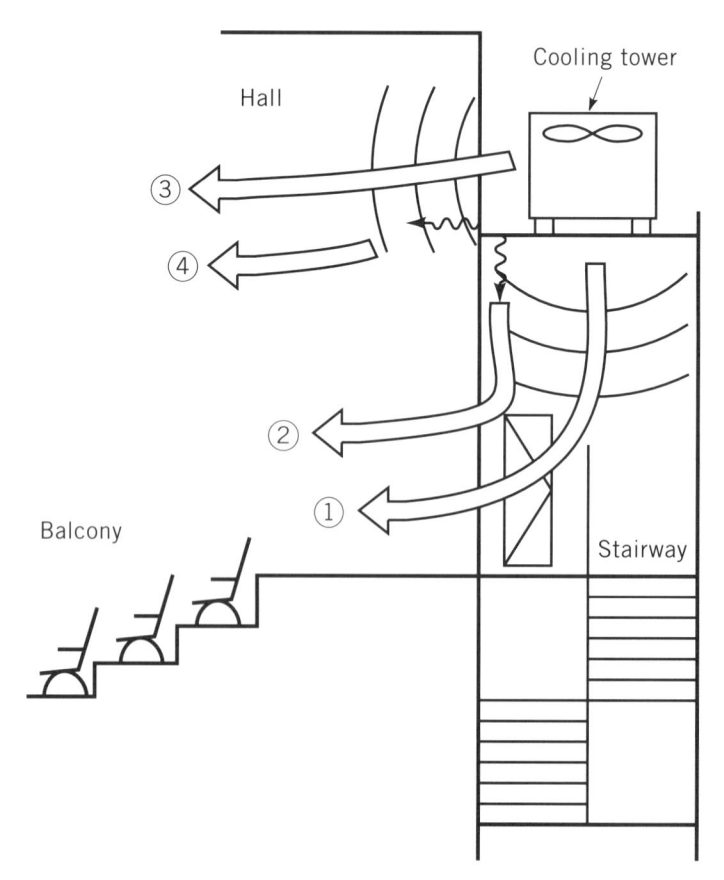

Figure 3.33 Section through Mechanics Hall illustrating four paths of airborne and structureborne sound transmission from a cooling tower located over a stairway into the hall. (Section A-A of Figure 3.32)

4

Acoustical Design: Places for Listening

L. Gerald Marshall
and David L. Klepper

4.1 INTRODUCTION

Within the world of architectural acoustics, the 20th century is often referred to as the "Sabine century" because the development of the reverberation equation by Wallace C. Sabine in 1898 initiated an era of outstanding progress toward improved understanding of the behavior of sound in buildings. (See case study in Chapter 1 of Harvard University's Fogg Art Museum lecture room, where Sabine initiated his original studies.) While significant advances in reverberation control and certain kinds of noise control in buildings occurred in the first half of the century, it was not until the second half that great strides were made in the art and science of designing places for critical listening. Although many made important contributions to the present level of understanding, Leo L. Beranek must be credited as one of the most important contributors during the last half of this century. In recent years, the digital revolution has accelerated the acquisition of knowledge about the behavior of sound in enclosed spaces by greatly simplifying and speeding the gathering and processing of information, and significant advances in knowledge are continuing (see Chapter 6).

In prior centuries, and well on into the 20th century, acoustics in places of listening were largely a matter of either happenstance or a result of reproducing buildings with known characteristics. Occasionally, an intuitive "science of acoustics" was applied to a building, but the effect was frequently either meaningless or harmful. Historically, the monumental architecture of churches produced an acoustic character within which a musical liturgy was more successful than a spoken liturgy, and that had a profound influence on the early development of music. Relatedly, secular music developed in rooms not designed specifically for music performance (palace ballrooms and courtyard theaters, for example), and as rooms specifically designed as places for musical performance evolved from those, the dominant influences were construction methods, esthetics, and the ability for patrons to see and be seen.

The scope of the topic—places for listening—is clearly much too comprehensive to cover in complete detail in a single chapter, or for that matter, in a single text. The material in this chapter is intended as a digest of theory and practice regarding acoustics in widely diverse places for listening.

4.2 SOUND PROPAGATION IN LISTENING PLACES—OUTDOORS AND INDOORS

Outdoor Sound Propagation

The transmission of sound in a free-space environment involves the fall-off of sound intensity by 6 dB, or to 25% of its intensity, for every doubling of distance. A further decrease in sound energy is evident in an outdoor facility because of the sound absorption by the audience as sound travels directly over people's heads. The well-known Greek and Roman outdoor theaters countered this problem by arranging the audience in steeply tiered fashion and by arranging them as close to the performance as possible, reducing the distance sound was required to travel and ensuring direct line-of-sight transmission from the performers to all members of the audience without much audience absorption. These two measures contributed to the acoustical success of those outdoor theaters. A third, and perhaps more important contributing factor, was the location of the theaters at quiet sites. Late afternoon performances required sites with the setting sun behind the audience, with strong implications for prevailing wind direction and sound refraction toward the audience. Finally, in Roman theaters in particular, a wall behind the performers added to the direct sound, with reflected energy arriving at listeners' ears within a short enough time interval (less than 30 msec) after the direct sound to reinforce both the clarity and loudness of the direct sound. These sound-reflecting walls may be thought of as an early first step toward creative room acoustics, with subsequent evolution leading to designing ceiling and wall surfaces of concert halls, opera houses, theaters, and worship spaces so that all surfaces contribute constructively to the listening conditions of the people assembled within them. Before moving on to the design of indoor spaces, however, some principles will be listed for outdoor listening spaces.

Amphitheater Design

A good deal of listening occurs outdoors with no sound-containing enclosure surrounding the audience. Except in the most small-scale and intimate of settings, electronic sound reinforcement is used to supplement the live sound. The audience may number in tens of thousands, it may be seated on flat or gently sloped lawns, and ambient noise from traffic and other sources may be present. Such conditions call for a concert enclosure for symphony orchestras, and an audience-coverage sound system.

Acoustical considerations for amphitheaters include (1) environmental noise levels at the site, (2) effect of activity noise from the site on surrounding areas, (3) conditions for the performers, (4) hearing/viewing conditions for the audience and, (5) amplification system design, including microphone pickup conditions on stage and sound system operator position.

With respect to items 1 and 2 consideration should be given to the use of barriers (berms, walls) to shield the audience and/or surrounding areas from noise intrusion if there is a likelihood of disturbance.

For item 3 concerning performing conditions, a performing area surrounded on three sides and overhead by sound-reflecting surfaces is most desirable for musical groups. These surfaces need not form a total enclosure, and some percentage of openness may be desired to maintain a view of the surrounding landscape. The sound-reflecting enclosure may be fabricated from solid materials or from a fabric membrane. A well-designed enclosure provides good instrumental blending and balancing, and provides the on-stage hearing/playing conditions desired for good ensemble playing and good microphone pickup conditions. Also, weather protection is a major consideration for musical instruments.

With regard to item 4 the ability to hear well is directly related to having unobstructed line of sight to the performers and loudspeakers. From this standpoint, a steeply sloped area for the audience is best. (A good example is Red Rocks Amphitheater in Denver, Colorado.) With a flat audience area, the performers must be elevated to be seen, and then little can be seen upstage beyond the first row of musicians. Loudspeakers can be elevated to be seen from a flat listening area (up to a height of no more than about 45 ft without causing echo-delay problems), but the problem of providing adequate sound levels at a large distance from the stage without "blasting" listeners close to the stage is more severe with sound grazing across a flat area than if distant listeners were elevated.

A summary of principles may be outlined as follows:

Listening Environment
- Provide good sightlines to performers.
- Select a quiet site or, in marginal cases, reduce noise with high barriers around listening area.
- Make certain on-site activities do not disturb neighbors.
- Provide a high-quality sound amplification system and locate the console in the audience area

Performing Environment
- Design for good response and communication for musicians. A full concert shell should be the goal. (Performance enclosures are discussed later in this chapter.)
- Design for good microphone pickup.

Transition to Indoor Acoustics

Figure 4.1 illustrates the transition from outdoor acoustics to indoor acoustics. Figure 4.1a shows an audience outdoors on a flat plane, with both inverse square law (whereby the intensity of sound diminishes inversely with the square of the distance traveled) and audience attenuation reducing sound levels for listeners at the rear. Figure 4.1b shows the sloped seating, which reduces audience absorption attenuation. Figure 4.1c shows the reflector behind the performers, adding reinforcement by reflected sound energy. In Figure 4.1d, the reflector is developed into a full stage shell with reflections both from behind and from above. Finally, in Figure 4.1e, shell surfaces are extended to envelop the audience, providing additional reinforcement by reflected sound energy, and the transition to an indoor theater is complete. In this example, the shaping of walls and ceiling assist both clarity and loudness, but sound-reflecting surfaces can also impair clarity by creating effects such as reverberation, echoes, flutter (rapidly repeating echoes), and focusing capable of interfering with good hearing conditions if not properly controlled. So, in comparing the indoor facility with the outdoor, an enclosure provides positive possibilities for improving hearing conditions but also numerous potential problems.

154 • *Acoustical Design: Places for Listening*

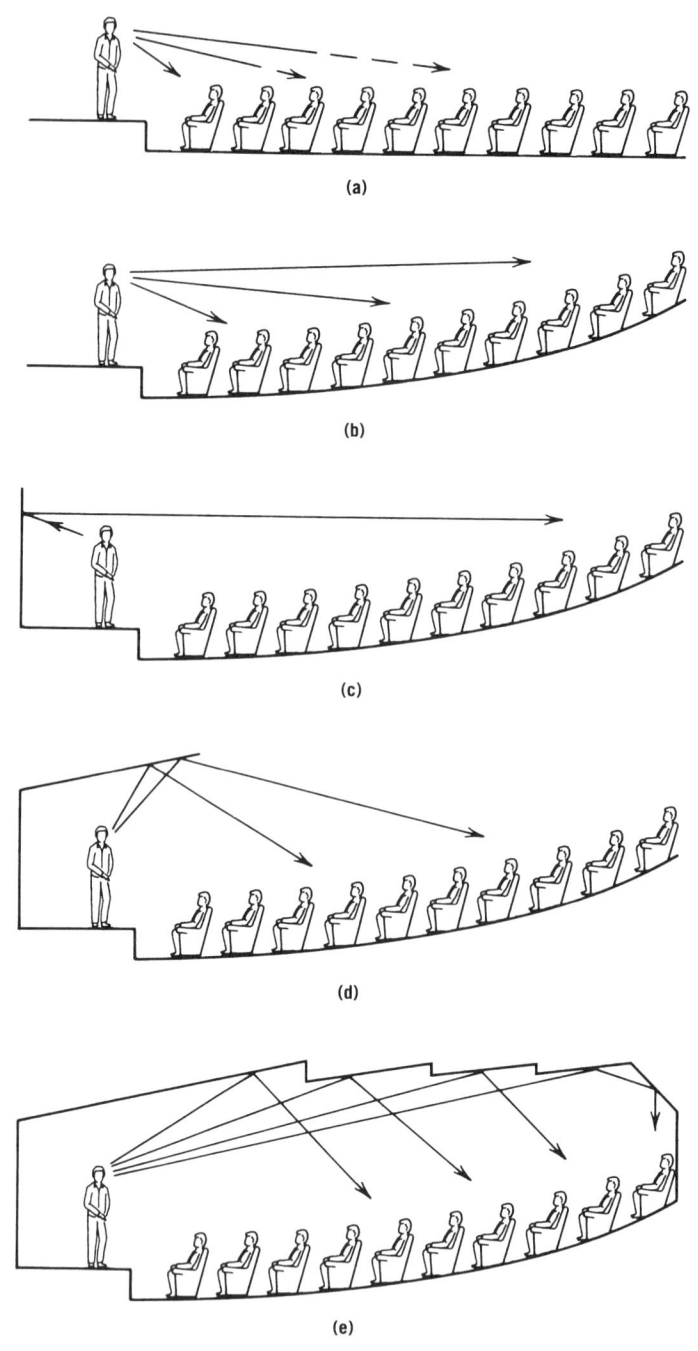

Figure 4.1 Stepped transition from outdoor to indoor conditions: (a) outdoors with no slope, (b) outdoors with sloped seating, (c) reflector behind performer added; (d) complete stage shell added, and (e) indoor auditorium. (From David Klepper, "Acoustical Design—Places of Assembly," Encyclopedia of Architecture: Design, Engineering and Construction, *Vol. 1, Joseph A. Wilkes and Robert T. Packard, Eds. Copyright © 1988 John Wiley & Sons. Reprinted by permission of John Wiley & Sons, Inc.)*

Noise control measures (including sound isolation and machinery noise and vibration control) and sound amplification concerns are common to both indoor and outdoor architectural acoustics projects dedicated to listening and are discussed in other chapters. As this chapter proceeds with discussions of acoustic behavior in various kinds of enclosed spaces, the reader must not overlook these other aspects of architectural acoustics.

Reverberation

The acoustical phenomenon of reverberation is inherent in every indoor space. If the audience and seating were removed from Figure 4.1e, if air absorption were eliminated (a physical impossibility), and if all interior surfaces, including the floor, were made perfectly sound reflecting (also a physical impossibility), a sound once initiated in the space would continue forever, being perfectly reflected between the wall, ceiling, and floor surfaces. This lingering of sound energy after the sound source has ceased is what is meant by reverberation. Of course, in a real space, some sound energy is lost with each reflection from a wall and ceiling surface, and even more is lost upon reflection from the audience and seats. Air absorption, which is usually a small additional effect, can become one that is not so small at high frequencies in very large spaces. The effect of all of this sound absorption is to give a finite length of time in which lingering sound energy can be heard. This is usually defined as the reverberation time, the time required for sound energy to decay 60 dB, or to one-millionth of its initial intensity. The concept of reverberation time and an associated calculation procedure date from the beginning of the 20th century, having been developed by Wallace C. Sabine. This development is credited with initiating architectural acoustics as a science, and to this day reverberation time remains as a fundamental design parameter. In any real indoor space, reverberation time will vary with frequency, sometimes only slightly, but sometimes considerably. When one refers to a single number reverberation time, the reference is usually to the time at 500 or 1000 Hz, or an average of the two.

For many years, achieving the proper reverberation time was considered a prime goal in the design of concert halls, opera houses, theaters, and worship spaces, along with such important matters as freedom from echo, flutter, focusing effects, and unwanted noise. Today, reverberation time is considered just one of several important parameters, and some even dispute its importance. Indeed, in experiments with electronically synthesized sound fields, reverberation time is not judged to be nearly as important as a number of other measures, particularly those dealing with early-reflection behavior (such as the early to late ratio and early decay time, described below). In real spaces, however, as opposed to simulated spaces where parameters can be altered and judged independently, these qualities are necessarily interrelated with reverberation time by geometry, and it happens that reverberation time retains considerable significance as an acoustic measure.

An ideal value for reverberation time is usually assigned on the basis of a room's function. Figure 4.2 is an example of one such set of criteria, developed primarily by Russell Johnson. The variation in ideal reverberation times (RT) from less than 1 sec to well over 3 secs is striking.

4.3 CONCERT HALLS AND RECITAL HALLS

Concert Halls

Concert halls represent one of the most interesting types of interior places of assembly to acousticians and may have benefited more from acoustical research than any other

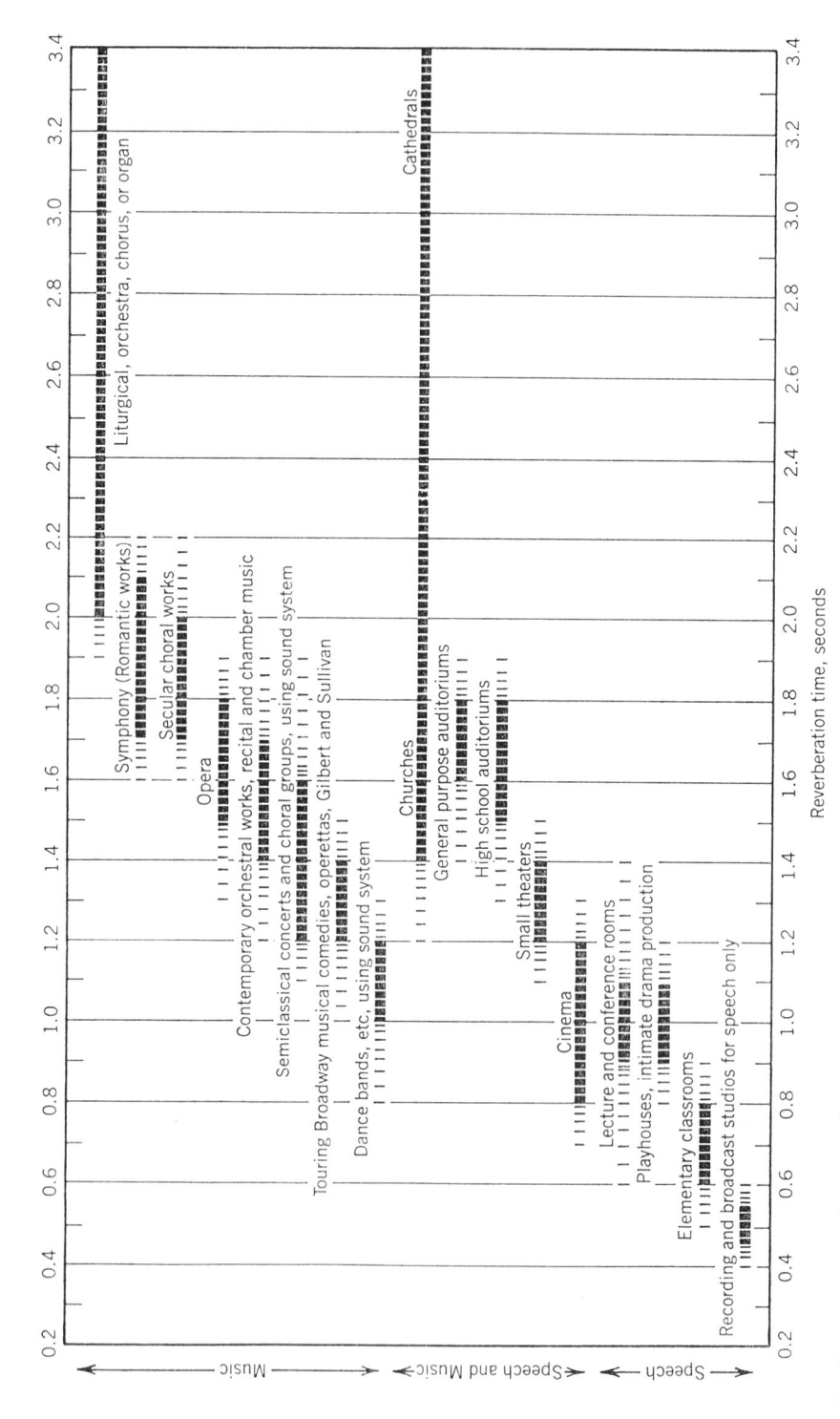

Figure 4.2 Optimum reverberation (500–1000 Hz) for auditoriums and similar facilities. (*Courtesy of Russell Johnson and Bolt Beranek and Newman, Inc.*)

building type. This research has reduced the element of chance in design, but not eliminated it completely. In 1962, after research based on the evaluation of 54 concert halls, Leo Beranek identified 18 characteristics of halls. Today, it is not certain that this particular list of terms would be the one used for characterizing a concert hall, but review of these attributes is useful, and the reader is referred to Beranek's *Music, Acoustics and Architecture* (1962)[1] as an important early influence on contemporary concert hall design, and to the more-recent (1996) *Concert and Opera Halls: How They Sound*,[2] which is an expanded and thoroughly updated revision of the former.

Subjective Attributes and Objective Measures

In his 1962 book, Beranek placed considerable importance on the attribute of intimacy (or presence) and related this quality (which should be largely self-explanatory) to the length of time between the arrival of the direct sound at the listener's ear and the first reflected sound, with 20 msec or less as optimal. He labeled this characteristic "initial time delay gap" (t_I). Since that time, more complex criteria have been developed, but, interestingly, t_I remains a good measure in spite of its simplicity because of several related physical implications that are of importance. For example, the narrow shoebox hall discussed below as one ideal configuration inherently produces a short t_I.

In a 1992 *Journal of the Acoustical Society of America* tutorial paper,[3] Beranek presented a condensed list of seven essential attributes of concert hall acoustics consisting of (1) reverberance, (2) loudness, (3) spaciousness, (4) clarity, (5) intimacy, (6) warmth, and, (7) hearing on stage. He noted that t_I relates closely to both intimacy and spaciousness, in part because of the effect of the quick reinforcing reflection and also because of the associated physical attributes from a geometry that produces a quick initial reflection.

The many acoustic parameters associated with concert halls (including those just listed) might be further reduced to just 3 basic groups: (1) those relating to clarity (intelligibility, articulation, definition), (2) those relating to audible room effects or ambience (sound quality, spaciousness, enhancement), and (3) those relating to loudness. Loudness is largely a function of hall size and seating capacity, and, therefore, is largely self-determining in an otherwise well-designed hall since the sound power output of the source (orchestra) is a somewhat fixed quantity. Occasionally, a designer must be concerned with avoiding excessive loudness in a small hall, or, on the other hand, with attempting to overcome weakness in a very large hall. Loudness cannot be totally separated from items 1 and 2 in this group since the audibility of spatial and reverberant effects changes with loudness, and also because clarity is influenced by the relative strengths of signal and noise. Loudness is rated by the objective measure, G, called strength factor or loudness index.

Both clarity and ambience are determined by a sequence of events beginning with the arrival of the initial signal, followed by a continuing series of room reflections that gradually dissipate because of losses experienced at the boundaries and in the air. The strength, time, and direction-of-arrival characteristics of these reflections in connection with the relative strength of the direct signal determine subjective impression. Strength–time characteristics can be examined on the energy–time curve (ETC) of Figure 4.3. The details of the early portion (consisting of discrete early reflections) are especially important, and this portion can be further divided at roughly 80 msec into early-early and late-early categories, with the former being the more significant of the two categories (see Figure 4.4). A hall's sound quality is largely established in the early-reflection period, with the effect of subsequent sound decay primarily being an audible persistence

158 • *Acoustical Design: Places for Listening*

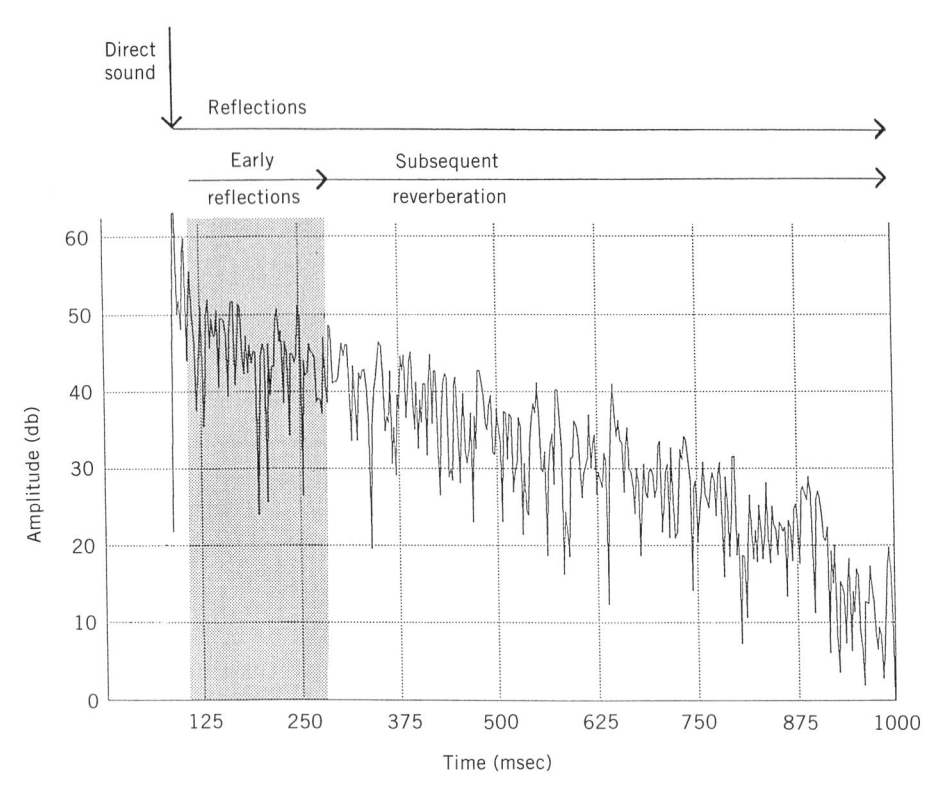

Figure 4.3 Energy–time curve with principal components.

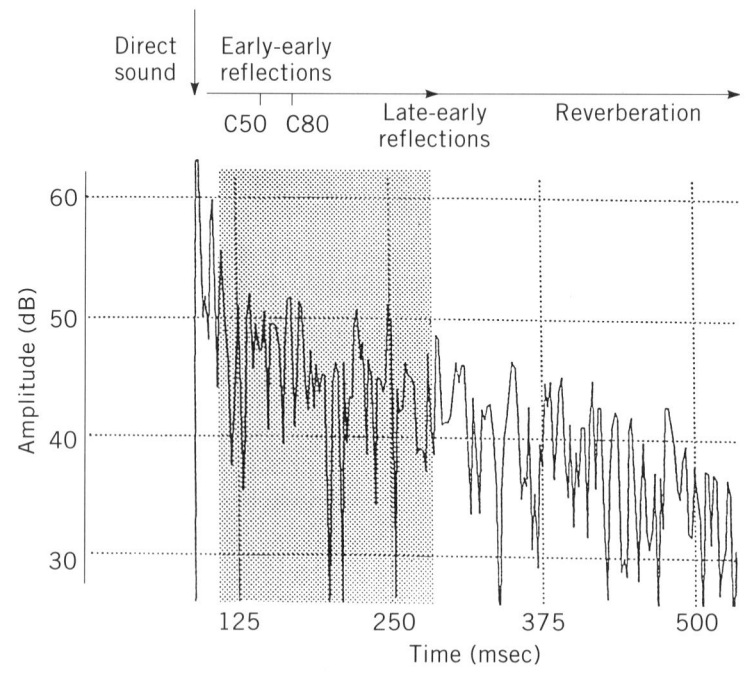

Figure 4.4 Early-reflection portion of energy–time curve (shaded).

during rests, separations between notes, and the like. The outlined early and late categories cannot be defined with precision since, among other things, they vary with room size, reverberation time, strength, spacing and direction of reflections, and type of activity (e.g., speech versus music or organ versus piano). In simple terms, reflections arriving within the first 50 msec will usually contribute beneficially to speech clarity; those within 50 to 100 msec may or may not be beneficial; and those arriving subsequently will probably not contribute to clarity and may instead be harmful (though still useful for ambience). For music, these useful/harmful times can be extended somewhat, as a generalization, although this is a variable that depends on type of instrument, tempo, scoring, compositional style, and so forth. This explains the different early-late ratio integration times used for the music and speech clarity ratios, C_{50} and C_{80}, described below.

A number of objective measures have been developed for rating clarity. Among the most widely used are the speech transmission index (STI) and its abbreviated derivative, rapid STI (RASTI), and the articulation loss of consonants (%Alcons). The simple early-late ratio (ELR) seems to function about as reliably as the other measures, and for speech, a 50-ms ratio dividing time is generally used, while for music an 80-msec dividing time is used. These measures are known as C_{50} and C_{80}, respectively. Energy preceding the dividing time is considered beneficial to clarity, and later energy is considered harmful, although this represents an oversimplification. Figure 4.5 equates C_{50} to STI, %Alcons, and a corresponding subjective rating scale. This figure also shows an ELR rating scale developed for music, using C_{80}. Frequency is an ever-present variable in acoustics and must be fully considered in any objective measure. For example, the 2-kHz octave band is by far the most important for speech intelligibility.

Room ambience is solely an effect of enclosed space and, depending on the details of the enclosure's response, can in the case of speech, affect both clarity and naturalness, and in the case of music, affect clarity, tonal quality, and spaciousness. A number of objective measures deal with the two parts of ambience, which may be described as reverberance and spaciousness. For reverberance (which follows from the lingering or time stretching of sound within an enclosure), RT, EDT (early decay time, the time expended for the initial 10 dB of reverberant decay multiplied by 6 for a 60 dB decay time extrapolation), and ELR are widely used measures. The ELR in this context views reverberance as a reciprocal quality to clarity, which is a commonly understood principle

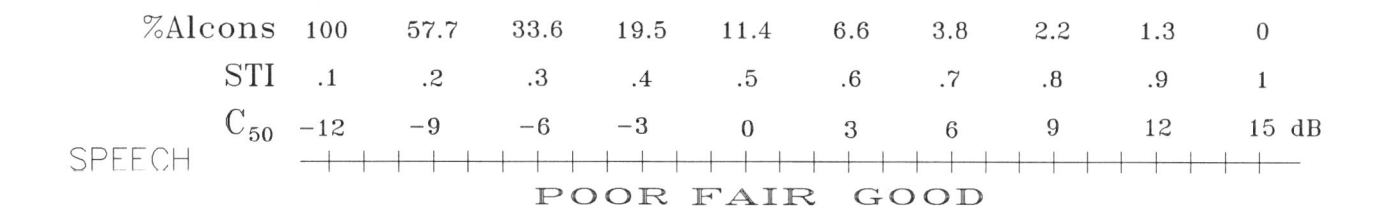

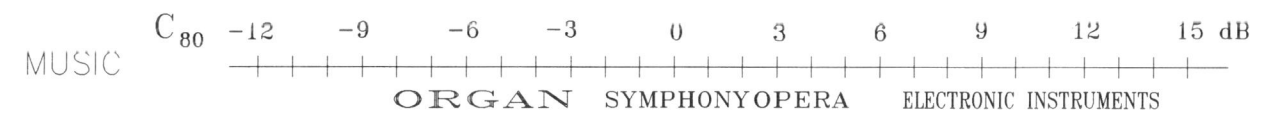

Figure 4.5 Speech and music objective measures vs. subjective rating scales.

in acoustics. Conventional wisdom has long held that good speech acoustics equate to poor music acoustics, and vice versa. For spaciousness, the interaural cross-correlation coefficient (IACC), a measure of the dissimilarity of sounds arriving at the left and right ears and the lateral fraction (LF), a ratio of lateral-to-omnidirectional energy arrivals, usually integrated over the first 80 msec are the widely used measures, though neither seems to be perfectly reliable. Two spatial qualities are often described: the apparent source width (ASW) being related to stereo effect or enlargement of the sound source, and the listener envelopment (LEV) related more to surround-sound effect or "room impression." Recent research indicates that ASW is strongly influenced by the strength of lateral reflections in the early-early reflection period, while the strength of later arriving lateral energy seems most influential for LEV, most particularly that of later-arriving lateral energy below 1,000 Hz.

Similarly, two types of reverberance are observed, one relating to the sense of singing tone or liveliness during running music and the other relating to the persistence of sound heard after a stopped tone. The EDT is used to define the former (commonly referred to as liveliness or running liveness), while RT is used for the latter (often called late or audible reverberation). Since EDT comes from the early portion of the decay process, it correlates well with ELR, and either measure may be used for rating this aspect of reverberance.

Even solid agreement in the acoustical consulting profession on the truly critical attributes would still leave the designer with the complicated problem of design. In the absence of solid agreement on the various attributes and their relative importance, creation of a universal set of design principles is still in the future. How, then, does one go about designing a concert hall with a good chance of success?

Shoebox Approach

The traditional, or European, shoebox-shaped concert hall and Western music grew up together and influenced one another. Middle Ages and Renaissance music performances often took place in spaces called oratories, which were rectangular-shaped rooms, and in similarly shaped palace ballrooms. When concerts moved out of palaces and were attended by the newly emerging middle class, the design of these oratories was copied in the design of the first concert halls. When Sabine, a pioneer in architectural acoustics, assisted in the design of the now highly regarded Boston Symphony Hall, he based the design on the already existing rectangular shoebox Boston Music Hall, and on such successful European halls as Leipzig's Neues Gewandhaus (destroyed in World War II). One very successful modern shoebox concert hall is Salt Lake City's Symphony Hall (Cyril Harris, acoustical consultant). (See Figure 4.6.)

Meyerson Hall in Dallas, Texas (Artec, acoustic consultant) is another successful modern shoebox hall (see case study in Chapter 6). This hall incorporates considerable acoustic adjustability, one significant part of which is the large sound-reflecting surface suspended over the performing area. This reflector, which is adjustable in height, improves the ability of musicians on stage to hear one another. Shoebox halls without such reflectors often have less-than-ideal conditions on stage because of high stage ceilings.

The shoebox approach to concert hall design is analogous to musical instrument design. There is a considerable amount of information about violin acoustics, but most successful new violins have designs based on old violins. Similarly, new concert halls often copy old successful ones. Note that the surfaces of these shoeboxes are not smooth and slick, either in the old halls or in the new halls. Side and rear balconies break up the smooth geometry, and detailed modulation is presented by niches and statues in the older halls and by deliberate surface modulations in both ceiling and wall areas in the new halls.

Acoustical success is not guaranteed by using a shoebox form, however, particularly if insufficient diffusion is employed. Also, a number of shoebox halls that are highly regarded for audience acoustics have on-stage player communication difficulties because of high stage ceilings, and, of course, performance quality must be negatively affected in these halls. The shoebox may be a good beginning because of several inherent desirable qualities, but other aspects of acoustical design, such as performing area configuration, proper volume, echo control, and good sightlines still need to be considered. Adequate diffusion is also important. (Figure 4.7 is an example of a sound-diffusing surface.) In any hall, a good on-stage environment must be produced.

If the hall is not too large, the shoebox approach agrees with the importance that many modern acoustical consultants attach to lateral reflections. In general, the modern shoebox geometry seems applicable for concert halls seating up to 2000 people, or perhaps 2500 at the most. If the hall is to remain narrow for producing early-arriving reflections, increasing the seating capacity too much can place the rear of the hall too far from the stage for any visual intimacy, regardless of acoustical excellence. This objection to the shoebox, plus the architect's natural inclination to be creative and not to copy old forms, has led to considerable experimentation with other shapes. Certainly, the shoebox does not have a monopoly on acoustical success.

Panel-Array Halls

The use of sound-reflecting panels in concert halls is often misunderstood. Their use was blamed for the acoustical deficiencies of Philharmonic Hall at New York's Lincoln Center, but undoubtedly it was quantity and arrangement of panels that was at fault, rather than the use by itself. Many acoustical consultants first used sound-reflecting panels to cure focusing effects and improve sound distribution in domed auditoriums, such as the Aula Magna in Caracas, Venezuela, MIT's Kresge Auditorium in Cambridge, Massachusetts; and the Reorganized Church of Jesus Christ of Latter-Day Saints Auditorium in Independence, Missouri. Randomly arranged square plaster panels were successfully used to interrupt the focusing effects of a multibarrel vaulted ceiling at Spaulding Concert Hall in the Hopkins Center at Dartmouth College. These allow the clear space above the panels to add to the room's reverberant volume, while assuring good distribution of early reflected sound. This favorable combination of early reflected sound with ample reverberation led several acoustical consultants to extend the use of sound-reflecting panels to auditoriums without focusing effects. Simply put, sound reflecting panels can be used to help shape a room's interior to provide early reflections without sacrificing the high volume required for reverberation. (See Figure 4.8 for an uninhibited use of sound-reflecting panels.) With sound-reflecting panels, one is not limited to the narrow shoebox shape, and configurations such as the large fan plan of Tanglewood can be made to work well with panels. Indeed, the shoebox hall can benefit from the use of sound-reflecting panels on stage for the performers and over the front portion of the audience where early reflections may not arrive as quickly as desired. Many very fine concert halls may be classified as panel-array halls.

Concert Hall In-the-Round

The desire to bring the audience close to the stage has extended beyond wide fan-plan halls to nearly circular halls with the concert stage near the center. One of the first important halls of this type was the Philharmonie in Berlin (Hans Scharoun, architect, and Lothar Cremer, acoustical consultant). The successful Christchurch Town Hall in

BUILDING CROSS SECTION A-A

(a)

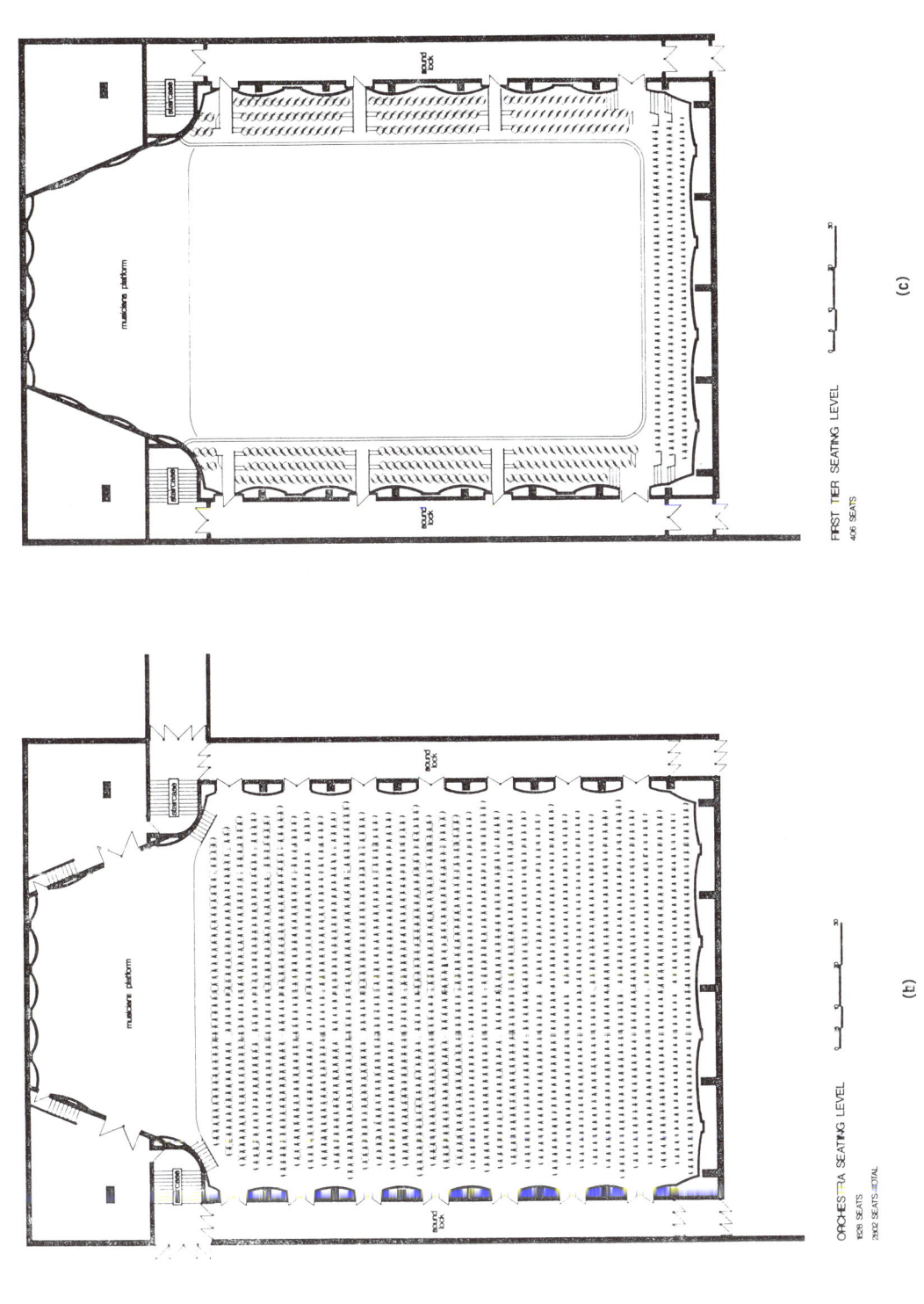

Figure 4.6 (a) Symphony Hall in Salt Lake City, Utah, a modern example of a shoebox concert hall; (b) main floor; (c) first balcony. C. Harris, Acoustical Consultant. (Courtesy of FFKR, Architects.)

164 • *Acoustical Design: Places for Listening*

Figure 4.7 Example of on-stage wall diffusion and open walls around the audience at the Finger Lakes Performing Arts Center, Canandaigua, New York. KMK Associates, Acoustical Consultant. (Courtesy of Handler/Grosso, Architects.)

Figure 4.8 An uninhibited and successful use of sound-reflecting panels. Lincoln Theater, Miami Beach, Florida. G. DeMarco Architect; P. Frink, Associated Theater Architects; KMK Associates, Acoustical Consultant. (Courtesy of Peter Frink.)

New Zealand (A. Harold Marshall, acoustic consultant) came soon after and was somewhat pioneering in its emphasis on lateral reflections from large sloping side wall panels. Another hall of the same type is Boettcher Concert Hall in Denver, Colorado (Christopher Jaffe, acoustical consultant). The Berlin and Denver halls both use panel arrays to complement irregularly shaped wall and ceiling surfaces to distribute sound energy. The experience of several acoustical consultants indicates that the "concert hall in-the-round" is probably not a best choice for a basic design, unless sophisticated use of electronics supplements the natural acoustics of the hall or unless a relatively small portion of the seating is behind the orchestra. Good acoustics may be achieved in the main part of the audience facing the orchestra, but achieving an equally good listening environment in a large area behind the orchestra is difficult indeed. Musical instruments and human voices are directional, and the traditional audience–performer relationship takes this into account. When seating is placed behind an orchestra, horns are pointed toward the audience, a piano lid obstructs the instrument's strings, singer's heads do not emit sound well to the rear, and so forth. Acoustics aside, though, in-the-round configurations have a good deal of audience appeal.

Partially Enclosed Concert Hall

For mild-weather concert going, the partially enclosed concert hall has proven to be a popular venue. Most major orchestras now have summer homes, and there are many very fine acoustic examples. Since these are commonly sized to accommodate large informal summertime audiences, the well-regarded examples often seem to defy the traditional criteria for good concert hall acoustics. For example, the open sides can be equated to absorptive boundaries in fully enclosed halls, and, in contrast, concert hall designers strain to obtain boundaries that are as massive and sound reflecting as possible in fully enclosed halls (see Figure 4.7). Further, designers of enclosed halls struggle to keep seating capacities in the ideal range of 1800 to 2200 seats, while partially enclosed halls often have considerably larger capacities. Yet, excellent results can be obtained when the hall is designed to produce the proper balance of clarity and reverberation by observing with particular care the requirements of early reflections, high volume, and good sightlines. A special consideration for this type of hall is the absolute need for a quiet site.

The music shed at Tanglewood in Lenox, Massachusetts, is a familiar example that has served as a prototype for many subsequent facilities. Tanglewood is blessed with a quiet site, but it must accommodate 5000 indoor listeners (6000 until wider seats were installed). Prior to 1959, the acoustics were less than ideal; there was ample reverberation but poor definition. The remodeling efforts designed by architect Eero Saarinen and acoustical consultants Bolt, Beranek and Newman in 1959 and 1960 gave the hall its present configuration, including the installation of sound-reflecting panels on the upper rear wall. These elements and the hall's sufficient volume assist in effectively resolving the acoustical problems associated with large size and missing side wall area. More recent studies[4] suggest that the triangular panel shape used at Tanglewood is optimal and that the panels evenly send strong reflections to the audience. The sound quality for full orchestra or orchestra with chorus is excellent, and better than one would expect from such a large, simple structure.

Yet, certain deficiencies are inherent in such a large, widely fan-shaped, flat-floored facility, the most serious being poor sightlines. Many of the facilities built later have solved the sightlines problem by either reducing seating capacity, sharply raking the seating, adding a balcony, or some combination of these. Several are quite successful in combining improved sightlines with the good acoustical features of Tanglewood, and the results almost equal the best of "winter" concert halls.

One other characteristic of most semienclosed halls is the use of hard, utilitarian seats instead of upholstered seats as normally used in indoor halls. Empty seats are sound reflecting instead of sound absorbing, making the empty hall excessively reverberant and the acoustics poor for rehearsals. Two solutions are possible—either rehearse elsewhere or incorporate large areas of sound-absorbing draperies that can be extended within the hall for rehearsals and completely withdrawn for performances.

Outdoors, uniform reinforcement of sound for lawn seating is typically accomplished by distributing loudspeakers along the upper rear and side walls of the auditorium. The signal to these loudspeakers must be delayed in time by a sufficient amount to integrate properly with the live sound from the stage. An earlier section in this chapter on amphitheater design contains a brief list of design considerations relevant to these partially enclosed outdoor performance facilities.

Concert Hall's Other Uses

The concert halls discussed thus far were designed specifically for symphony and choral concerts. Some have adjustability to optimize the acoustics for varying sizes of performing groups and/or varying periods of music (the Romantic period requiring more reverberation and less definition, according to many, than the classical and modern periods). Yet even the most single-purpose concert hall will generally be used for lectures, popular music concerts, film viewing, and conventions. A good electronic sound system can go a long way toward adapting a concert hall to these uses, but a sound system combined with adjustable sound absorption to reduce reverberation can be even more effective in assuring that all uses of the room will benefit from good hearing conditions. This need for adjustability will be explored further when multipurpose auditoriums are discussed. Economics dictates that nearly every concert hall, theater, or opera house will experience multiple uses.

Electronic Concert Hall

When the Royal Festival Hall in London opened in 1951, it was criticized for the lack of adequate bass response and lack of reverberation. The causes of those defects were several, including too small a room volume and the use of sound-absorbing elements. Instead of major architectural surgery, the hall was improved by a subtle and sophisticated electronic system. P. H. Parkin developed a system called assisted resonance, using multiple amplification channels, each having a rather narrow bandwidth, and easily adjusted to have a gain just under feedback. Together, when properly adjusted, these amplification channels provided the hall with additional reverberation and additional bass energy without being obviously artificial. Almost simultaneously, John Ditamore was experimenting with broadband amplification and reverberation at Purdue University, using reverberation chambers and tape recordings into tape-loop delay devices. In recent years, C. Jaffe has been a vocal proponent of "electronic architecture" and can certainly share in being credited for the increasing acceptance of electronically synthesized concert hall sound. More recently, fine systems have been produced by Lexicon Co. (LARES) and by Acoustic Control Systems (ACS).

Thanks to digital technology, the state of the art has progressed to the point that electronically synthesized acoustics can be of exceptionally high quality, with the only major criticism being a philosophical one—should electronic acoustics be substituted for natural acoustics? In existing facilities that have hopelessly poor acoustics with no reasonable hope for improvement by practical cost effective means, the case for electronic acoustics can be argued with considerable justification. Beyond that, the authors will leave the judgment to others.

Recital Halls

Halls for chamber music are an especially attractive form of concert hall to acoustic consultants since some of the constraints associated with larger halls are not present. It is not uncommon, though, for an extremely wide range of musical organization sizes and types to use a recital hall (organ, symphony orchestra, and chorus at one extreme and solo recital at the other, e.g.), and this means that a large range of sound power outputs, optimal reverberant characteristics, and performing area sizes should be accommodated by the hall. Adjustable absorption (usually in the form of tracked draperies) can be useful for controlling both reverberation time and loudness.

Whenever a hall accommodates many sizes of performing groups, physical adjustability at the platform is desired to allow each activity to occur within an ideal stage setting. With performing area adjustability, a soloist need not perform on a large, bare stage. Yet, the stage may expand to handle activities requiring more space. Movable stage wall elements and batten-hung ceiling elements may be used to accomplish performing area adjustability, as shown in the hall in Figure 4.9.

If there is an organ, it should be placed on the upstage wall, and the mouths of the pipes should be high enough so that choir members, when present, are not overwhelmed by organ sound. This means a high stage ceiling will be required in such cases, and any suspended sound-reflecting elements over the stage provided for on-stage communication and response for instrumentalists should be mecha-

Figure 4.9 Track-suspended stage enclosure units that can be easily moved to surround any size performing group. The black reverse sides of the panels serve as masking legs for other kinds of events. Wellin Hall at Hamilton College, Clinton, NY Ewing Cole Cherry, Architects; KMK Associates, Acoustical Consultant. (Courtesy of Tom Bernard.)

nized to move out of the way for organ recitals if otherwise acoustically or visually obstructing.

4.4 OPERA HOUSES, THEATERS, GENERAL-PURPOSE AUDITORIA, AND WORSHIP SPACES

Opera Houses

From the standpoint of acoustics, an opera house may be described as a concert hall designed to accommodate singers on stage and an orchestra in a pit. Consequently, the hall should produce beauty of tone for both the singers and the orchestra, maintain an acceptable balance of audibility between the two, and produce a sense of responsiveness to the singers and the instrumentalists.

Acoustic enhancement of musical sources requires a room that produces both reverberance and clarity, and in a balance appropriate to the specific activity. Though simplistic, there is a rule-of-thumb that says symphony halls, opera houses, and drama theaters should have RT characteristics of about 2, 1.5, and 1 sec, respectively. Referring back to Figure 4.2, it is seen that the RT considered ideal for opera is about 1.3 to 1.8 sec, and is lower than that considered ideal for orchestral music. As noted earlier, clarity and reverberance are largely reciprocal qualities. Since opera involves language, both sung and spoken, plus orchestral music, it should not be surprising that clarity–reverberance balance requirements for opera fall between those for symphony and drama, though normally closer to those for symphony (see Figure 4.5). In houses where opera is performed exclusively in the vernacular, those balance requirements might tilt a little more toward clarity.

Both clarity and sound enhancement effects are influenced by the details of a room's geometry, finishes, and furnishings. For clarity, emphasis should be placed on natural reinforcement of singers' voices, but the opportunity for natural reinforcement from room boundaries is minimized by the considerable forward directivity of the voice. So, good line of sight to the stage from all seats is needed for both good acoustics and the visual requirements of musical theater. Orchestral sound from the pit is the other acoustic factor in opera houses, and this aspect is discussed in the next section.

One defining difference between the typical opera house and concert hall is that the opera house has a proscenium arch separating the performers from the audience, and a tall stage tower behind the proscenium to permit rapid changes of scenery by flying scenery in the stagehouse. Another major difference is that the concert hall developed its traditional shoebox shape from the Renaissance oratory, while the horseshoe-shaped opera house developed from the typical court theater in which the audience was arranged in boxes at the rear and both sides, with many levels from floor to ceiling. (See Figure 4.10 for a horseshoe design in the grand tradition.) For opera, the goal was to minimize the distance from the audience to the stage, and a limitation was the degree of balcony cantilever possible at the time. Modern use of the horseshoe is evident in such halls as the Metropolitan Opera House and New York State Theater, both at New York's Lincoln Center (C. Harris and Vilhelm Jordan, acoustical consultants). Classic opera houses, such as the Vienna Staatsoper (reconstructed after World War II with 1700 seats) and Milan's La Scala, usually seated less than 2200 people and would seat considerably less with up-to-date seating standards. In contrast, the Metropolitan Opera House seats 3900 with modern seating standards and is acoustically successful. Of course, voices of international stature are essential to fill this house.

Figure 4.10 Fulton Opera House, Lancaster, Pennsylvania. Elegance and grand scale are commonly associated with horseshoe-design opera houses. Kessler Associates, Inc., Renovation Architect; KMK Associates, Acoustical Consultant. (Courtesy of David Rose.)

Orchestra Pits

In opera houses and multiuse auditoria, a properly designed pit is a basic need. The design approach should be to produce a pit having a large opening in front of the stage, a stage undercut with clearance for standing musicians, and, for multiuse auditoria in particular, a powered lift. A lift is extremely important to the practical use of a multiform forestage area. With a powered lift, the changeover from pit use, to audience seating, to extended forestage is accomplished with a minimum of effort, and all these uses are important in a multipurpose facility. Manual staging elements (similar to tables with folding legs) can be installed instead, but, in fact, these tend to be unused because of the major effort involved in accomplishing changeover.

The provision of more than a single lift can be helpful in tailoring the size of the pit opening to match the size of the pit orchestra used for a particular production. Often a primary lift will exist directly in front of the stage, with a front–back dimension corresponding to about three rows of audience seats; and a secondary lift will exist in front of this, with a front–back dimension of about two rows of seats.

The more spacious a pit is, the better the sound produced within it. A stage undercut is very helpful in achieving adequate floor area and roominess without removing an inordinately large number of audience seats. With any alternative, a removable, solid-paneled pit railing should separate the pit area from the first row of seats. In plan, a pit should not be exceptionally wide and shallow. The arrangement of instrumentalists and conductor must always be given consideration.

Pits are often constructed too small for their purpose in modern halls because of the desire to bring the audience as close to the stage as possible. A minimum design figure for pits is 16 ft^2 (1.5 m^2) per musician. Logically, this is about the same as the on-stage per-musician design figure since the size of instruments, players, and needed clearances are the same in a pit as on a stage. For small-capacity pits, a larger per-musician figure should be used since the area used up by percussion and keyboard instruments is very often not much different for small orchestras than for large.

Examples of open pits include most European opera houses, and the Metropolitan Opera House and New York State Theater at New York's Lincoln Center. Pits where Wagner's operas predominate should have a large covered area to develop a more blended and distant sound from the orchestra and to allow better balance with the singers. The prime example is the Bayreuth Festspielhaus, of course. Most modern pits are a compromise between the two types, but with the larger portion of the orchestra in the open area.

Adjustable sound-absorbing treatment is desirable in most cases for controlling sound levels (i.e., pit–singer balance) in certain kinds of events. Usually, this treatment is in the form of curtains on the pit's rear wall. Also, pit railings can be reversible, with one side sound reflecting and the other sound absorbing. Sometimes, loose sections of carpet are brought in for some varieties of high-powered pops events.

Theaters

In any theater, the goal of effortless perception of speech overshadows any other acoustical design consideration. Other acoustical goals are speech naturalness and wide dynamic range, from a whisper to a yell. Of course, these goals may be achieved by proper room acoustics design in moderate-sized theaters, but sound reinforcement is essential in large theaters. Proscenium theaters with capacities up to and even exceeding 1000 seats are possible without electronic amplification, but theaters-in-the-round require amplification above about a 600-seat capacity because of the difficulty in providing reinforcement behind the actor's back through room acoustics.

In contrast to a concert hall, a proscenium theater is probably best served by the moderate fan plan, which puts a larger number of the audience closer to the stage than does a rectangular plan. For the same reason, a plan with balconies is better than single-level seating, except in the smallest of theaters. Thrust-stage and in-the-round theaters are complicated by the fact that, in the speech range where much intelligibility lies (from f's, s's, and t's), voice levels are roughly 10 dB lower behind the speaker's head than in front of it. The best thrust-stage and in-the-round theaters are, therefore, quite small, and the unidirectional theater with a mild fan plan has an acoustical advantage when compared to other theaters of the same audience size. Monumental designs with high ceilings, useful in assuring adequate reverberation from high volume in concert halls, are not practical in speech theaters where the goal of high intelligibility demands a low reverberation time and strong reinforcement. This is best met by low ceilings that reflect sound quickly to the audience no more than 30 msec after the direct sound from the actor. Ensuring such quick reflections for all members of the audience can be difficult because the ceiling and wall sound-reflecting surfaces must be coordinated with stage lighting slots and visual design requirements for the theater. Note that in some theaters where monumentality and consequent high ceilings are essential, specially shaped side balconies can help provide short-delayed sound reflections for the central main floor, underbalcony soffits can provide reflections for the seats under them, and the main ceiling can be shaped to reflect sound energy quickly to the balconies. This is especially so for the popular "courtyard" form, which replicates the classic horseshoe opera house in form but at a smaller scale.

All wall and ceiling surfaces that do not produce quick reflections (reflections arriving at the listener's ear within 30 msec of the direct sound) should be treated with efficient sound-absorbing material to ensure adequate control of echo and reverberation. Reverberation can be a positive contribution in music spaces, but not in speech theaters. Successful theaters have been built with measured reverberation times ranging from less than 0.5 sec to over 1.5 sec, all having good speech intelligibility without electronic reinforcement. The lower reverberation times are essential for thrust-stage and in-the-round theaters because there is very little direct energy behind the actor's back to begin with. The higher reverberation times are applicable only to the larger proscenium and other unidirectional fan-plan theaters, particularly those using hard box sets that can efficiently reinforce an actor's voice. Today, the calculation of reverberation time should be viewed as an important check on the acoustical design of theaters, rather than the basis of design.

The Alley Theater in Houston, Texas, is a good example of theater acoustic design (see Figure 4.11). It is a mild fan-shaped plan, with a steep rake and a relatively low ceiling shaped to reflect sound from the stage directly to the audience with minimum delay. The rear wall surfaces have an efficient sound-absorbing treatment of carpet over glass fiber. The reverberation time is 1 sec, unoccupied, and only slightly lower when occupied since upholstered seats compensate for the missing audience in the empty theater (Ulrich Franzen, architect).

Multipurpose Auditoria

While concert halls seem to command the most interest and research for listening places within the architectural acoustics profession, the reality is that most auditoriums must accommodate nearly every sort of activity, including concerts. How well the multiuse function is satisfied depends on (1) design concept, (2) budget, and (3) auditorium staffing. These 3 items are interrelated, and each is influenced by the others.

A fundamental nonacoustic design decision that, nonetheless, is of major acoustic significance concerns whether or not a stagehouse is to be provided. The acoustic design philosophy employed for the auditorium's design will follow from that decision.

With a stagehouse, a full orchestra enclosure is required for the sake of good music acoustics, and one of the necessary functions of the enclosure in this alternative is to seal off the wing and loft volumes to prevent the loss of sound into these spaces.

Without a stagehouse, a facility may be designed that is geometrically more closely related to a pure concert hall, but still appears as a traditional proscenium auditorium. With this alternative design philosophy, there is almost no physical separation between the stage and audience volumes, although there may be an appropriately proportioned proscenium frame in place that gives the appearance of separation. On stage, orchestra enclosure surfaces are still required to provide player communication, sectional blend and balance, and distribution of sound to the audience, but these surfaces will not fully seal off surrounding backstage areas, and the entire stage space may remain in communication acoustically with the total volume of the auditorium. In addition, the orchestra enclosure elements can be designed to function as masking legs and borders for many nonconcert uses (see Figure 4.9).

An acoustic environment that is ideal for one type of activity can be unsatisfactory for another (drama versus music, e.g.), and a good multipurpose auditorium design accommodates these differences by incorporating acoustic adjustability. The orchestra enclosure constitutes one form of adjustability. Another form of acoustic adjustability effects changes in reverberation time and is usually implemented in the form of large areas of tracked sound-absorbing curtains installed along the boundaries. When retracted into storage pockets, maximum reverberation is achieved for the benefit of concert hall acoustics. When extended, reverberation is reduced to optimize

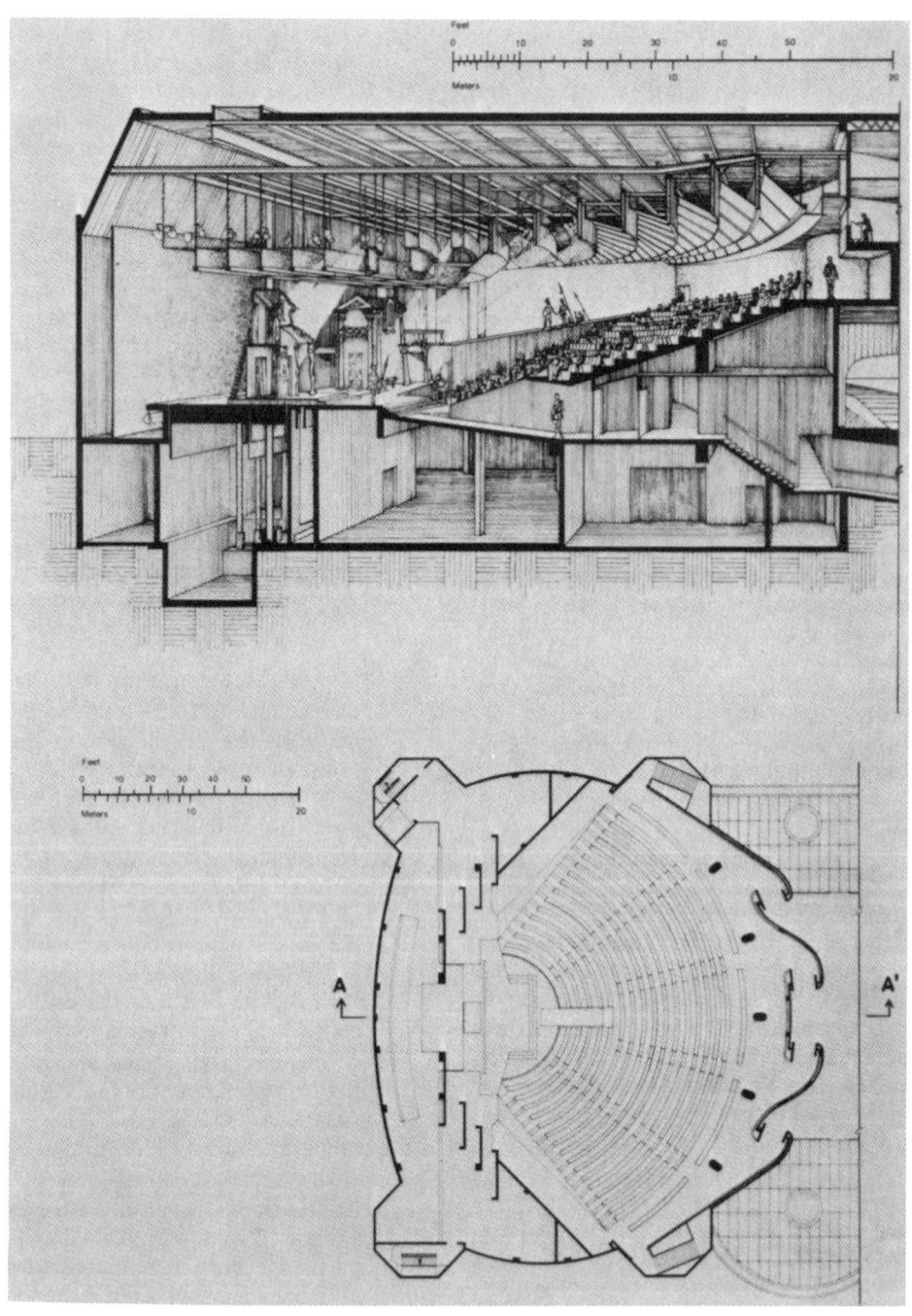

Figure 4.11 The Alley Theater, Houston, Texas. Sound-reflecting surfaces provide early reflections to the listeners. Bolt, Beranek and Newman, Inc., Acoustical Consultant; U. Franzen and Associates, Architect. (By permission of the George C. Izenour Archive.)

speech intelligibility and/or to reduce the loudness of high-sound-output groups on stage.

Movable Stage Enclosures

As just pointed out, a movable stage enclosure serves the important function in a multiuse auditorium of converting the sending end of a basically theatrical stage designed for scenery handling into an efficient sending end for concert hall use.

The stage enclosures that are among the simplest and most economical are commercially available modular devices that are usually fairly low, allow considerable "leakage" into the backstage area, and are relatively lightweight and imperfectly sound reflecting at lower frequencies. Though widely used, these enclosures are lowest in efficiency, and whenever the project budget allows, custom enclosures of heavier materials and more complex design can be much more effective.
An enclosure should be designed with the following acoustic goals in mind:

1. Prevent the loss of sound backstage.
2. Efficiently transfer and distribute full-range sound to the audience.
3. Accomplish good sectional balance.
4. Provide good on-stage communication for the musicians.

Item 1 requires an enclosure with heavy, tight-fitting components.

Item 2 is best accomplished with a high, unrestricted opening between the stage and audience so the orchestra and audience will be in a single space, acoustically. Full-range reflectivity requires the use of heavy materials, and recent research has demonstrated that low-frequency reflections from the enclosure's ceiling and upper walls are of greatest importance since low-frequency reflections from lower wall areas are severely attenuated by grazing across the audience area.[5] Often, an enclosure's ceiling will be its lightest and "leakiest" surface.

Item 3 requires a geometry that provides maximum reinforcement of the string sections and deemphasizes the inherent strength of the percussion and brasses and the extra reinforcement they receive from the upstage wall and rear corners.

Item 4 requires "feedback" reflections that are not excessively delayed before reaching the musicians. This requirement can be in conflict with the high-ceiling criterion of item 1 and a number of major halls are deficient in this respect. Long ago, for example, a famous concert hall installed suspended canvas flat reflectors well below its approximately 45-ft high ceiling. During a more recent restoration, however, those flats were removed and this caused a strong negative reaction from many long-time users. Panel arrays, as discussed in the prior section, represent one good option for obtaining feedback reflections and sectional balancing.

The concept of suspending a panel array subceiling within a high-volume enclosure leads to the "double-enclosure" concept, which can have several attractive features. Once a large high-volume outer enclosure is in place to produce a spacious sound quality and to couple well with the audience, movable inner elements can be used to accommodate any size performing group with surrounding surfaces, both in plan and section. For example, portable "recital screen" elements might be used to define the performing area in plan, and a winched panel array would provide overhead surfaces at a height compatible with the needs of the performing group. Many manifestations of this concept are possible. A straightforward early example exists at Uihlein Hall, Milwaukee, Wisconsin (acoustics by Bolt, Beranek and Newman). This enclosure is constructed primarily of heavy damped steel and forms a large outer enclosure some 45 ft tall, within which a series of clear sound-reflecting panels are suspended. With

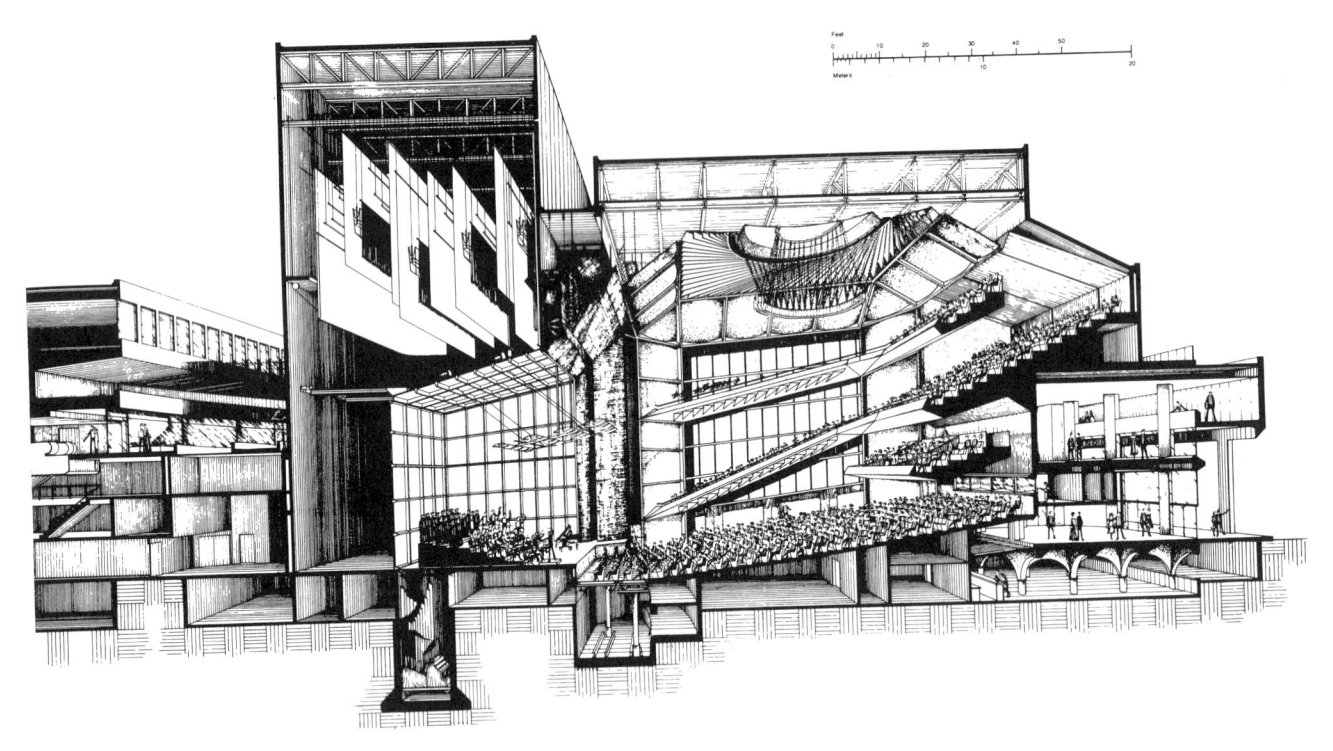

Figure 4.12 *Uihlein Hall at the Performing Arts Center, Milwaukee, Wisconsin. The stage enclosure ceiling is high to develop reverberation near the source, while articulated sound-reflecting panels aid definition and on-stage communication. H. Weese and Associates, Architects; Bolt, Beranek and Newman, Inc., Acoustical Consultant. (By permission of the George C. Izenour Archive.)*

this enclosure, seen in Figure 4.12, the stage area and hall are truly a single space. Other examples are seen in Figures 4.8 and 4.9.

Motion Picture Theaters

As opposed to a theater for unamplified drama, a motion picture theater is designed to accommodate amplified sound, rather then to reinforce live voices. Therefore, room shaping is not an acoustic consideration, while room boundary reflectivity is a dominant consideration. Ideally, motion picture theaters will receive large quantities of wall- and ceiling-mounted broadband sound-absorbing material to control reverberation. At least 50% of the walls should be treated, and the ceiling should be 50 to 100% treated.

The goal in motion picture theater acoustic design is to achieve high intelligibility and accurate localization of both stationary and moving sound sources from a multichannel sound reproduction system that extends down to at least 25 Hz.

Lucasfilm has championed motion picture sound system and acoustic standards through its THX group, which has a certification program for theaters meeting its requirements. In addition to sound system performance and room noise control requirements, reverberation criteria are provided, and those are shown on Figure 4.13. For the sake of good response from the behind-the-screen loudspeakers, the mounting details prescribe a fully absorptive behind-the-screen environment, plus baffling to capture radiation to the rear.

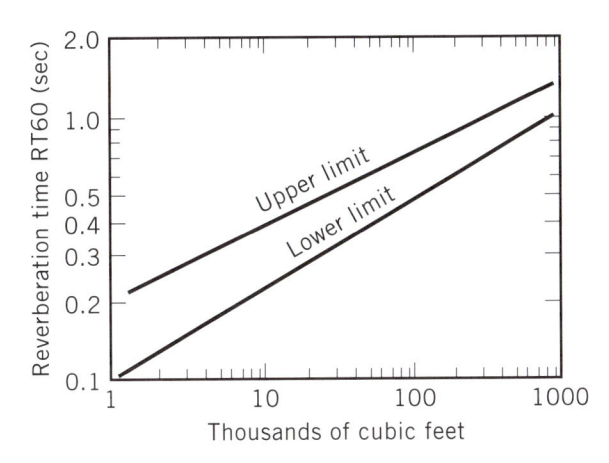

Figure 4.13 *500-Hz reverberation time vs. room volume, recommended for THX-qualified theaters. (Courtesy of THX division of Lucasfilm, Ltd.)*

Worship Spaces

Churches and synagogues resemble multiuse auditoriums in that both speech and music are expected to have good acoustics. In these spaces, though, it is impractical to constantly change movable acoustical panels and draperies during a service since speech acoustics and music acoustics must coexist at all times. Speech intelligibility is important in almost every worship space, but the proper environment for worship music will depend on the type of music performed. An amplified gospel choir requires a relatively low reverberation time and high definition, while traditional church music, and particularly choirs of men and boys performing the music of English cathedrals, requires a relatively long reverberation time, often even longer than 3 sec. Size is also an important consideration. Cathedral acoustics in a small wood church would sound unnatural, as would dry, lecture hall acoustics in a large cathedral. A description of four very different architectural styles may point out the degree of variation in the acoustical designs of worship spaces.

Cathedrals

Large cathedral-type spaces usually house large pipe organs and large choirs, and the music appropriate to such spaces was composed with large, reverberant spaces in mind. The accession in the 4th century of the first Christian emperor, Constantine, brought dramatic change in the construction of church buildings. Large public structures, the majority of which were adaptations of the Roman civic basilica, were built to replace private house churches as places of worship. The reverberant acoustics associated with the increasingly monumental church architecture that evolved from the basilica church could literally turn speech into music, and those highly reverberant environments were responsible for the adoption of a musical liturgy as a necessary alternative to a spoken liturgy. In turn, the musical liturgy that developed from monophonic chant (as a logical replacement for speech) and rhythm (from the Latin text), and came to include harmony and polyphony, all of which was facilitated by the creation of a system of notation, established the course of development of the early history of Western music, and there should be no wonder why music created in response to the reverberant acoustics of cathedrals sounds so wonderful within those spaces.

Today, of course, speech must also be intelligible, but long reverberation times are required for the musical portion of the service. The usual modern approach in reconciling these seeming opposites is to (1) design a spacious room with hard, sound-reflecting surfaces that effectively distribute the sound of music throughout the space, allow the congregation to hear themselves sing, and produce ample reverberation and (2) design a sophisticated sound amplification system that can place amplified speech energy into the sound-absorbing congregational seating area without directing large amounts of amplified energy at the wall and ceiling surfaces. This concept is easy enough to state, but its implementation requires close cooperation between the architect, acoustical consultant, and sound system contractor. Detailed descriptions of the sound systems needed to assure high intelligibility in this kind of space are not within the scope of this chapter, but there are a variety of sound systems in the repertory of the acoustical designer. They are (1) the large central cluster of directional horn loudspeakers over the usual speaking positions; (2) delay-synchronized distributed loudspeakers pointing down and usually located 12 to 15 ft above the floor in chandeliers; (3) column loudspeakers mounted on side-aisle columns, facing away from the speaking position and receiving a signal progressively delayed with distance from the speaking positions, with each loudspeaker covering a relatively small area; and (4) pew-back loudspeakers, also using signal delay, with one small loudspeaker for every three or four people. Variations and combinations of these basic types are possible, of course.

In years past, attempts were made to make large cathedral spaces good for speech by massive applications of sound-absorbing treatment. Such churches became acoustically dead, very poor for music, and not much good for speech either since they were too large for unaided voices and the sound-absorbing walls and ceiling did not distribute sound energy. So, complex and expensive sound systems were needed. Once the decision to use sound systems is made, there is no reason to make such spaces acoustically dead. As long as microphones can be located close to the person who is speaking, a sound system can provide high intelligibility if the amplified sound energy is concentrated on the sound-absorbing congregation (see second case study in Chapter 2).

From digital technology, an option for producing convincing "cathedral" acoustics in nonreverberant spaces through the use of electronics has become commercially available in recent years. With such a system, natural-sounding acoustics are obtained by electronically simulating the pattern of early reflections and reverberation of a stereotypical cathedral.

Small Churches

An opposite of the cathedral is the small, low, meeting-house style church, which may have a medium-sized pipe organ or an electronic instrument. Music may be important, but the style of architecture precludes a long reverberation time. Speech can be intelligible without electronic amplification because of short distances, and the music program should feature music that has emotional impact in an intimate acoustical environment. The electronic reverberation and surround systems referred to in the preceding paragraph are also being successfully used to extend the music acoustics ranges of small and medium-sized churches.

Medium-Sized Churches

Many churches will fall halfway between the extremes just described. Cathedral acoustics may not be possible in such spaces, but midsized churches can possess good concert hall acoustics, with reverberation times in the 1.8- to 2.0-sec range. Surfaces should be hard and sound reflecting. In certain cases, pew cushions may be applied to control the difference between full and empty acoustical conditions. A moderate- to large-size pipe organ will usually be present or planned for. A relatively simple sound system, usually of the central cluster variety, can ensure good intelligibility, even for weak-voiced speakers. Most cathedral-style music can still sound good in such a space, and the acoustics will be appropriate to the size. Synagogues and mosques have acoustical design characteristics related to this type of church.

Evangelical Churches

The large evangelical church will be quite different from the 3 described above. An electronic organ, a piano, and an amplified choir are the main music sound sources, and the music is more similar to contemporary popular music than to traditional liturgical music. Preachers employ a wide dynamic range of voice levels. The service is often televised, and the entire church can be considered a large TV sound studio.

The acoustical design of these large evangelical churches is basically similar to that of a large speech auditorium, with reverberation times in the 1.0- to 1.5-sec range. Rear walls should usually be sound absorbing, and ceilings arranged so that reflections arrive at the listeners' ears within 30 msec of the direct sound. The sound amplification systems often resemble the portable systems used for contemporary popular music concerts.

Organ–Choir Arrangement

In any of the four varieties of worship spaces, the organ, organ console, piano (if any), and choir should be located within a 40-ft radius to avoid sound delay problems. The organist-choirmaster must hear the organ and the choir without excessive delay. If the organist and choir director are two individuals, they must also be close to each other for proper coordination, as should the pianist, if any.

An exception to this placement rule is the antiphonal organ, which may be at the opposite end of the church and which is intended for special effects and for keeping the congregation synchronized along the length of the church. Antiphonal choirs are also special and require considerable skill on the part of the choir director and singers.

Choirs and organs are best located either behind the altar on the chancel wall facing the congregation or in the balcony or gallery at the rear of the church. The organ and choir should face down the main axis of the church with good line of sight to the director and congregation. Divided chancels and transept placement are less satisfactory.

Both the choir and the organ should be placed somewhat higher than the congregation. In small churches, platforms may be used; in larger churches, some sort of balcony is preferable. Choir members should not stand under very low ceilings. Ceiling height above choir platforms should be in the 20 to 30 ft range. In cathedral-like spaces, nearby side walls can provide the reflections necessary to help communication among choir members. The mouths of organ pipes should not be located directly behind choristers' ears.

Materials

Finish materials for worship spaces should usually be hard and sound reflecting, as for concert halls. Wood paneling absorbs low-frequency energy unless it is very thick or bonded to something more massive. Brick, stone, and concrete are all appropriate materials for church interiors. Plaster is also good but should be relatively thick. Carpet should not be used, except in evangelical churches, and should particularly be avoided near the choir and organ. When sound-absorbing treatment must be used to control reverberation or echo, it should not be placed on the ceiling, which should instead be hard and sound reflecting to distribute sound energy. Pew cushions reduce the quality of stand-up congregational singing, but they are useful and sometimes necessary for controlling reverberation to permit good understanding during those times when only a portion of the congregation is present. Upholstered pews are normal for evangelical churches with amplified music.

4.5 OTHER PLACES FOR SPEECH AND MUSIC ACTIVITIES

Sports Facilities

This category of enclosed assembly spaces includes coliseums and sports activity arenas and stadiums that are also usable for convocational and large popular music events. On occasion, these facilities may house a series of cultural activities, including orchestral concerts and plays. Sometimes curtains are used to divide off a portion of the seating area. These are large facilities, with seating capacities of well over 5000. The largest examples are the various domed stadiums, such as Houston's Astrodome and Seattle's King Dome, which can enclose entire baseball fields and their surrounding seating stands (see the case study in Chapter 5).

Sound Absorption

The basic problem with all of these facilities is that the surfaces are too far from the listener and/or performers to provide useful reinforcement by reflected sound energy. Instead, reflections from the boundary surfaces are heard as echoes. The basic acoustical design consists of making these surfaces sound absorbing by means of massive applications of acoustical treatment. The acoustical materials used must have high absorption coefficients in the speech frequency range, particularly 250 to 4000 Hz. Concrete surfaces may employ thick sound-absorbing formboards that are left in place on the underside of the concrete. Steel surfaces may employ perforated metal roof decks, consisting of a sandwich of perforated metal on the bottom, glass fiber in the middle, and sheet metal above. Inflatable domes can employ special sound-absorbing fabrics installed to sag beneath the actual dome material. The airspace between the sound-absorbing fabric and the dome material above is necessary for efficient sound absorption. Many, many alternatives exist for treating ceilings with sound-absorbing materials.

Wall surfaces are usually also treated. Hard seats may be essential in some projects for reasons other than acoustics, but upholstered seats are better, and sometimes perforated seat bottoms (which expose sound-absorbing material within the seats) are used to control the buildup of reverberation that accrues when occupancy is low.

The concept of reverberation time is somewhat meaningless in regard to such large spaces. Even with all significant areas treated efficiently, reverberation times can run to 4 or 5 sec because of their tremendous volume. (If room surfaces are left untreated, reverberation times as high as 10 sec have been recorded.) The important goal is to treat any potential echo-producing surface and not to aim for a particular reverberation time.

A final reason for the massive application of sound-absorbing treatment is to control crowd noise. In large, untreated sports spaces, crowd noise can build-up to the point where even massive, high-level sound systems cannot communicate emergency information. With treatment, crowd noise during an exciting part of a sports event can usually be held to broadband levels of approximately 95 dB (flat), and the sound system can be designed to override such levels without causing hearing damage.

Cultural Events

The accommodation of cultural events in these large spaces should, after proper treatment of boundary surfaces, be similar to that for outdoor facilities. Stage enclosures or concert shells are useful for symphony orchestras and choruses. Beyond that point, the problem faced is basically one of proper sound system design, with the emphasis on achievement of adequate levels, proper coordination in time with sound from the live source, and uniformity of coverage. Some acoustical consultants would also include provision of electronic concert hall sound.

Music Buildings

Music buildings contain a variety of spaces with special room acoustics needs. Included are recital halls, large rehearsal rooms, ensemble rooms, faculty studios, music classrooms, practice rooms, and recording facilities, among others. Covering all of these in detail is beyond the scope of this chapter, but the principles may be outlined. (Recital halls are discussed earlier in this chapter.)

Musical instruments produce as much sound power in small rooms as in large auditoriums and can be uncomfortably loud. For reduced sound intensity, sound-absorbing materials are used extensively in music buildings. For ensemble and rehearsal rooms, ample spaciousness is helpful, too. For example, large rehearsal rooms are ideally 2 stories high, and in no case should they be less than $1\frac{1}{2}$ stories high. With 2-story-high ceilings, suspended sound-reflecting panels are often incorporated to assist definition and player communication.

Rehearsal rooms also require a good deal of installed sound-absorbing material for the sake of reverberation control, loudness reduction, and, in particular instances, elimination of flutter-echo paths between parallel walls. The first two of these items relate to quantity of material, while the last relates to distribution. Because of the greater amount of sound power produced in an instrumental room than in a choral room, the former needs more treatment than the latter. Adjustable-absorption devices can be used to tailor a room's response to better match the size and composition of a particular performing group, and also to better suit users' varying tastes.

With respect to distribution of treatment, one consideration is to make certain that the material exists on walls between seated and standing head height, at a minimum. This statement applies to most kinds of spaces, both large and small. The desired quantity of material in high-ceiling rooms means that more than just this region will be covered, but in smaller rooms (practice rooms, e.g.), wall-mounted treatment need not extend beyond that limited region.

Still another general consideration is to make certain that absorption exists in each of the 3 dimensions—lateral, longitudinal, and vertical. In heavily treated rooms, discrete "live" sound paths resulting from poor distribution of treatment can produce highly noticeable "flutter" or "ring."

Music classrooms experience playback of recorded material, among other things, and can be viewed as combination lecture-plus-listening rooms, both of which are discussed in this chapter.

Conference Rooms

Conference rooms should allow effortless conversing either with or without the assistance of a sound system. Acoustically, this implies reverberation control and freedom from defects such as flutter (between parallel hard surfaces) and focusing from concave hard surfaces. Applied absorbing materials might be placed most strategically on walls between seated and standing head height. On the ceiling, absorption might be limited to the perimeter so the central portion will serve to reinforce and distribute sound.

A special form of conference room requiring careful acoustic design is the teleconference or videoconference room. Teleconference rooms are linked through interactive video networks, and they can be said to have a broadcast studio function since successful use requires site-to-site conversing that is as comfortable as face-to-face conversing. Reverberation and room mode control are obtained through the proper application of efficient broadband sound-absorbing materials. These should be applied so they exist in each of the room's principal axes. Unequal distribution of sound-absorbing material can result in nonuniform sound decay causing "coloration." Thus, the average absorption coefficient of the end wall surfaces should be similar to the average absorption coefficient of the side walls, and to the average absorption coefficient of the floor and ceiling, the two surfaces of a pair taken together. This condition is preferably not accomplished by applying acoustic treatment on only one of each pair of opposing surfaces. Instead, treatment should be distributed on as many surfaces as feasible. Because the walls serve as a stage set or "backdrop" in videoconferencing, wall surfaces must be simple and uniform. Consequently, the acoustic treatment and its facing should be uniform over any wall to be acceptable as a background for digitized video transmission. To increase sound absorption at lower frequencies, supplemental treatment in the form of special low-frequency absorbers may be considered. Examples include resonant-panel and Helmholtz absorbers designed for peak absorption values at lower frequencies. Such treatment would be necessary if relatively thin sound-absorbing materials were used elsewhere, especially in small rooms.

Home Listening Rooms

Small rooms can produce significant coloration and poor spatial imaging of reproduced sound if they are not treated extensively and uniformly with broadband sound-absorbing materials. In other words, the sound of the room itself (due to reflections and resonances) should be substantially suppressed, though not totally, and this requires large areas of effective sound-absorbing and sound-diffusing treatments on all surfaces. Absorptive treatments should be applied so that average coefficients in each of the room's principal axes are about the same. (See also the discussion of teleconference rooms above.) For example, if each side wall were 50% absorptive, and the wall behind the front loudspeakers were 100% absorptive (to minimize phase cancellation), the rear wall might then be 100% hard and diffusing.

For accurate imaging, a listening room must be totally symmetrical about the vertical plane along the principal listening axis with respect to both treatment and geometry. In other words, the left and right playback channels should see identical sound transmission paths.

Surround channels must also be given consideration with respect to loudspeaker placement and transmission. In Lucasfilm's Home THX Audio System (which was developed to better reproduce motion picture program material in the home environment), the ambience loudspeakers are placed directly to the sides of the listener, but these are loudspeakers with specially defined radiation characteristics. More commonly, surround loudspeakers are moved further to the rear.

Monitor room acoustic requirements parallel those of the home listening room. A widely used approach to monitor room acoustic design is called live-end–dead-end, or LEDE (developed by Chips Davis and first described by Don and Carolyn Davis). In an LEDE room, the end containing the monitor loudspeakers is made nonreflective to increase the delay of the first reflections to optimize imaging and minimize phase interference. The opposite end is made both sound reflecting and sound diffusing to provide a sense of room response. Whenever sound-absorbing treatment is concentrated in this manner, rather than distributed, it is important to provide both a large room volume and efficient sound-diffusing geometry exist in the untreated portion to avoid severe modal coloration. Established criteria must be met for a room to be LEDE qualified.

Lecture Rooms

Large lecture room acoustic requirements may be equated to those of small theaters. The boundaries should be shaped for good natural reinforcement of the speaking voice, and usually this applies most particularly to the ceiling.

Applied sound-absorbing treatment may be useful for controlling reverberation, echo, and flutter. Likely areas to receive treatment include the rear wall, the perimeter of the ceiling, and side wall areas between seated and standing head height. The latter is especially helpful in rooms with parallel side walls. The stock solution of acoustic tile ceiling and hard walls should be avoided.

Classrooms linked through interactive video networks—known as "distance learning" classrooms—are becoming widely used. The acoustic design principles for teleconferencing described in a preceding section are largely applicable since ease of communication both within and between rooms is essential to the distance learning function.

REFERENCES

1. Beranek, L. L., *Concert and Opera Halls,* Acoustical Society of America, New York, 1996.
2. Beranek, L. L., *Music, Acoustics, and Architecture,* Wiley, New York, 1962; reprinted by Krieger, Huntington, NY, 1979.
3. Beranek, L. L., "Concert Hall Acoustics—1992," *J. Acoust. Soc. Am.,* **92,** 1–39 (1992).
4. Nakajima, T., Ando, Y., and Fujita, K., "Strong Lateral Low-Frequency Reflected Components from Canopy Complex with Triangle Plates in Concert Halls," *J. Acoust. Soc. Am. Suppl.* **1, 88,** S185 (1990).
5. Bradley, J. S., "Some Effects of Orchestra Shells," *J. Acoust. Soc. Am.,* **100,** 889–898 (1996).

FURTHER READING

Abdou, A., and Guy, R. W., "Spatial Information of Sound Fields for Room—Acoustics Evaluation and Diagnoses," *J. Acoust. Soc. Am.,* **100,** 3225 (1996).

Barron, M., *Auditorium Acoustics and Architectural Design,* (E & FN Spon) Chapman & Hall, London & New York, 1993.

Bradley, J. S., "Relationships among Measures of Speech Intelligibility in Rooms," *J. Audio Eng. Soc.,* **46,** 396–405 (1998).

Bradley, J. S., *Proceedings of the 16th International Congress on Acoustics and 135th Meeting of the Acoustical Society of America,* 1661–1662 (1998).

Cremer, L., and Mueller, H., *Principles and Applications of Room Acoustics,* Vol. 1, English translation with additions by T. J. Schultz, Applied Science Publishers, Essex, England; in USA and Canada, Elsevier, New York. Originally published in German (1978) by Hirzel, Stuttgart, 1982.

Crocker, M. J., *Encyclopedia of Acoustics,* Vol. 3, Wiley, New York, 1997.

Egan, M. D., *Architectural Acoustics,* McGraw-Hill, New York, 1988.

Forsyth, M., *Auditoria: Designing for the Performing Arts,* Mitchell, London, 1987.

Forsyth, M., *Buildings for Music,* MIT Press, Cambridge, Massachusetts, 1985.

Izenour, G. C., *Theatre Design,* Yale University Press, New Haven, 1997. (originally published, McGraw-Hill, New York, 1977.)

Klepper, D. L. "Sound Systems in Reverberant Rooms for Worship," *J. Audio Eng. Soc.,* **18,** 391 (1970).

Lubman, D., and Wetherill, E. A., *Acoustics of Worship Spaces,* Acoustical Society of America, New York, 1985.

Marshall, L. G., "An Analysis Procedure for Room Acoustics and Sound Amplification Systems Based on the Early-to-Late Sound Energy Ratio," *J. Audio Eng. Soc.,* **44,** 373–381 (1996), and in *Second Sound Reinforcement Anthology,* D. L. Klepper, ed., Audio Eng. Soc., 1996, pp. 480–488.

Talaske, R. H. and Boner, R. E., *Theatres for Drama Performance,* Acoustical Society of America, New York, 1985.

Talaske, R. H., Wetherill, E. A., and Cavanaugh, W. J., *Halls for Music Performance,* Acoustical Society of America, New York, 1982.

Case Study

HITCHCOCK PRESBYTERIAN CHURCH, SCARSDALE, NEW YORK

Architect: Peter L. Gluck & Partners
Acoustical consultant: KMK Associates

The sanctuary for Hitchcock Presbyterian Church in Scarsdale, New York, is a $6.25 million replacement for one destroyed by fire. For the rebuilding program, the architect presented five possible designs to the church, only two of which were very good acoustically, and one of those was ruled out because of zoning restrictions.

The chosen scheme, shown in Figure 4.14, is the one preferred by the acoustic consultant. The narrow axial plan (2:1 length-to-width ratio) with the organ and choir at front center, with no interference from a crossing or chancel arch, plus a high ceiling and totally hard interior all add up to a nearly perfect setting for the musical portion of the service in a small protestant church. High speech intelligibility is obtained with the assistance of a distributed sound system having loudspeakers contained within the chandeliers. Synchronization of the multiple loudspeaker signals and the live sound is obtained by means of a digital delay unit to maximize realism and intelligibility.

An interesting aspect of the renovation scheme is that the original roof line could not be raised because of zoning prohibitions, so the old sanctuary floor was ripped out to obtain greater room volume by dropping to the basement. The square windows along each upper side wall are in openings that were at people level before the floor was removed. (See also exterior view in Figure 4.15.)

An important feature is that all items on the platform, including the lectern, pulpit, and railing, are removable so a pure concert platform can be produced. Note that two rows of choir seating are placed between the tracker organ's console and case so the organist can face the choir.

Figure 4.14 Interior view of Hitchcock Presbyterian Church, looking toward chancel. Loudspeakers for the speech reinforcement system can be seen in the center of the chandeliers.

Figure 4.15 Street-side view of Hitchcock Presbyterian Church. The line of small windows close to ground level are those at clerestory level in the interior view.

Case Study

LENNA HALL AT CHAUTAUQUA INSTITUTION, CHAUTAUQUA, NEW YORK

Architect and theater consultant: Assembly Places International
Acoustical consultant: KMK Associates

The Chautauqua Institution in Chautauqua, New York, desired a new dual-use, simply structured facility to serve as (1) a large orchestra rehearsal room for the Chautauqua Symphony Orchestra and other organizations of the institute and (2) a recital hall. Acoustic requirements for the first function are very different from the second in a small hall because of the considerably greater sound power output of a full symphony orchestra. A small hall can be wonderful for recitals but can be uncomfortably loud for a full symphony orchestra. Therefore, loudness control by means of operable sound-absorbing draperies was incorporated into the design (see Figure 4.16). Sound enhancement (or tonal beauty) is influenced favorably by reverberation, and, especially in a small hall, by diffusion. Reverberation is affected by hall volume and boundary reflectivity, so the operable draperies also provide reverberation control by altering boundary reflectivity. Clarity is assisted by strong early-arriving sound reflections from nearby surfaces, and an array of suspended sound-reflecting panels was used for this purpose. Clarity can be further improved by extending the operable draperies to reduce reverberation.

The hall's volume was chosen to obtain an upper limit reverberation time of 1.8 sec for recital uses. Full exposure of the curtains can reduce this to 0.9 sec. Wall panels were designed to provide highly diffused reflections from a diffraction-grating profile defined by a number theory regulating the depths and widths of adjacent channels.

Some of the adjustable draperies are installed on tracks around the perimeter of the room. Additional sets are hung vertically from the roof along four sides. All curtains retract into storage pockets for the maximum reverberation condition. All draperies can be moved quickly to any desired position (the upper-volume draperies are motorized) so that performers may quickly tune the hall to best suit their requirements.

In the recital mode, the program called for an audience of 500 and a maximum of 30 performers (see Figure 4.17). In the rehearsal mode, the program required sufficient floor area for a full symphony orchestra. The schedule of activities at the institution requires transformation from one mode to the other in minimal time. The logistic problem of transforming the flat floor for rehearsal to a tiered-seating recital configuration is resolved by having a number of seating rows located on tiers of telescoping retractable platforms, and additional rows located on adjustable-height platforms within a terraced pit.

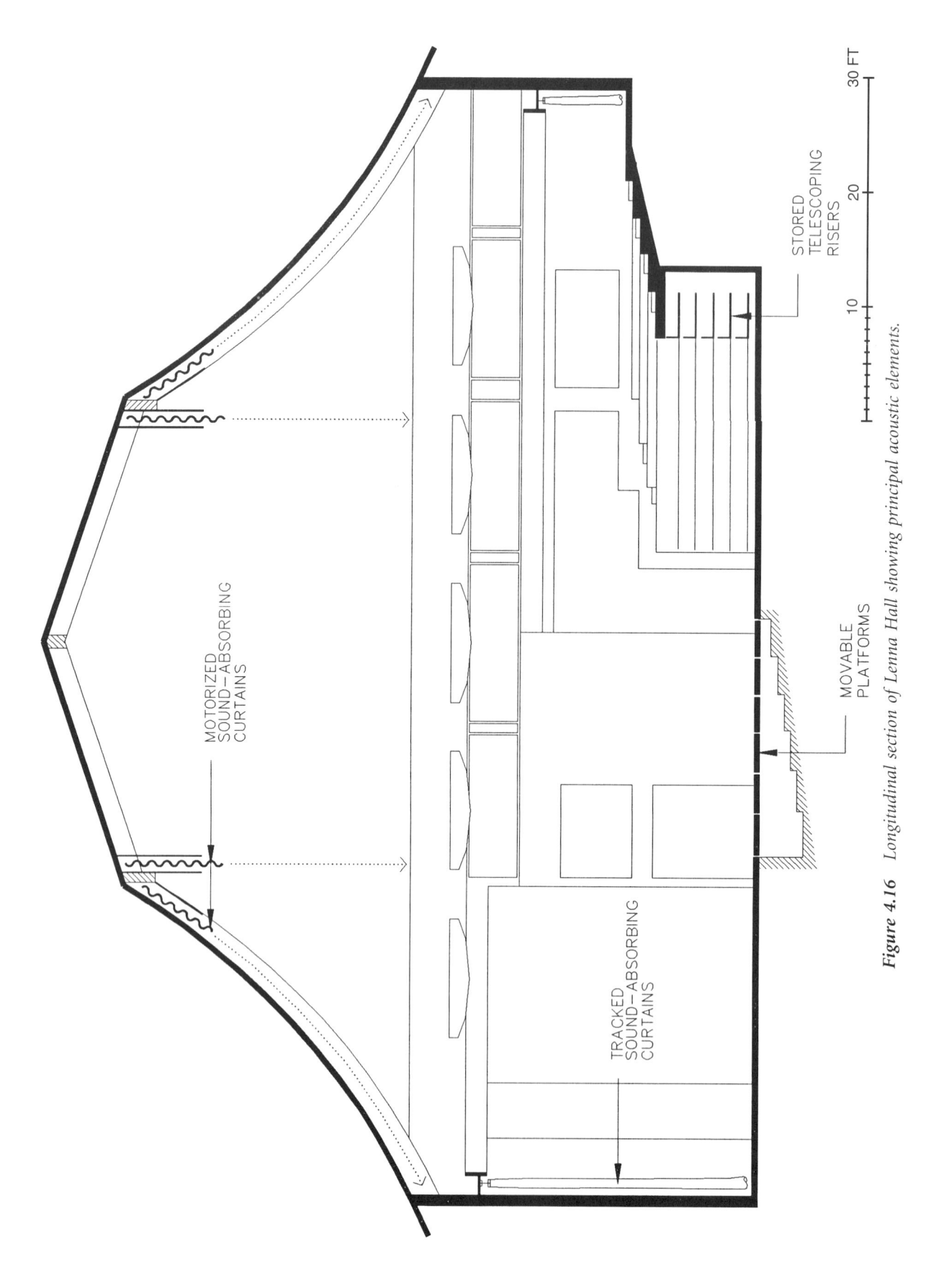

Figure 4.16 Longitudinal section of Lenna Hall showing principal acoustic elements.

Figure 4.17 Lenna Hall in recital configuration.

Case Study

BEN FRANKLIN HALL, AMERICAN PHILOSOPHICAL SOCIETY, PHILADELPHIA, PENNSYLVANIA

Architect: Bower Lewis Thrower
Acoustical consultant: KMK Associates

Benjamin Franklin founded the American Philosophical Society in 1743 for the promotion of useful knowledge. Past and present membership includes 12 American presidents and over 200 Nobel prize recipients.

The 400-seat auditorium created within a space carved out of a 19th-century bank building's interior is a special-purpose facility designed to satisfy the society's principal requirement for a high-quality lecture space. Two secondary functions are also of importance, however, and these are occasional banquets and music recitals. The challenge to the acoustic designer, then, was to reconcile clarity requirements for speech, reverberance and liveliness requirements for music, and suppression of activity noise during banquets. The three-story-high ceiling and shoebox geometry provided the opportunity for achieving superb music acoustics. Elegantly

Figure 4.18 Reversible hard/soft acoustic panels used to alter acoustic characteristics for lecture and concert uses. (Photo courtesy of the American Philosophical Society)

Figure 4.19 Interior view of Ben Franklin Hall. Horizontal line-source loudspeaker system is recessed into lintel at top of columns. (Photo courtesy of Gregory Benson)

designed wall-mounted pivoting panels that are sound absorbing on one side and sound reflective on the other were created as the means for radically altering the reverberation and loudness characteristics of the space to allow optimal acoustical environments for the 3 principal uses (see Figure 4.18).

The electronic speech reinforcement system consists of a horizontal line-source loudspeaker system at the front of the room and an audience response system consisting of distributed ceiling-mounted loudspeakers (see Figure 4.19).

5

Sound Reinforcement Systems

J. Jacek Figwer

5.1 INTRODUCTION

Most sources of natural sound, such as human voice or musical instruments, produce sound levels that are adequate for listening at relatively short distances in rooms with small volume. It is generally accepted that acoustically well designed auditoriums with a volume in excess of 1500 m^3, where the voice must travel more than 15 m, will require a sound reinforcement system to provide adequate loudness and uniform distribution of sound. Even in smaller rooms, such as meeting rooms with a seating capacity for 100 people or more, where strong-voiced speakers can be heard clearly, the weaker voices must be reinforced. Often, there is the need to reproduce a program from a disk, tape, or motion picture sound track. Outdoors, sound reinforcement may be required for a talker whose audience is at a distance of more than 7 m. In rooms with long reverberation, which is not favorable for speech, the intelligibility can be improved with a sound reinforcement system, which increases the amount of direct sound at the listener's position. For satisfactory intelligibility, in order to avoid masking, the voice should be at least 25 dB above the level of ambient noise. In many instances, this will require reinforcing of the useful sounds. In theaters and opera houses, sound reinforcement systems are needed to amplify the voices of the performers so that they will not be overpowered by the sounds of the orchestra.

The issue of sound reinforcement in opera is a subject of musical concern because many people prefer to hear the performance without artificial amplification. This is not always possible, especially with singers who have less than perfectly trained voices. The need for amplification has become more pronounced in recent times, when reasons of profitability led to the construction of lyric theaters with seating capacities larger than before. In other instances, loudspeakers are used for reproduction of organ sound in halls not equipped with this instrument. In the Barbican Theatre in London, a sound system is used to feed into the hall the sounds of an orchestra playing in another room.

The human voice has a sound power spectrum with a maximum between 500 and 600 Hz, and the sound power decreases above 1000 Hz at a rate of about 8 dB per octave. The vowel sounds are not as critical to the speech intelligibility as the consonant sounds. The consonant sounds are relatively weak and therefore easily masked by noise. Most of the sound energy in consonants is in the high-frequency range. At high frequencies voice is directional and the intelligibility of speech drops off rapidly when the speaker turns away from the listeners. This poses problems during conferences and theatrical plays, when the speaker or the actors do not always face the audience. The intelligibility of speech can be restored through the amplification of speech sounds at frequencies above 500 Hz. However, reinforced speech, with the bandwidth limited to high frequencies, does not sound natural. For that reason high-quality sound reinforcement systems include frequencies starting at approximately 125 Hz to ensure a pleasant and natural quality of speech.

Sound systems intended for the reinforcement or reproduction of music must cover a wider frequency range. The fundamental frequency of organ pipes starts at 16 Hz and that of the bass drums starts between 16 and 20 Hz. Few sound systems are capable of reproducing such low frequencies. A normal frequency range for high-quality reinforcement systems extends from 40 to 12,000 Hz.

Sound reinforcement systems must be designed to produce adequate sound pressure levels without distortion or other signs of overload. Typically, the systems intended for reinforcement of speech should be capable of delivering sound levels of between 85 and 90 dB. Systems for reinforcement or reproduction of concert music should be designed to deliver sound pressure levels of between 100 and 105 dB. Sound reinforcement systems used by rock bands can develop sound levels of 110 dB or more. The exposure to such high sound levels for extended periods of time can lead to hearing damage.

The localization of a sound source by the listener occurs with the arrival of the first sound impulse from that source. Successive repetitions of the same signal, either in the form of natural sound reflections from hard surfaces in the room or in the form of sound impulses from loudspeakers, are not registered by the listener as separate sounds until the delay reaches relatively large values of more than 50 msec. The suppression of echoes arriving with short delays is called the Haas effect, after the discoverer of this phenomenon. Such repeated impulses contribute to the loudness of the primary signal, and under some circumstances increase the intelligibility. The Haas effect is widely used in the design of sound systems when the directional realism in reinforcement must be preserved. This requires placement of the loudspeakers in such a way that the reinforced sound arrives at the listener's ears with a slight delay following the sound from the natural source. In this arrangement, the listener will localize the sound as arriving from the direction of the source, not from the loudspeaker. The largest increase in the intelligibility of speech due to the operation of a sound system will occur if most of the sound from the loudspeakers reaches the listeners directly, without reflections from the walls or from the ceiling. This requires that the loudspeakers either be placed close to the listeners or that they radiate the sound in a highly directional way, or both.

The size of the room in which a sound system operates has an effect on the performance of the system and will influence the choice of loudspeakers most suitable in a given situation. In small rooms, for example, such as domestic living rooms, all sound reflections from the room boundaries arrive at the listener in very quick succession, following the direct sound within a few thousandths of a second, and are easily integrated with the direct sound. In small rooms, therefore, loudspeakers with wide dispersion of sound will perform satisfactorily, and there is no need to use highly directional loudspeakers. In very large rooms, sound reflections from the walls arrive following the direct sound with very long delays, sometimes in tens or hundreds of milliseconds. Such long-delayed reflections do not integrate with the direct sound and are generally harmful to the intelligibility and clarity of sound. Therefore, in large rooms, better results are generally obtained through use of loudspeakers that radiate sound within a narrow angle and are pointed toward the audience. Their directivity tends to reduce the amount of harmful sound reflections from the walls.

In rooms with small volume, intended for private listening, background noise can be maintained at very low levels. Consequently, the efficiency of the employed loudspeaker systems is of little consequence, and they are being selected primarily from the standpoint of delivering the sound with highest possible fidelity over the widest possible frequency range. In very large spaces, such as sports arenas, where the background noise due to the presence of audience cannot be effectively controlled, the ability of a loudspeaker system to deliver sound at a very high output with very high efficiency is of great importance.

5.2 LOUDSPEAKER SYSTEMS

There are two basic types of loudspeaker arrangement used in sound reinforcement systems: the central and the distributed loudspeaker systems. In a central system, the loudspeakers are normally located above the source of live sound. In medium- or large-size auditoriums, a central loudspeaker system is the preferred type in most situations, because it preserves the directional realism in sound reinforcement. In instances when the arrangement of the audience areas is such that the distance from the central loudspeaker system to the nearest listeners is much shorter than the distance to the most remote seats, it may not be possible to obtain a satisfactorily uniform coverage of the audience area with the sound from a single source, without the use of supplemental loudspeaker systems. Typically, supplemental loudspeakers may provide the coverage for listeners in deep balcony or underbalcony areas, or for the listeners sitting in the first rows of auditoriums that are wide in the front. Supplemental loudspeakers must operate with properly delayed signals, so that the sound from different loudspeakers is received in a close sequence to preserve the directional realism and to avoid artificial echoes. A section through an auditorium with a central loudspeaker system and with supplemental loudspeakers is shown in Figure 5.1. In supplemental or distributed systems that use directional loudspeakers operating with delayed sound, the loudspeakers must be oriented away from the source of the nondelayed sound in order to avoid artificial echoes.

The other basic arrangement of loudspeakers in a room is a distributed system that uses a large number of loudspeakers distributed uniformly over the audience area. An example of such a loudspeaker arrangement is shown in Figure 5.2. Usually, the loudspeakers in distributed systems are recessed in the ceiling. However, in rooms with a fixed seating, they can be installed in the backs of the seats or, in church sound systems, in the backs of the pews. With the loudspeakers located near the audience, each listener receives the sound primarily from the closest loudspeakers. Although the directional realism in reinforcement using a distributed system cannot be maintained, the clarity and intelligibility of sound is usually very good. Distributed systems are used in situations where a low ceiling does not allow for a central system or where not all the listeners can have unobstructed lines of sight to a central system. The distributed systems are also used in large meeting, convention, or exhibiting rooms, which must have a flexible arrangement of seats or require the reinforcement of sound from any position in the room, or when the room is divisible by movable partitions.

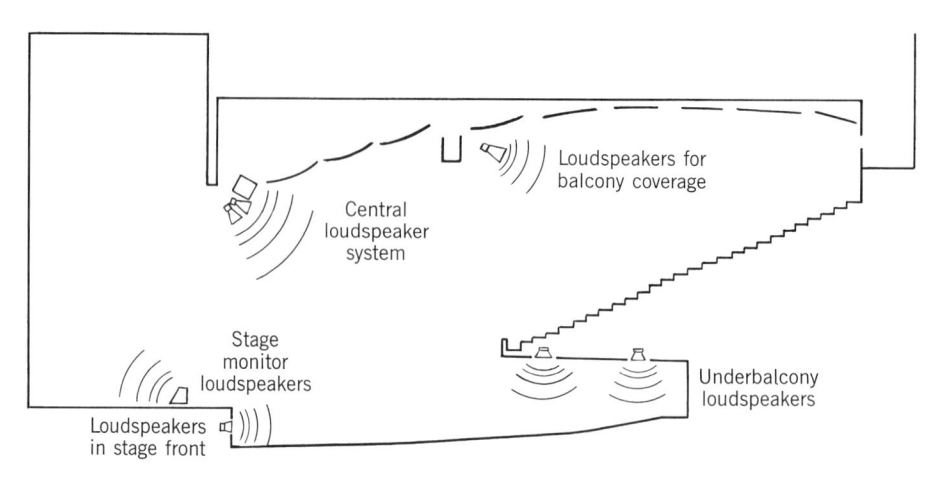

Figure 5.1 Auditorium with central loudspeaker system and supplemental loudspeakers. (From Jacek Figwer, "Sound Reinforcement Systems" Encyclopedia of Architecture: Design, Engineering, and Construction, Vol. 4, Joseph A. Wilkes and Robert T. Packard, Eds. Copyright © 1989 John Wiley & Sons, Inc. Reprinted by permission of John Wiley & Sons.)

The distributed loudspeaker systems with loudspeakers mounted in the ceiling should not be used in rooms where the ceiling is very high.

A well-designed, carefully installed, and properly operated sound system will provide a pleasant, natural quality of sound without distortion or coloration. The distortion of sound occurs when the system, for one reason or another, is not capable of delivering at its output an acoustical signal that has the same shape as the signal at the input of the system. Common causes of distortion are an improper selection of components, which are undersized for the required output; an incorrect gain structure, with a resulting overload of intermediate stages of amplification; or an input overload, when a signal of excessive amplitude is applied to the input of the system. Undersized

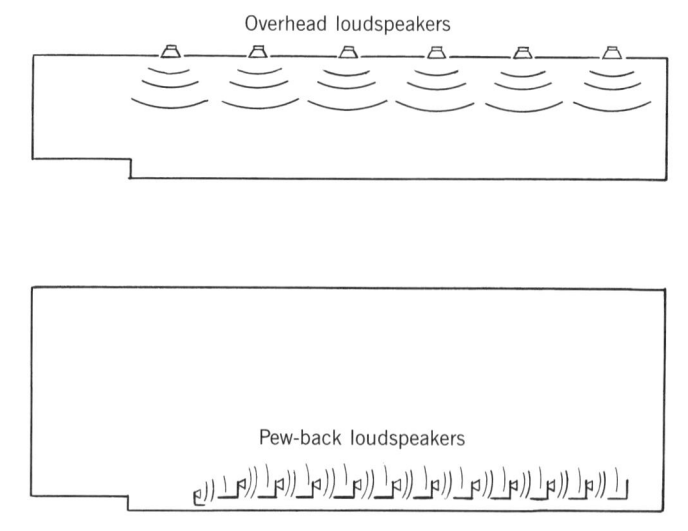

Figure 5.2 Distributed loudspeaker systems. (From Jacek Figwer, "Sound Reinforcement Systems" Encyclopedia of Architecture: Design, Engineering, and Construction, Vol. 4, Joseph A. Wilkes and Robert T. Packard, Eds. Copyright © 1989 John Wiley & Sons, Inc. Reprinted by permission of John Wiley & Sons.)

components are largely a matter of an incorrect design and cannot be easily corrected at a later stage. The incorrect gain structure may be a matter of an incorrect design or incorrect adjustments; the first may not be easy to rectify; the second can be easily readjusted by instruments and tests. An input overload is always a matter of the incorrect operation, easy to avoid and correct. The sound coloration is a state in which the sound system loses the naturalness of sound reproduction and acquires an unpleasant, ringing quality. This may be caused by a nonuniform frequency response, when certain frequencies are amplified more than others, or it may be the result of operating the system with an excessive gain, near the point of an acoustic feedback or howling. Both conditions can be corrected, although the corrections of the overall frequency response may require changes in the equipment, in its adjustments, or in the system design.

5.3 EQUIPMENT

Microphones

Microphones are electroacoustic transducers that convert acoustic energy into electric energy. They convert sound waves into electric signals, which are further amplified, transmitted, or processed as required in a sound system installation. The microphones most commonly used in sound systems are the electrodynamic transducers. An electrodynamic microphone is based on the principle of the sound pressure acting on an electrical conductor, either a specially folded metal ribbon (ribbon microphones) or a coil (moving-coil microphones), causing it to move in a magnetic field and to have current induced in it, at a frqeuency corresponding to the fluctuations of the sound wave. The moving-coil microphone is currently the most popular type. Its advantages are an excellent quality of reproduction; rugged construction; directional sensitivity, which can be varied through simple means; and the fact that it needs no supply of power.

From the standpoint of directivity, microphones can be omnidirectional, used mostly for recording or for close pickup in sound reinforcement, and directional, with maximum pickup from the direction of their main axis. Directional microphones, with cardioid or supercardioid pickup characteristics, are most widely used in sound systems because they help reduce the acoustic interaction between the loudspeaker and the microphone that causes the acoustic feedback, the interference caused by reverberation, and the effect of ambient noise. In order to use all of these advantages, the directional microphone should be placed close to the source of sound, pointing away from the loudspeakers. A variety of microphones used in sound reinforcement systems is shown in Figure 5.3.

In addition to the conventional microphones, hand-held or on stands, there are miniature (lavaliere) microphones, which can be worn to allow more freedom of movement. Other special microphone types include the ultradirectional, narrow-beam, "shot-gun" microphones, used for a distant pickup in theatrical reinforcement. The condenser, also called capacitor, microphones are less rugged than the electrodynamic microphones and require a power supply. They are used in sound systems for their very small size and exceptionally smooth frequency response. Microphones of any type can be used with small, wireless transmitters to allow full freedom of motion of the speaker. The wireless microphones are especially useful in the reinforcement of theatrical plays.

Preamplifiers

The function of a preamplifier is to reinforce the signal from the microphone to a level suitable for further processing and feeding other components of the sound system.

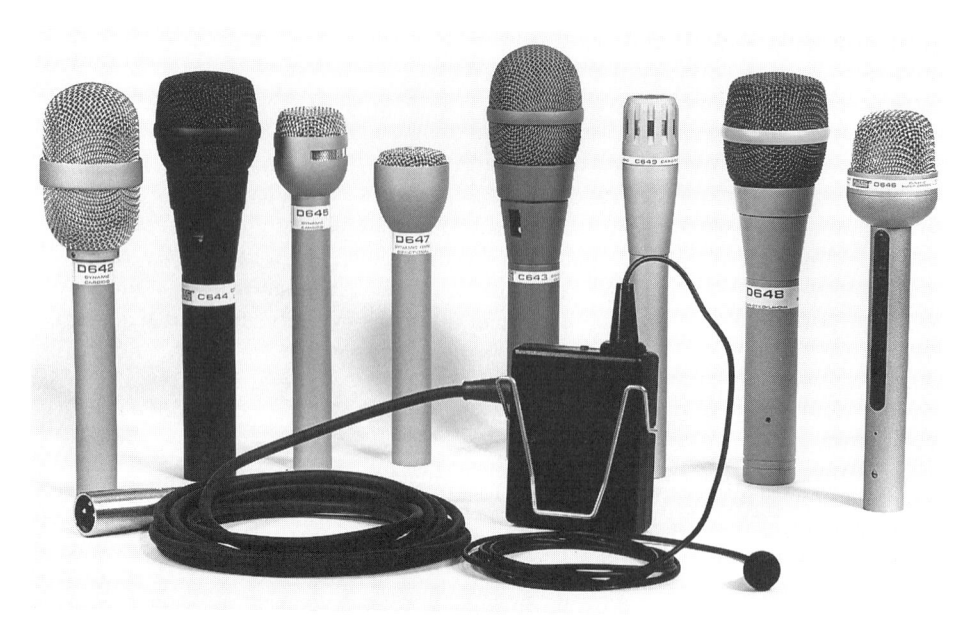

Figure 5.3 *Microphones used in sound reinforcement systems. (Courtesy of Altec Lansing Corp.)*

In addition to the inputs for microphones, the preamplifiers may have inputs especially designed to accept signals from phono cartridges, tape, film, or other signal sources. Preamplifiers are normally provided with several inputs allowing the mixing (combining) of signals from several sources. Preamplifiers may be equipped with tone controls, allowing tonal corrections to the signal.

Control Consoles

The control consoles fulfill the same role as the preamplifiers, except that they are much more complex and offer more flexibility in signal processing. Typical control consoles have 12, 16, 24, 32, or 40 inputs and two or four main outputs, depending on the requirements. Control consoles often include elaborate frequency equalization in each channel, allowing the introduction of peaks or dips at selected frequency bands. Some consoles include matrix circuitry at the outputs, which allow the delivery of separate and different mixtures of sound (mixes) as needed for simultaneous sound reinforcement, recording, broadcasting, and monitoring on the stage. A mixing console with 12 input channels, used in small to medium size theater and auditorium sound systems, is shown in Figure 5.4.

Signal Processing Equipment

There is a large variety of signal processing equipment that is used in sound reinforcement systems. Equalizers are used to alter the frequency response of the system. In their simplest form, they act as tone controls; in a more elaborate version, they are used to increase or decrease the gain in the individual frequency bands with an octave or $\frac{1}{3}$-octave bandwidth. The purpose of an equalizer is to introduce corrections in the electrical frequency response of the system to compensate for irregularities in the frequency response of transducers (microphones and loudspeakers) combined with the effects of room acoustics. The detailed equalization of a sound system or a smooth

Figure 5.4 Mixing console. (Courtesy of Rupert Neve, Inc.)

overall frequency response is invaluable for obtaining a natural quality of sound and for increasing the gain of the system before feedback.

Limiters restrict the level of the signal that can be applied to the inputs of the electronic equipment, thus protecting the amplifiers and loudspeakers from an overload. Limiters are particularly useful in sound systems serving large arenas or outdoor stadiums, where the components, in order to deliver the high sound pressure levels required, are likely to operate near their power handling capacity.

Electronic delays are used to retard the signal to the amplifiers serving the supplemental loudspeakers located closer to the listeners than the main loudspeaker system. The purpose of the delay is to compensate for the difference in time it takes the sound to travel from the main loudspeaker system and from the supplemental loudspeakers. Electronic delays typically employ digital delay lines; they have one input and one or more outputs, each with a different delay, as required.

Electronic delays may also be used with distributed loudspeaker systems to compensate for the difference in time required for sound to travel from the live source and nearby loudspeakers. A typical application is for distributed systems with loudspeakers in pew backs in churches. Caution is needed when this approach is used in large, acoustically live halls, especially when the installation involves loudspeakers mounted in the ceiling and operated with relatively high sound levels, so that the sound from the loudspeakers in the rear of the hall can be heard in the front. Under such circumstances, delaying the sound to the rear loudspeakers may adversely affect hearing conditions near the live source located in the front of the hall. The delay in hearing the sound from distant loudspeakers, added to the electronic delay, may create conditions of artificially increased reverberation.

Feedback suppressors are used in systems that must operate with weak-voiced speakers or with a long-range microphone pickup, such as in theatrical reinforcement. Some types of feedback suppressors employ the method of frequency shifting. This method consists of shifting by a few cycles per second the frequency of sound radiated from the loudspeaker with respect to the frequency picked up by the microphone. Consequently, the peaks in the point-to-point acoustical frequency response in the room are shifted into the valleys of the response, allowing an increase of the gain before feedback by several decibels. Newly developed feedback suppressors operate on the principle of adaptive filtering. Such feedback suppressors seek out the feedback frequency and automatically introduce a narrow-band filter at the offending frequency, which cancels the feedback. Feedback suppressors that simply reduce the overall broadband gain of the system, when the threshold of howling is reached, are the least effective type of suppressors.

Distribution amplifiers are used in large sound systems consisting of several independent amplification tracks, where it is necessary to feed a number of signal lines from one source without affecting that source through an interaction from the lines or from the increased load. This is best accomplished through the use of distribution amplifiers, which provide the division of signals while maintaining the isolation between the source and the receivers.

Power Amplifiers

Power amplifiers are designed to provide a signal output with sufficient power (voltage and electric current output) to feed the loudspeakers connected to the system. Typically, in a large sound system, in order to reduce the losses due to the resistance in the loudspeaker lines, the distribution of loudspeaker power is via 70- or 100-V lines. The size of the power amplifiers is selected to allow connecting the loudspeaker loads provided in the design plus at least 10% reserve capacity. Typically, power amplifiers are sized in 3-dB increments of power, delivering output at 50, 100, or 200 W.

Loudspeakers

Electromagnetic, direct-radiator loudspeakers are the most commonly used in sound reinforcement. The task of a loudspeaker is to convert the electric signal supplied from the power amplifier into an air vibration that the ear perceives as sound. An electromagnetic loudspeaker consists of a magnet with a round air gap and a voice coil, which moves inside the gap. The voice coil is attached to a paper cone or diaphragm, which is designed to radiate the vibration. The cone-type loudspeakers vary in size from less than 7 cm in diameter, for loudspeakers intended for reproduction of high frequencies, to over 45 cm in diameter for low-frequency loudspeakers. Direct-radiator loudspeakers must be housed in enclosures to cancel the effect of radiation from the back of the cone. Small-diameter loudspeakers are not as efficient as large-diameter ones in radiation of low frequencies. The efficiency of conversion of the electric energy into the acoustic energy is relatively low for the direct-radiator loudspeakers. The typical efficiency of such loudspeakers is between 1 and 4%.

One of the shortcomings of a simple, direct-radiator loudspeaker is its directional characteristic, which tends to focus the high-frequency sounds into a narrow beam. An improved radiation characteristic, with a wider radiation angle at high frequencies, is possible with a coaxial design, in which the radiation of low frequencies originates from a large-diameter diaphragm, and the radiation of high frequencies originates from a separate, small radiator, installed concentrically inside the large unit. Coaxial loudspeakers show a uniform dispersion angle of at least 60° up to a frqeuency of 4 KHz. They are the preferred loudspeaker type in installations that use distributed loudspeakers.

The efficiency of radiation of loudspeakers can be dramatically improved when an electromagnetic driver is coupled to the outside atmosphere via a horn. The horn acts in this arrangement as an impedance transformer, decreasing the losses due to the impedance mismatch between the driver diaphragm and the air. In the middle- and high-frequency ranges, horn-type loudspeakers with compression drivers can have radiation efficiencies of up to 50%, making them the most efficient loudspeaker systems currently available. A high-frequency, horn-type loudspeaker with constant directivity radiation over a wide range of frequencies is shown in Figure 5.5. The nominal coverage angle of this loudspeaker is 90° in the horizontal plane and 40° in the vertical plane.

Equipment • 195

Figure 5.5 High-frequency constant directivity horn. (Courtesy of Electro-Voice, Inc.)

Figure 5.6 shows a compression-type, high-frequency reproducer used with the high-frequency horn shown in Figure 5.5.

Due to physical limitations, it is not possible to produce a simple loudspeaker capable of satisfactory reproduction of the entire audio-frequency range from one unit. The accepted method is to divide the audio-frequency range into two, three, or four frequency bands, assigned for reproduction by separate loudspeakers. The performance of the individual units is optimized from the standpoint of directivity, frequency response, efficiency, and the power-handling capacity in their respective frequency bands. Large sound reinforcement systems frequently use two- or three-way loudspeaker

Figure 5.6 High-frequency reproducer. (Courtesy of Electro-Voice, Inc.)

systems with the audio-frequency range divided at 500 or 800 Hz and 6000 Hz crossover frequencies. A large low-frequency louspeaker used in bass enclosures is shown in Figure 5.7.

Manufacturers of loudspeaker equipment offer packaged modular loudspeaker systems designed to meet the requirements of a range of typical, frequently encountered applications. Such systems allow simplified installation and easy integration with the interior design. One such loudspeaker system is shown in Figure 5.8. It consists of two loudspeaker arrays with controlled directivity, a bass enclosure, and an electronic controller that shapes the frequency response of the system. If only speech reinforcement is required, the system can be limited to two loudspeaker arrays. For reproduction or reinforcement of music, the bass enclosure must be added.

Sound Control

No sound reinforcement system will operate satisfactorily unless it is properly controlled. The optimum location for a sound control room or position is in the rear of the auditorium, in an arrangement in which the operator can directly hear the sound he or she is controlling and can follow the activity. The sound control position should be either fully open to the volume of the auditorium, which is practical only for small systems, not involving much control equipment, or it should be located in a special sound control room provided with a large, operable sound control window. Controlling the sound during the reinforcement of live events while listening to a monitor loudspeaker, or from a control room located remotely, without lines of sight to the activity being reinforced, will invariably lead to unsatisfactory results.

Many performing arts sound reinforcement systems now use control facilities located entirely within the audience area. These may be dedicated locations where all control equipment and all activity occurs, or auxiliary locations intended only for events during which their use is feasible.

Figure 5.7 Low frequency loudspeaker. (Courtesy of Altec Lansing Corp.)

Figure 5.8 Compact modular loudspeaker system. (Courtesy of Bose Corporation.)

5.4 EXAMPLES OF SOUND REINFORCEMENT AND REPRODUCTION SYSTEMS

There is an almost infinite number of variations in the design of sound systems that are applicable to rooms of various uses. It is essential that many aspects concerning sound systems be considered in early phases of facility planning. These include a clear understanding of the functions that the sound system must perform, the size and the placement of loudspeakers, the location of a sound control position, and the provision of adequate space and ventilation for the signal processing and the amplification equipment. Electrical conduit of adequate capacity must be provided in the structure a long time before the sound system equipment can be installed. All of these factors require careful coordination with the architects and engineers.

Laying out of conventional distributed loudspeaker systems may be simple and may require only coordination of the loudspeaker placement on reflected ceiling plans, with the locations of lights, sprinklers, and air-handling system diffusers taken into consideration. Design of central loudspeaker systems often involves application of large loudspeakers, which require a design of suitable architectural enclosures and adequate supports. To give an idea of the size of various loudspeaker system components, a typical small, high-frequency horn loudspeaker is roughly 45 cm wide, 35 cm high, and 50 cm long. A typical full-size, high-frequency horn is approximately 75 cm wide, 85 cm high, and 100 cm long. While a typical vented-box, low-frequency enclosure may be 100 cm high, 60 cm wide, and 45 cm deep and have an internal volume of 270 l, a large horn-loaded, bass-reflex, low-frequency loudspeaker system may be 85 cm high, 200 cm wide, and 75 cm deep and have a gross volume of roughly 1275 l.

The design of larger sound systems is a complex task that requires knowledge of several areas of acoustics including room acoustics, psychoacoustics, and electroacoustics. In many instances, therefore, the design requires services of an expert who, in addition to a thorough understanding of the field, has sufficient experience in practical applications. Many specialists who design sound systems can be found among members of the National Council of Acoustical Consultants; some others operate independently. Before engaging a consultant, it is recommended to check his or her performance on projects completed in the past.

It should be emphasized that even properly designed sound systems that use the best components may not perform satisfactorily if they have been installed in spaces with unsuitable acoustics. Factors of room acoustics, such as reverberation, distribution of sound-reflecting and sound-absorbing surfaces, and a level of ambient noise from interior and exterior sources, must be considered together in connection with the sound system, for they will influence the overall results.

Theaters and Opera Houses

The critical needs of theatrical reinforcement, where the directional realism plays an important role, require the sound systems in theaters and opera houses to use central loudspeaker systems located at the proscenium. Depending on the shape of the hall, the system may include supplemental loudspeakers.

Receptacles for microphones are provided at the stage front for use with microphones located at the footlights. Additional microphone receptacles are normally available in the stage wings. If the acting area is deep, microphone pickup from the footlight locations becomes unsatisfactory due to the large distance from the actors. Because the use of microphones suspended overhead generally does not provide a usable pickup and they often interfere with the movements of stage equipment and scenery, the best recourse is to use wireless systems with miniature microphone capsules concealed in the costumes of the performers.

Signal pickup for recording or broadcasting of live performances with audience is normally accomplished by means of microphones suspended in front of the proscenium. Halls designed for opera or for musical comedy are equipped with orchestra pits with one or more floor platforms on lifts. Microphone receptacles are provided in the pit floor because a performance of ballet frequently requires amplification of the orchestra sound from the pit onto the stage.

Sound effects during theatrical plays are reproduced by means of portable loudspeakers located offstage or in the scenery. There are large collections of prerecorded sound effects available. For use in stage productions, these effects are dubbed from disk to tape and provided with identification cues marked on the inserts of leader tape. The playback of sound effects is performed from the sound control room or from the stage wing, following visual or aural cue signals.

Halls that are designed for multipurpose uses, including performances of popular music with vocalists, require more microphones than those restricted to the performances of opera or drama. Microphones used by popular music groups are connected via multipair microphone cables called "snakes," the use of which reduces the equipment clutter on the stage floor. Performances of popular music also require the use of a multitude of loudspeakers on the stage for monitoring of sound by the performers. Full-size monitoring systems are independent from the sound reinforcement systems serving the house and employ separate sound control consoles, equalizers, power amplifiers, and so forth; this equipment is normally located in one of the stage wings.

The transmission of information to the cast offstage is accomplished by means of a program monitoring and stage manager's paging system covering the backstage areas. The loudspeakers in this system are distributed in the dressing rooms, waiting areas, and access corridors. The house sound, as heard by the audience, is derived from the sound reinforcement system during amplified events, or during events without amplification, from a house sound pickup microphone suspended in the hall. The loudspeaker circuits in the dressing rooms normally include individual volume controls. The equipment allows defeating the settings of the local controls to ensure the priority of the stage manager's announcements. The program monitoring and stage manager's paging system is independent from the sound reinforcement system serving the house proper, and its equipment is normally located in one of the stage wings, near the stage manager's position, where it can be used without the help of a sound system operator during events not requiring amplification.

The staff positions are interlinked by a production communications system. This is an intercom system of special design that allows a hands-free operation and voice communication in environments with high levels of ambient noise. The production intercom system normally includes a master station located at the stage manager's position and a number of fixed and portable remote stations. These are located in the light and sound control rooms, projection rooms, at follow spot locations, on lighting catwalks, and the like. The fixed stations can be equipped with loudspeakers or with handsets. Portable remote stations are constructed in form of beltpacks connected to outlets through extension cables. Remote stations with wireless transmission provide the utmost mobility. The stations allow the use of single- or double-muff headsets with boom microphones. The production communication systems typically have at least two independent channels; one used by the lighting staff and one used by the sound crew, for example. More elaborate systems have four or eight channels. Even more flexibility in communications can be provided with a matrix arrangement for connecting the channel outputs with the lines to the remote intercom stations. The system allows listening to the house sound program. All channels are accessible for announcements by the stage manager. Outlets for production intercom are provided at the play director's and lighting director's positions in the audience area for use during rehearsals and setting up of the performance. During rehearsals, the play director makes announcements to the actors and to the conductor through a separate system with fixed loudspeakers at the proscenium and portable loudspeaker extensions that can be placed where required.

Concert Halls

Although the performances of classical music and vocal and instrumental recitals are normally not amplified, modern concert halls are equipped with high-quality, full-range sound reinforcement systems. These systems are needed to meet the requirements of multipurpose uses of the concert halls, for announcements, the reinforcement of speech during concerts with narration, and for use by an increasing number of vocalists and instrumentalists who use reinforced sound as an integral part of their performances. The challenge in designing sound systems for concert halls is in reinforcing speech with high intelligibility in spaces that have long reverberation. Typically, concert halls use central loudspeaker systems with high directivity, similar to the systems in theaters and opera houses. Due to the lack of a proscenium wall, those systems are often suspended in free space or integrated architecturally with sound reflectors or canopies used above the orchestral platform. Many concert halls are used for mixed theatrical and concert productions, ballet, and the like, and have orchestra pits. Receptacles for microphones are provided in the front of the orchestra platform, in the orchestra pit, and at the ceiling for suspended microphones. If the concert hall is used for performances of popular music, the sound system will include, in addition to the central loudspeaker system, loudspeakers for reinforcement directly from the stage, and stage monitor loudspeakers. Concert halls require systems for performance monitoring and paging backstage, as well as production intercom systems.

Auditoriums and Lecture Halls

Most auditoriums and lecture halls are equipped with central loudspeaker systems for full-range reinforcement of speech and reproduction of motion picture sound. Microphone receptacles are provided in the front of the platform and in locations designed for portable lecterns. For economy of space, the sound, projection, and lighting control are frequently located in one room. A desirable feature is the provision, at the lecturer's position, of remote controls for projection, volume control of sound, and dimming of the lights.

Meeting and Conference Rooms

Meeting and conference rooms often have systems of distributed loudspeakers recessed in the ceiling. Audio facilities may include microphone preamplifiers designed for automatic operation and mixing of sound. Depending on the requirements, these rooms may include specialized equipment for teleconferencing and equipment for projection of images from slides, film, videotape recorders, or television. Front projection is used when a complete blackout of the room is possible; it is not advisable for functions where note taking is essential. Rear projection provides satisfactory results in rooms that must remain partially lit during the event.

Motion Picture Theaters

The loudspeakers for playback of motion picture sound are located in the front of the room, behind a perforated screen. The high-frequency loudspeakers in the system are placed approximately $\frac{1}{3}$ of the screen height below its top. The number of independent sound channels may vary from one, for films with single-channel sound tracks, to four, for Cinema Scope sound reproduction, with three loudspeaker systems serving the main channels, located behind the screen, and the sound effects loudspeakers serving the fourth channel, distributed on the walls of the hall in a surround fashion. Figure 5.9 shows two-way loudspeaker systems used in medium-size motion picture theaters. The unit on the right is a compact loudspeaker with shallow depth, used for reproduction of sound effects in surround fashion. Motion picture films that rely heavily on the use of sound effects may require the installation of special sound systems including ultralow-frequency loudspeakers, called subwoofers, used to simulate spectacular events, such as earth tremors, explosions, and the like. Many sound systems in motion picture theaters have inputs for playback of music during intermissions and for use with microphones for anouncements or during events when the hall is used as an auditorium.

Office Spaces

Except for meeting and conference rooms, the functions of the sound systems covering offices are limited to announcements or to the playback of background music.

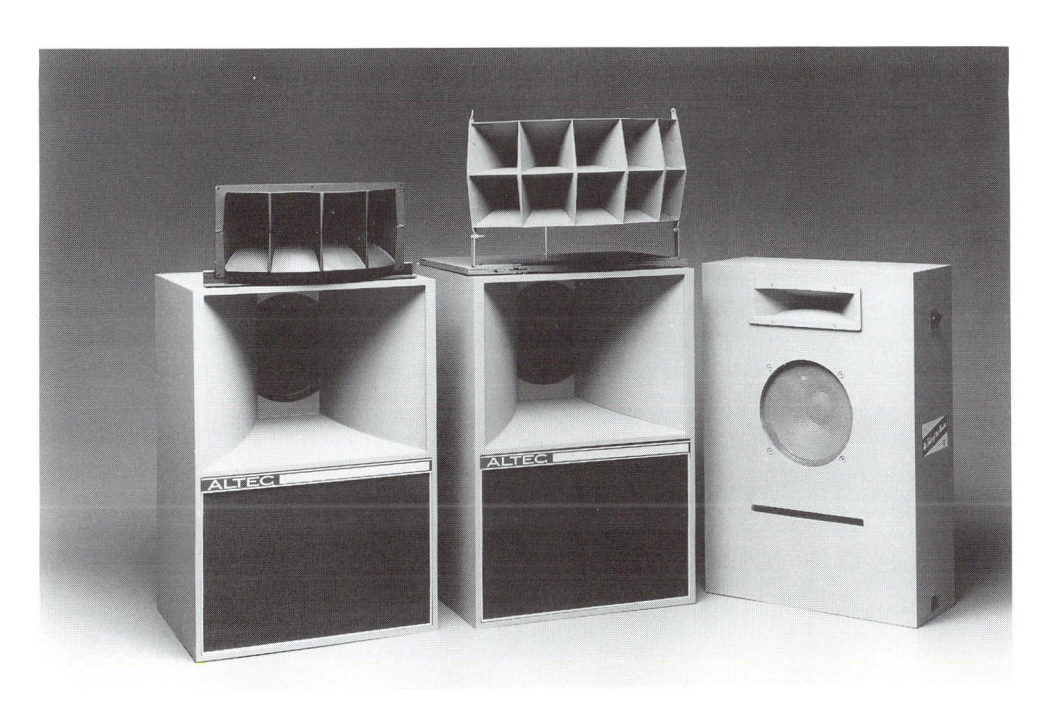

Figure 5.9 Loudspeakers used in motion picture theaters. (Courtesy of Altec Lansing Corp.)

Open-plan offices are frequently equipped with systems emitting a low-level, bland-noise, which helps to mask speech sounds intruding from adjacent workstations and improves the speech privacy. Such systems can combine both the masking noise and the distribution of announcements. Offices, as a rule, employ systems of distributed loudspeakers either recessed in the ceiling or concealed in the ceiling plenum behind sound transparent ceiling panels. Depending on the building code requirements, the high-rise office towers may include sound systems designed for life safety announcements and for the transmission of warning signals. Such systems are normally separate from other systems and use fire-resistant equipment, wiring, and installation materials.

Houses of Worship

Houses of worship generally require acoustics suitable for music, that is, conditions with long reverberation. The problems facing the designer of sound systems for houses of worship are similar to those with concert halls, except that the requirement of the high intelligibility of speech is of primary importance, with the requirement of directional realism in reinforcement playing a lesser role. Under these conditions, distributed systems with loudspeakers in the backs of the pews are sometimes a solution. The use of wireless microphones for mobility and of automatic microphone mixers for simplicity of operation are common in the present-day installations in houses of worship.

Exhibition Halls

Sound reinforcement systems in exhibition halls are designed for reproduction of announcements and of background music. If an exhibition hall includes a permanent or demountable platform allowing some form of stage presentations, the design normally includes a central loudspeaker system located above the front of the platform to provide sound reinforcement with directional realism.

Hotel Ballrooms and Banquet Halls

Hotel ballrooms and banquet halls are often divisible spaces with microphone pickup required from any location on the floor. Like exhibition halls, hotel ballrooms and banquet halls use distributed systems of overhead loudspeakers. If a stage platform or a motion picture screen is provided, the design includes a central loudspeaker system. Hotel ballrooms and banquet halls often feature such functions as floor shows and fashion shows, in many respects resembling theatrical performances. For this reason, production communications facilities serving lighting, projection, and sound control are often essential.

Assembly Halls

Sound reinforcement systems for assembly halls may use either central or distributed loudspeaker systems, depending on the configuration of the space and the requirements of the users. Microphone arrangements may vary from simple to the most complex, when the participation of the audience or of the delegates from the floor is required. Assembly halls for multinational conferences must be equipped with systems for simultaneous translation.

Sports Facilities

Typically, indoor arenas, designed strictly for sports events, with seating capacities of up to 15,000 spectators have directional loudspeaker systems located in the center of the arena. Often, arenas are also used for other types of performance staged in the center or at the end of the floor. In such instances, more loudspeaker systems are required, one for each performing platform location. In multipurpose arenas, the provision of performance monitoring and paging systems in dressing rooms and locker rooms and the provision of production intercom system for stage management, lighting, and sound is required. In extremely large indoor arenas, such as the Houston Astrodome or the Louisiana Superdome, it is not possible to provide the sound coverage with adequate sound level and with a sufficient ratio of direct to reflected sound employing only one central loudspeaker system. These huge spaces have sound systems using many powerful, directional loudspeaker arrays distributed above the audience areas.

In such large systems, as well as in other distributed systems, the loudspeaker arrays or loudspeaker groups should be provided with cut-off switches to limit sound coverage to the occupied areas. This measure helps reduce reverberation. In cases when the microphones are located in the areas covered by loudspeakers, muting certain loudspeakers can reduce the danger of feedback.

Some large arenas, in order to save construction cost, use acoustical treatments that have depth insufficient to control reverberation at low frequencies. The low-frequency reveberation time in such spaces may be several times longer than the reverberation time at higher frequencies. Operation in such spaces of a full-range sound system causes almost uninterrupted persistence of a low-frequency sound, which is harmful to the clarity and intelligibility of speech. Because the reproduction of low frequencies is not essential for maintaining intelligibility, this situation can be remedied by limiting the bandwidth of the transmitted sound to higher frequencies only. Typically, the low-frequency roll-off is selected at 200 or 300 Hz. Higher roll-off frequencies will impair the naturalness of speech.

The selection of the configuration of a sound system for an outdoor stadium depends entirely on the stadium uses, the seating configuration, and the availability of supporting structures for the loudspeakers. A major problem in many outdoor stadium sites is echo from large, hard surfaces such as scoreboards, unoccupied bleachers, nearby housing structures, and so forth.

Sound systems in natatoriums are designed primarily for announcements to the spectators. The loudspeaker coverage is normally restricted to the bleachers, using systems of distributed, directional loudspeakers. In some instances, underwater sound is provided for swimming instruction and coaching of the swimmers.

5.5 SPECIAL SOUND SYSTEM INSTALLATIONS

Paging and Voice-Alarm Systems for Power Plants and Other Industrial Facilities

Power plants and large manufacturing facilities create difficult conditions for voice communications through public address systems that use loudspeakers. This is due to high levels of ambient noise normally present in such facilities and to their typically long reverberation time caused by large volume of rooms and by scarcity of sound-absorbing materials. Both high background noise and long reverberation time are harmful to speech intelligibility.

Intelligibility decreases rapidly if the average levels of reproduced speech are less than 20 dB above the A-weighted levels of ambient noise. Because listening to public address announcements at speech levels of more than 95 to 100 dB creates discomfort, industrial plants in which the average ambient noise levels exceed 75 to 80 dBA are generally not suitable for use of public address systems. Another aspect is that a provision in large industrial plants of public address systems capable of delivering a uniform coverage with very high sound levels requires application of large quantities of loudspeakers and amplification equipment, the cost of which may be prohibitive. In plants with high noise levels, it may be necessary to use lights or sirens to communicate simple messages. If reliable transmission of voice messages to selected individuals is indispensable, it can be, in case of high ambient noise, accomplished through headsets equipped with effective earmuffs.

"Hard-of-Hearing" Systems

Systems for people with hearing impediments are required in many public auditoriums. Often, the hard-of-hearing stations are provided at selected seats in the auditorium. Each such station is equipped with an outlet jack for headsets and with an individual volume control. A typical location for a jack and a volume control is under the armrest of an auditorium chair. If the coverage of all audience areas with a hard of hearing system is required, a wireless system is used with the transmission via an induction loop or frequency-modulated infrared radiation.

Simultaneous Interpretation Systems

Auditoriums designed for international conferences are equipped with multilanguage simultaneous translation systems. In such systems, the signal from the speaker's microphone (floor language) is distributed to the interpreters' booths. The interpreters listen to the floor language via headsets and translate the speech into other languages. The translated language signals are reinforced and distributed to the delegates' seats in the hall. The delegates listen to the translation via headsets, having selected the language channel with a switch. Each listening station is provided with an individual volume control. Another method of distribution of translated speech is via wireless transmission using radio waves, induction loops, or frequency-modulated infrared radiation. In a wireless transmission system, each delegate carries his or her own receiver and headset and is not restricted by a connecting cord in his or her freedom of movement. Wireless systems are required in halls with movable seats and flexible seating arrangements. The interpreters' booths must have satisfactory sound isolation between each other and must be located and oriented so that the interpreters see the lips of the person whose language they translate. Each booth must accommodate two interpreters sitting side by side.

REFERENCES AND FURTHER READING

AES Recommended Practice Specification of Loudspeaker Components Used in Professional Audio and Sound Reinforcement, AES2–1984 (ANSI S4.26–1984), Special Publications Office, Audio Engineering Society, New York.

Ballou, G. M., *Handbooks for Sound Engineers, the New Audio Cyclopedia,* Howard Sams, Indianapolis IN, 1987.

Doelle, L. L., *Environmental Acoustics,* McGraw-Hill, New York, 1972.

Figwer, J. J., "The Louisiana Superdome Sound System," *JAES,* **24**(7), 554 (1976).

Harris, C. M., "Sound and Sound Levels," in *Handbook of Noise Control,* 2nd ed., C. M. Harris, ed., McGraw-Hill, New York, 1979, Chap. 2.

Journal of the Audio Engineering Society, selected papers.

Klepper, D. L., ed., *Sound Reinforcement* (*Anthology of Papers*), Audio Engineering Society, New York, 1978.

Kuttruff, H., "Electroacoustic Installations in Rooms," in *Room Acoustics,* 2nd ed., Applied Science, London, 1979, Chap. 10.

Rettinger, M., *Practical Electroacoustics,* Chemical Publishing Company, New York, 1955.

Schroeder, M. R., "Improvement of Acoustic Feedback Stability by Frequency Shifting," *JASA,* **36,** 1718 (1964).

Sprinkle, M. C., "Environmental Control—Sound Systems," in *Time Saver Standards for Architectural Design Data,* 5th ed., J. H. Callender, Ed., McGraw-Hill, New York, 1974, Chap. 4, p. 1013.

Webster, J. C., "Effects of Noise on Speech," in *Handbook of Noise Control,* 2nd ed., C. M. Harris, Ed., McGraw-Hill, New York, 1979, Chap. 14.

Case Study

CONCORD-CARLISLE REGIONAL HIGH SCHOOL, AUDITORIUM SOUND SYSTEM, CONCORD, MASSACHUSETTS

The Concord-Carlisle High School was completed in 1960 to provide educational facilities for a growing number of children from the "baby-boom" generation. Financial constraints dictated the construction of a school without frills. The auditorium (Figure 5.10) is a multipurpose space, with a fan-shaped floor plan. It has capacity for an audience of up to 650 people.

The ceiling of the auditorium is sound-reflecting and the rear wall has sound-absorbing treatment. Originally, the auditorium was not equipped with a sound system, but the existence of a 2.5-m-wide and 1.5-m-high, fabric-covered recess in the sloping front portion of the ceiling indicated that the architects considered the installation of a central loudspeaker system at some point in their design.

Over the years, the users of the auditorium experimented with a variety of temporary sound systems using portable loudspeakers and loudspeaker lecterns. None of these stop-gap solutions worked satisfactorily, and in 1986 the school turned to a local consultant for the design of a permanent sound system. The system was to provide reinforcement of speech, with multiple locations for microphones on the stage, reproduction of motion picture sound and other prerecorded material, instrumental and vocal reinforcement, basic stage monitoring, and an output for simple taperecording. The sound system was to be inexpensive and have a straightforward design to allow for use by untrained personnel.

The completed sound system provides for eight microphone inputs and one input for signals from a portable motion picture sound projector. All inputs are switchable to accept microphone and line level signals, which allows for the use of a variety of input equipment.

All control, signal processing and amplification equipment is located in an existing, open projection booth in the rear of the auditorium. The operator sees the action on the stage and hears directly the sound he or she is controlling. The equipment includes a moderately priced, reliable mixer/preamplifier with four inputs, supplemented by a five-input mixer extender. A frequency-shift feedback stabilizer, which increases the available gain before feedback during speech events, is also provided. The system has active $\frac{1}{3}$-octave-band equalization and a dual-channel power amplifier. One channel feeds the central loudspeaker cluster and a second channel is reserved for stage monitor loudspeakers. The block diagram of the system is shown in Figure 5.11.

The two-way central loudspeaker system consists of two small-format, high-frequency horns with compression drivers and one vented-box, low-frequency enclosure with two drivers. The loudspeaker crossover frequency is 800 Hz, employing a passive network. The loudspeaker system is partially recessed in the space that was provided for this purpose in the original design of the ceiling.

Because of a very limited budget, the installation of the sound system was implemented in stages. First, the school purchased most of the equipment. A year later, the equipment was installed by the school maintenance staff and wired by electricians. With all equipment in place, the consultant checked out the installation and performed the final adjustments and equalization of the system. After adjusting the system for best performance, the consultant provided the school with a summary report that included basic data on the system adjustments and the recommended typical settings of controls.

Figure 5.10 Concord-Carlisle Regional High School auditorium. (Photo by author.)

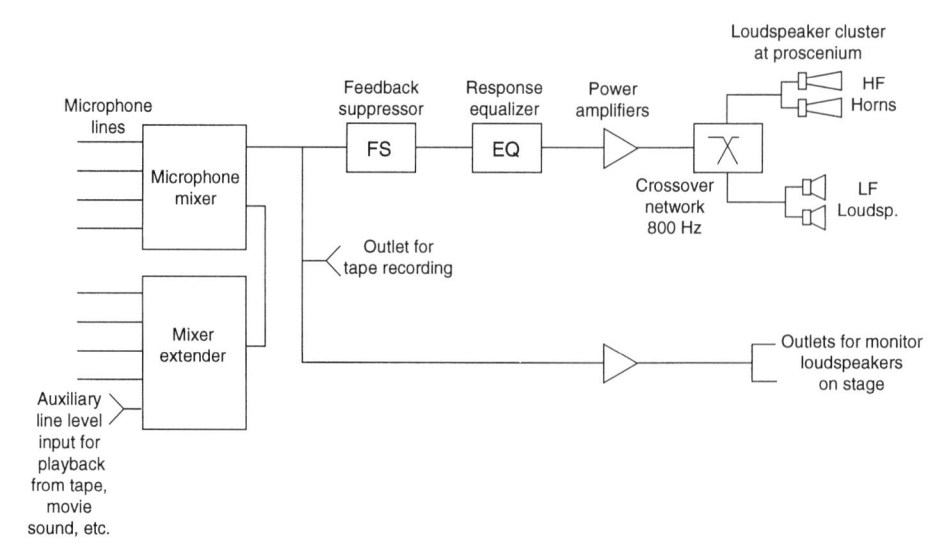

Figure 5.11 *Simplified block diagram of the Concord-Carlisle Regional High School auditorium sound system.*

The overall acoustical frequency response of this simple sound system, measured with ⅓-octave-band pink noise signals, is essentially flat, with a gradual roll-off below 80 Hz and above 5000 Hz. The uniformity of sound coverage, measured using an octave band of pink noise centered at 4000 Hz as a test signal, is within ±2 dB. The maximum sound levels for broadband program material reach up to 100 dB. If the adjustment of the system components is accidentally brought out of balance, the performance of the system can be restored in a matter of minutes by following the simple and easily understandable procedure outlined in the consultant's report.

Case Study

CARGILL 97 LECTURE AUDITORIUM SOUND SYSTEM, NORTHEASTERN UNIVERSITY SCHOOL OF LAW, BOSTON, MASSACHUSETTS

The sound system and audio-visual facilities in the Cargill 97 Lecture Auditorium were provided as part of a complete refurbishing and upgrading of the hall (Figure 5.12). Before the renovation, the auditorium was known for its unsatisfactory hearing conditions, and one of the major objectives of the upgrading was to improve the acoustics. Another requirement was to increase the seating capacity of the hall from 170 to 208 seats. This had to be accomplished without changes in the basic floor plan, since it was not possible to change the existing structure. The auditorium has a fan-shaped floor plan and a curved rear wall.

Before the renovation, the auditorium ceiling was made of sound-absorbing acoustical panels, the floor was carpeted, and the walls were finished with painted plasterboard. The mid-frequency reverberation time, measured in the hall without an audience, was 1.8 sec. The extended persistence of sound was mainly due to the presence of a strong flutter echo. Subjectively, the flutter echo was adding an unpleasant coloration to the sound. Flutter echo is typical for rooms that have sound-absorbing ceilings and floors and hard, sound-reflecting walls. The mechanical system noise was excessive, with the noise spectrum following the noise criterion curve NC-40. There was a strong intrusion of traffic noise from the street, particularly trolley car noise. A study of the existing roof construction revealed that the auditorium was covered with a thin sheet-metal roof that, together with the lightweight acoustic-panel ceiling, did not provide enough sound isolation for exterior noise.

Considering the relatively small size of the auditorium and the fact that classes held in the auditorium are conducted in an interactive way, which requires that the students engage in dialogue with the lecturer and among themselves, the goal of the acoustical redesign was to create hearing conditions that would allow good communication without the use of a speech reinforcement system. This was accomplished through the installation of a continuous, sound-reflecting, plasterboard ceiling, which provides useful sound reflections between speaking persons located anywhere in the room. The heavy construction of the new ceiling, combined with the effect of sound-absorbing glass-fiber blankets added in the void between the ceiling and the metal roof, greatly improving the sound isolation of the auditorium from exterior noise. Sound-absorbing panels were installed to control the rever-

Figure 5.12 Lecture Auditorium Cargill 97, Northeastern University School of Law. *(Photo by author.)*

beration and focusing of sound by the curved rear wall. As result of the changes to the room finishes made during the renovation, the mid-frequency reverberation time was shortened to 0.8 sec in the empty auditorium and to 0.6 sec with full audience.

The mechanical systems serving the auditorium were completely redesigned. With the improved isolation between the auditorium and the mechanical equipment room, with reduced air discharge and air return velocities, and with the provision of duct silencers and duct lining where required, the ambient noise in the room was reduced to meet the NC-30 criteria.

Even though the improved acoustical conditions in the auditorium after renovation allow free speech communication without electronic reinforcement, a loudspeaker system was required for playback of sound from audio-visual sources. The loudspeakers in this system can be used as an option for reinforcement of live speech to assist weak-voiced lecturers. At times, subtle reinforcement of speech may also help to boost the intelligibility when the lecturer is turned away from the audience, while writing on the blackboard or providing comments during a slide presentation. Another system was required to provide good-quality recording on standard cassettes of lecturers and questions and answers from the audience for those students who could not participate in the lecture.

Since the lecture hall has a relatively low ceiling compared with the depth and width of the audience area, a distributed loudspeaker system was chosen to ensure a good uniformity of sound coverage. With the intelligibility of speech being of primary importance, the system uses ceiling-mounted line-source loudspeakers with higher directivity than that of commonly used loudspeakers. The use of directional loudspeakers increases the ratio of direct-to-reverberant sound, which is favorable for intelligibility. The directivity of the loudspeakers also limits the amount of sound energy projected toward the front of the hall where microphones are located. This reduces the danger of acoustical feedback. The danger of feedback is further reduced through the use of a frequency-shift feedback stabilizer installed in the signal path for live sound reinforcement.

The ceiling loudspeakers are installed in two rows, with two loudspeakers in the first row and three loudspeakers in the second row. The signal to the first row of loudspeakers is delayed 15 msec and the signal to the second row of loudspeakers is delayed 30 msec. In this arrangement, the sound from the loudspeakers arrives to the listeners with a short delay following the natural voice of the lecturer. Delaying the signal preserves the directional realism, and the listeners perceive the speech as arriving from the direction of the lecturer, not from the loudspeakers. The function of the loudspeakers in this system is to enhance the sound by simulating ceiling reflections.

The lecture hall sound system is equipped with microphones intended to meet a wide range of pickup conditions that may be required by the lecturer. This includes fixed, unidirectional microphones for use at the lectern, wired lavaliere microphones for people who do not want to speak from a fixed position while delivering a lecture, and wireless microphones for use by lecturers or by the audience during question-and-answer sessions.

In order to minimize the need for adjustments to the system during lectures, the system is equipped with an automatic mixer. The signals from the microphones and the signals from the automatic mixer are available for recording with a standard audio cassette deck or with a VHS video tape recorder during events that are video recorded. During playback of prerecorded cassettes or videotapes, the system accepts

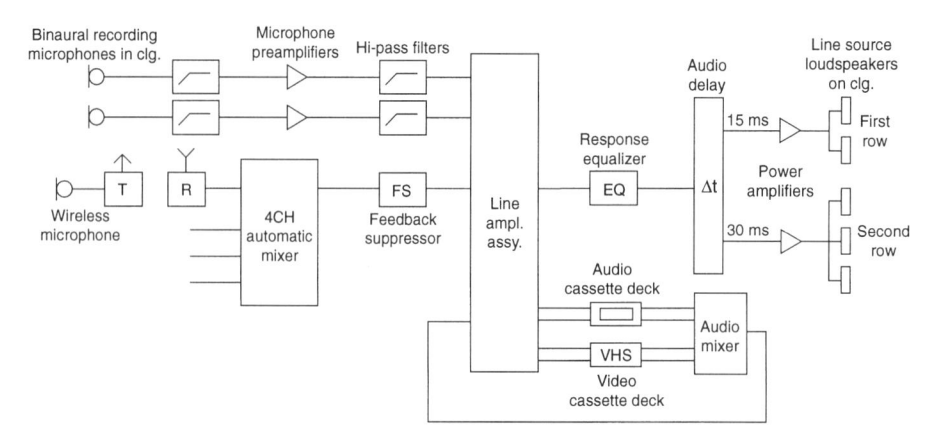

Figure 5.13 Simplified block diagram of the Cargill 97 Lecture Auditorium sound system.

audio signals via a four-input audio mixer, independent from the automatic microphone-signal mixer. The system is equipped with a ⅓-octave-band equalizer, installed ahead of the digital audio delays and power amplifiers. The purpose of the equalizer is to compensate for irregularities in the frequency response of the loudspeakers and to help achieve a smooth overall frequency response.

The lecture auditorium is equipped with a screen for front projection of images from a slide projector or from a video projector equipped with an interface that accepts video signals from a computer or VHS video player and video camera.

The part of the system that is designed to provide audio recording of lectures and of questions from the audience for use by absent students is separate from the sound reinforcement and playback system. This recording system employs the method of binaural recording by means of two ceiling-mounted microphones. In a binaural recording, the signal from one microphone is reproduced through one earpiece of a headset and the signal from the second microphone is reproduced through the other earpiece. A binaural recording offers a higher degree of directional perspective and provides increased intelligibility and better discrimination against ambient noise than a single-channel recording.

The pressure-zone-type microphones in the binaural recording system are flush-mounted on the ceiling. The signals from these microphones are passed through high-pass filters and a preamplifier, followed by second-stage high-pass filters. Because low frequencies are not needed in recording speech, the two stages of high-pass filtering eliminate the pickup of low-frequency noise or rumble.

The binaural signals are recorded on a standard audio cassette deck. They can be also routed to the audio input of a VHS video tape recorder. Binaural speech recordings can be made without using the system for reinforcement of live speech.

A simplified diagram of the sound system is shown in Figure 5.13.

Case Study

THE LOUISIANA SUPERDOME SOUND SYSTEM, NEW ORLEANS, LOUISIANA

The Louisiana Superdome stands 82.3 m high and has a dome that measures 210 m in diameter. The volume of the main bowl is approximately 1.7 million cubic meters. Used as an auditorium, it accommodates up to 95,400 spectators, although it is said that during one densely packed Rolling Stones concert, the audience was over 100,000. The Superdome is designed to accommodate sports events, including football, baseball, basketball, and to serve a large variety of uses as an auditorium.

The Louisiana Superdome is the second in the series of large-scale, domed stadiums constructed in the United States starting in the 1960s. Inaugurated in August 1975, the Louisiana Superdome followed the opening of the Houston Astrodome, an indoor stadium similar in concept. Using experience from the design and initial operation of the Houston Astrodome, the design of the Louisiana Superdome was set to avoid mistakes made in the design of the Astrodome and to improve upon it. Among other considerations, this concerned acoustics and the design of the sound system.

One of the design features of the Astrodome was the provision of a translucent roof intended to let daylight in to allow natural grass to grow on the playing field. This idea backfired because, in the relatively dark interior of the stadium, the translucent roof produced glare against which the players could not see the ball. This problem could not be overcome and ultimately the clear roof cover had to be painted blue. This solved the problem of glare, but the entire ceiling of the Astrodome was left hard and sound-reflecting, with all inherent problems of long reverberation, echoes, and focusing of sound.

In the Louisiana Superdome (Figure 5.14), the roof and the ceiling are two independent structures, separated by an airspace of varying depth. The separation reaches a depth of approximately 3 m in the center of the dome and decreases gradually toward its perimeter. The roof of the Superdome rests on a steel structure and consists of a composite system of 3.8-cm-deep, 20-gauge, fluted steel deck, covered with a 2.5-cm-thick layer of polyurethane foam and a 0.6-mm-thick waterproofing membrane made of Hypalon, sprayed over the polyurethane.

The interior ceiling of the dome is a 3.8-cm-deep, ribbed, perforated, acoustical steel deck with glass-fiber pads installed in the ribs. The space between the roof and the acoustic deck below it is used for catwalks, winching systems, lighting systems, and air return. It offers convenient access to all equipment that needs servicing in that area and allows access to the roof structure. In addition to absorbing sound, the double-layer roof and ceiling structure offers outstanding isolation from outside noise.

The sound-absorbing treatment in a domed stadium serves three purposes: to control the reverberation (persistence of sound), to counteract the sound-focusing effects of the dome, and to reduce the level of crowd noise. The huge volume of the Superdome and the large distances from the sources of sound to the boundaries of the space make any surface in the hall a potential source of echoes. Such conditions call for a massive application of sound-absorbing treatment.

In addition to the acoustic deck that forms the ceiling, sound-absorbing treatment was applied to the curved walls in the upper volume of the hall. This treatment consists of perforated metal panels filled with glass-fiber blankets, enclosed in polyethylene film bags.

The concourses and other open spaces that extend to the spectator areas have sound-absorbing ceilings consisting of suspended, perforated metal pans, filled with glass-fiber pads. Concerns about cost and maintenance problems precluded the use of upholstered seats in most of the audience areas. Upholstered seats would have helped stabilize the reverberation under conditions of varying occupancy.

The average absorption coefficient for the entire interior of the Superdome, calculated from the reverberation time measured in the unoccupied hall, is close to 0.5. The reverberation time in the unoccupied hall is approximately 7 sec at 125 Hz, 6 sec at 500 Hz, and 5 sec at 2000 Hz. The interior of the Superdome is shown in Figure 5.15.

In selecting the types of loudspeakers for a space as large as the Superdome, one must consider the levels of direct sound in the audience areas that must be produced during typical events. The crowd noise during sports events in an acoustically treated indoor stadium frequently reaches sound levels in excess of 95 dB. Under such circumstances, the sound system must be capable of delivering at the listeners' ears undistorted sound levels of at least 100 dB. The requirement for high sound levels precludes the use of relatively inefficient cone-type loudspeakers in those parts of the sound system that cover the audience in the open bowl of the arena. Horn-loaded loudspeakers with compression drivers, which are much more efficient than direct-radiator loudspeakers, are a better choice in this application. The distribution of such loudspeakers in the space depends on the radiation patterns of the loudspeakers and their distance from the audience areas. The distance must be chosen so that the loudspeakers can deliver the required sound levels without exceeding their rated power. In examining the coverage requirements for the Superdome, it became apparent that even with use of the most efficient loudspeakers, it was not possible to develop sufficient sound levels in the audience areas from one centrally located loudspeaker cluster. A more complex loudspeaker system, with multiple loudspeaker clusters located on the circle closer to the circumference of the hall, was required.

Figure 5.14 Outside view of the Louisiana Superdome. (Photo by author.)

Figure 5.15 Interior of the Louisiana Superdome. (Photo by author.)

Another aspect that must be considered is the distance between the source of live sound (the performer) and the main loudspeaker system. For listeners seated far away from the performer, the loudspeaker system becomes the principal source of sound. In the case of simple announcements, the direction from which the announcement arrives does not matter, as long as the message is loud enough, sounds natural, and is intelligible. In the case of stage performance, the main loudspeaker system must be located close to the performing position in order to preserve the directional realism. If the distance is too large, the performer hears the sound from the loudspeaker with an excessive delay. This is unpleasant and confusing. The effect of such a delay can be offset to some degree by using stage monitors (also called foldback loudspeakers) located close to the performer.

In large halls, the positioning of performing platforms with respect to the loudspeaker systems is critical. The platforms cannot be placed just anywhere on the field. They must be placed where the primary loudspeaker systems are located. If several performing stage locations are required, each location calls for its own primary loudspeaker system. Since there are practical limits of how many such loudspeaker systems can be provided, the number of workable stage locations is also limited. The above applies to events that are served by the permanently installed house sound system. If a touring group brings its own temporary sound system, the performing platform can be located almost anywhere to meet the specific requirements of the production. However, there are serious drawbacks connected with the use of temporary sound systems. Unless a temporary system is carefully and intelligently integrated with, and uses parts of, the house system, it rarely provides uniform sound coverage. Typically, with temporary systems, the listeners close to the stage are exposed to excessive sound levels, while listeners seated far away do not receive enough sound.

The design of a sound system must be preceded by a detailed study of the program for the facility. The Louisiana Superdome had a program that includes 18 types of events, each having its own seating plan. These could be served by four basic sound system configurations, each using a different primary loudspeaker system. The primary loudspeaker systems include the following three assemblies: (1) the Center Gondola Loudspeaker System, which consists of six loudspeaker clusters, serving for reinforcement of events staged in the center of the stadium; (2) the Auxiliary Gondola Loudspeaker System, formed by one large loudspeaker cluster, which can be operated from two locations, 22 or 38 m west of the center of the field (a system intended for use during basketball games or for reinforcement of performances from two off-center stage locations); and (3) the Circumferential Loudspeaker System, consisting of twelve loudspeaker clusters, suspended from the dome on a 122-m diameter circle (a system used for reinforcement of sound for spectators seated in areas exposed to the main volume of the hall).

The locations of the main loudspeaker systems and their operating positions are shown in Figure 5.16. When not in use, the center and auxiliary loudspeaker gondolas can be hoisted into their elevated storage positions to clear the space above the playing field.

The spectator areas and connected concourses, which have no direct sightlines to the main loudspeaker systems, are covered by distributed cone-type loudspeakers. The typical locations of these supplemental loudspeakers are shown in Figures 5.17 and 5.18. The distributed supplemental loud-

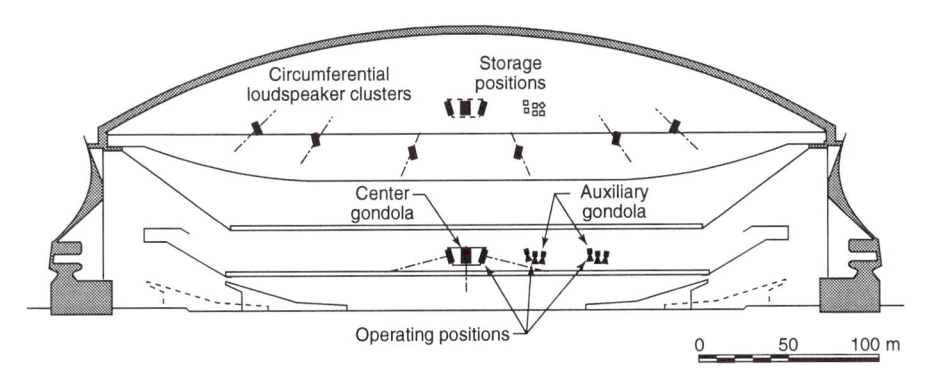

Figure 5.16 Locations and operating positions of main loudspeaker systems in the Louisiana Superdome.

speakers are divided into zones, based on their distance from the primary loudspeaker systems. The primary loudspeakers operate without signal delay, while the signal to individual zones of supplemental loudspeakers is delayed. The increments of signal delay between adjacent loudspeaker zones do not exceed 25 msec. The range of delays used in the system is 50 to 300 msec.

The primary sound control position is located next to the press area, in a space shared with the sports announcer. Both the system operator and the announcer have clear lines

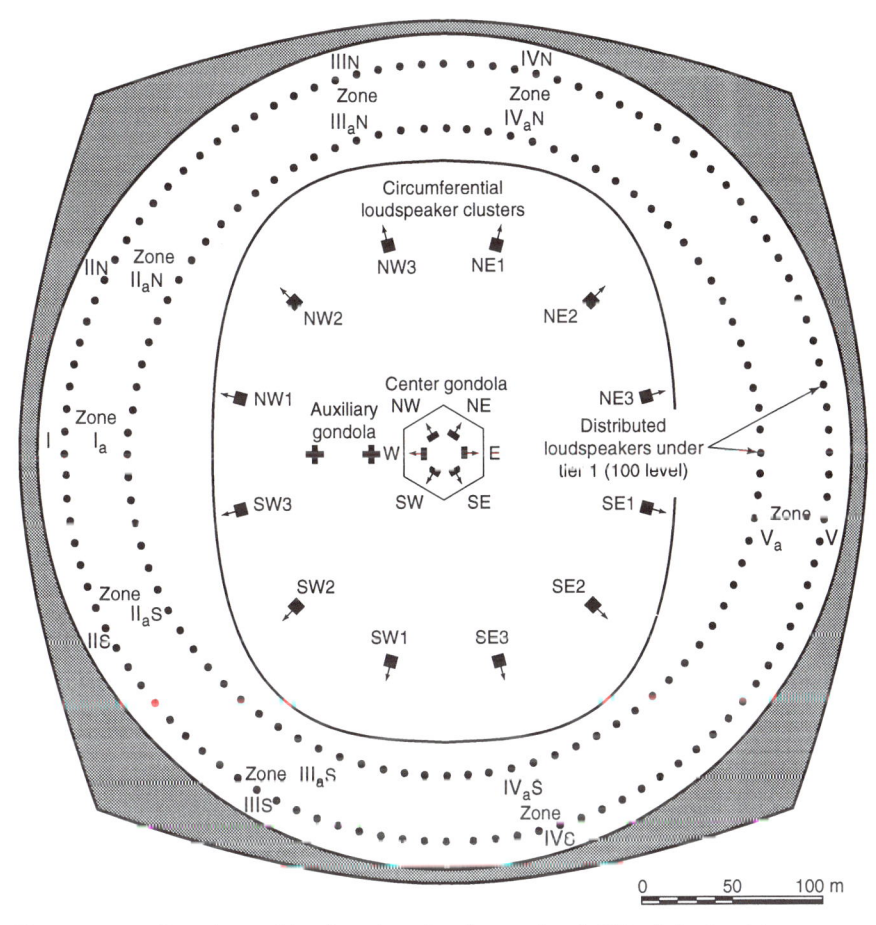

Figure 5.17 Locations of loudspeakers in plan at Level 100 of the Louisiana Superdome.

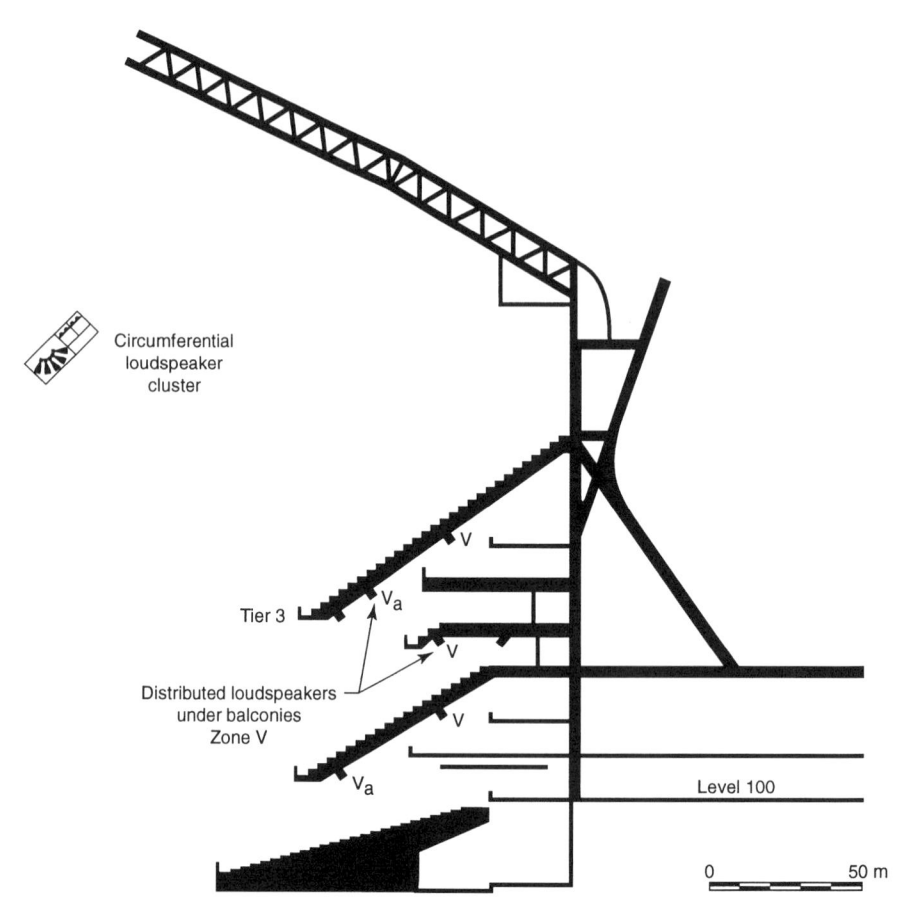

Figure 5.18 Locations of loudspeakers in Section at the Louisiana Superdome.

of sight to the entire playing field. The sound can also be controlled from remote locations on the field, which are close to the performing platform locations.

A simplified block diagram of the Superdome Sound System is shown in Figure 5.19. The input equipment consists of microphones, wireless microphones, music sources, and signal feeds from TV and giant-screen video projection systems.

The distribution of audio power to the loudspeakers is at 70 V and, in the case of long runs, at 140 V. Changing the system configuration depending on the event does not require patching. It is simplified through the use of a switching matrix with relays. The system is equalized in $\frac{1}{3}$-octave bands to obtain a smooth frequency response, which is essentially flat from 63 to 2500 Hz and descends gradually to a relative level of -10 dB at 10,000 Hz.

The Louisiana Superdome sound system has been used with success for many years. It has served as the prototype for a number of systems installed in large multipurpose sports arenas. Recently, the Superdome sound system underwent an expansion to improve the sound coverage in the spectator areas located under the balcony overhangs. In the original design, the audio power required for the individual loudspeaker zones was calculated with the assumption that crowd noise would be uniform in all spectator areas in the stadium. Accordingly, the systems were designed to deliver uniform sound levels in all areas. In reality, due to the sound reflected from the hard underbalcony soffits, the crowd noise in spectator areas located under the balcony overhangs proved to be higher than in the open seating. The problem of insufficient sound levels in the areas under the balcony overhangs were corrected through addition of loudspeakers operating with increased output. The system underwent further upgrading with the addition of a sound control console with increased input capacity and with the installation of a system that allows sequential monitoring of the condition of all compression drivers without the need to disconnect the loudspeaker lines. This further enhanced the reliability of the system and its ease of maintenance.

Case Study

THE COMMUNITY CHURCH OF VERO BEACH, FLORIDA, SOUND SYSTEM

When the Congregation of the Vero Beach Community Church outgrew the facility it had occupied for years, it de-

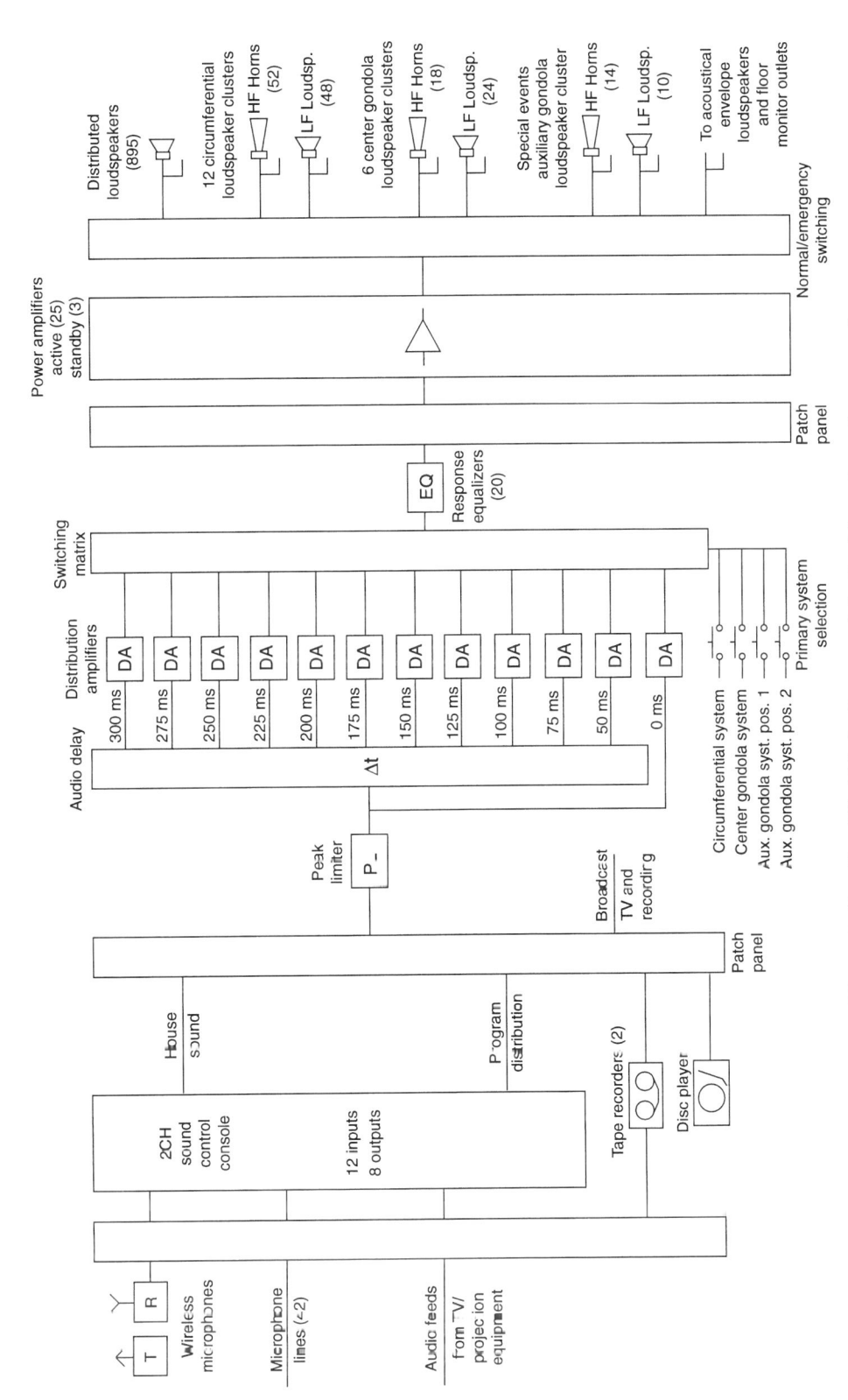

Figure 5.19 Simplified block diagram of the Louisiana Superdome sound system.

cided to build a new church that would meet all its needs for an extensive program of activities. These needs were outlined in discussions with the pastor, the music and choir director, the architect, and the acoustical consultant. The congregation is very much involved in church programs and it has young as well as elderly members. Preaching is very important, therefore the listening conditions and the sound system needed to provide excellent intelligibility. The church has a very active Music Department and an excellent choir, therefore music acoustics were equally important as speech. The existing pipe organ was to be relocated from the old facility to the new church. The preliminary design had a fan-shaped nave, with a volume of over 8500 m^3, and included a balcony extending above the narthex. Since the bids for construction in accordance with the preliminary design came above the budgeted amounts, and additional funds were not available, the design had to be revised to reduce the cost. The redesign eliminated the balcony, leaving it as an option for future expansion. In addition, the height of the building had to be reduced, which necessitated lowering the ceiling in the nave. This reduced the volume of the nave to approximately 7500 m^3.

The organ, the choir, and the bell choir surround the sanctuary. All surfaces in the front of the nave are hard and sound-reflecting. This includes the hardwood floor in the sanctuary, which extends up to the first row of pews in the nave. The ceiling in the sanctuary is acoustically hard and shaped to form a sound-reflecting shell for the organ and for the choir. The ceiling in the nave consists of pyramids made of gypsum board. The ceiling is sound-reflecting and strongly diffusing. There are sound-absorbing areas on the upper parts of the rear wall, where openings to the balcony were originally planned. The floor under the pews is carpeted.

The reverberation time in the unoccupied nave is 2.2 sec at 125 Hz, 2.0 sec at 500 Hz, and 1.6 sec at 2000 Hz. There is not much variation in reverberation with varying occupancy because the pews are upholstered with sound-absorbing pads and breathable fabric. The mechanical and air-handling noise is in the range of NC-25 to NC-30. A view of the nave, looking toward the sanctuary, is shown in Figure 5.20.

The sound system uses a single cluster of directional, high-frequency, horn-type loudspeakers and vented low-frequency enclosures suspended centrally in the sanctuary. This loudspeaker system delivers a maximum of direct sound energy to the audience and enables the reception of reinforced speech with high intelligibility and clarity. The sound system is equalized in $\frac{1}{3}$-octave bands for smooth, extended frequency response. This ensures a high gain of the system before feedback and allows reinforcement of soloists and instruments with a natural sound quality. The central loudspeaker cluster does not cover the seats for the choir or the bell choir. Instead, these seats are covered with supplemental systems of small loudspeakers installed in the guard rail. The signal to the supplemental loudspeakers is delayed. The coverage from local supplemental loudspeakers operating with low signal levels provides better results than projecting sound at a high level from a central loudspeaker cluster. With this approach, less reverberant sound is produced in the sanctuary, and the negative effects of loudspeaker sound being picked up by microphones are reduced. Listeners who are seated in the sanctuary area during sermons receive coverage from floor monitor loudspeakers. Floor monitor loudspeakers are also available for soloists. The pulpit and altar speaking positions are equipped with hypercardioid microphones. A wireless microphone is also provided. Cardioid microphones are used for soloists and instrumental reinforcement. Omnidi-

Figure 5.20 Community Church of Vero Beach. (Photo by author.)

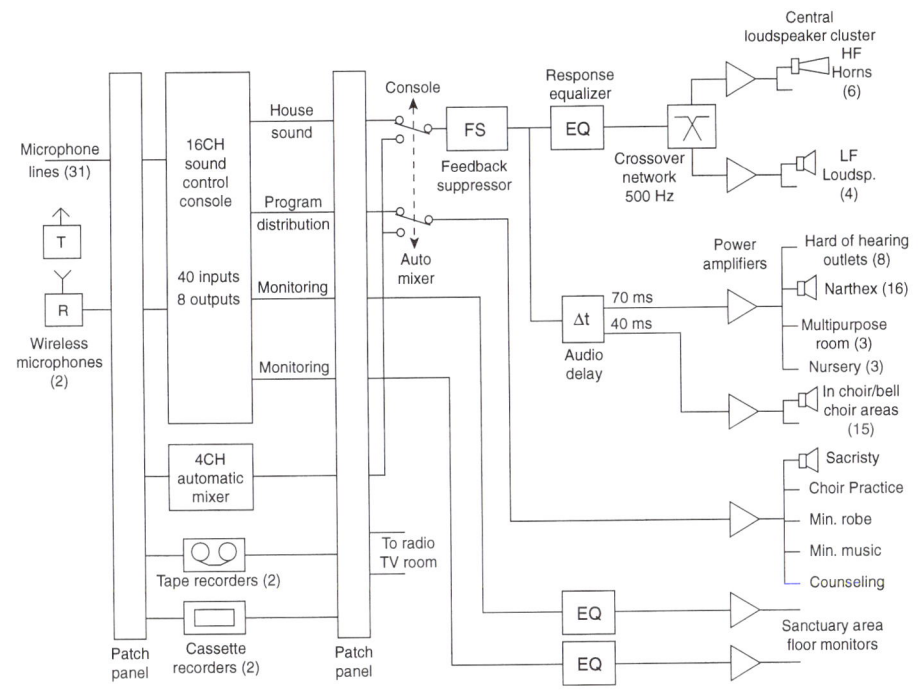

Figure 5.21 Simplified block diagram of the sound system at Community Church of Vero Beach.

rectional, suspended microphones are provided for sound recording and stereo signal transmission of services by a local broadcasting station. The control equipment consists of an automatic mixer. The number of microphone lines and the patching facilities provided in the system were designed to anticipate the future provision of a more elaborate control console for sound recording. A simplified block diagram for the system is shown in Figure 5.21. The coverage of the system extends to include the narthex, sacristy, choir practice, and nursery. Outside the nave, the system uses distributed loudspeakers recessed in the ceiling.

The sound control room is located in the rear of the nave and has an operable window. This allows the operator to follow the activities in the nave and hear directly the sound he or she is controlling. Since some of the events taking place in the church are complex and involve a number of people or performing groups, it was necessary to equip the facility with a professional-quality production intercom system. Such a system helps in controlling and coordinating activities. The intercom system has loudspeaker stations in the robing room, ministry of music, sacristy, choir practice, multipurpose room, and corridor leading to the baptistry. Fixed headset/handset stations are provided at the organ console, the pulpit, the narthex (two stations), bell choir, sound control, and lighting control. In addition, there are outlets for remote beltpack/headset stations installed in the sanctuary floor at the location where weddings take place and at the catwalks in the space above the ceiling.

Since its opening in 1991, the Community Church of Vero Beach has gained a reputation for its fine acoustics. It has become a place of choice for performances by local and visiting music groups.

Case Study

QUEEN SIRIKIT NATIONAL CONVENTION CENTER SOUND SYSTEM, BANGKOK, THAILAND

The Thai National Convention Center, dedicated to Queen Sirikit, was built to host the 1991 Conference of the World Bank and International Monetary Fund (Figure 5.22). As such, it had to meet the requirements for conference facilities dictated by the World Bank. The complex includes a large, divisible Plenary Hall and a set of divisible meeting rooms (all fully equipped for simultaneous translation), an entrance foyer, a delegate meeting plaza, a restaurant, and a cafeteria. The convention center was designed, built, and ready for operation in 18 months. In the following, we will review the characteristics of the individual spaces and the sound systems and supporting installations provided in the center.

216 • Sound Reinforcement Systems

Figure 5.22 Queen Sirikit National Convention Center, Bangkok. (Courtesy of Design 103, Architects.)

Plenary Hall

The name Plenary Hall was adopted for the largest space in the Convention Center because the program of the World Bank meeting called for holding the plenary session in this hall (Figure 5.23). In reality, it was known that after the World Bank conference was over, the hall would have to be capable of serving a large variety of events other than conferences. One of these uses, which is of great importance for the economy of Thailand, is the ability of the Convention Center to accomodate large-scale industrial exhibitions and trade shows. For that reason, the Plenary Hall was designed with a flat floor and with no permanent seating. The load-bearing capacity of the floor allows for the display of heavy machinery. The hall has an access ramp and oversized doors to allow large exhibits to be brought into the hall. The clearance between the floor and the bottom of the roof structure is 11 m.

To increase its flexibility, the Plenary Hall can be divided by means of operable partitions into three separate, acoustically isolated spaces, called Halls A, B, and C. The sound isolation between the divided spaces of the Plenary Hall, tested with $\frac{1}{3}$-octave-band filtered noise signals, meets the Noise Isolation Class NIC-42.

When undivided, the Plenary Hall has a length of 112.5 m and a width of 35 m. The roof structure is a space

Figure 5.23 Queen Sirikit National Convention Center, Plenary Hall. (Courtesy of Design 103, Architects.)

frame 3 m in height; it was lifted from the ground as one component. The overall volume of the hall is approximately 55,000 m^3.

The floor of the hall is covered with carpet. The hall has extensive sound-absorbing treatment on the entire area of the ceiling and on the walls. The hall is free from echoes and has a mid-frequency reverberation time of 1.7 sec unoccupied. Extensive noise control treatment of mechanical and air-handling systems keep the ambient noise below NC-35.

The Plenary Hall is equipped to accommodate a variety of events ranging from large meetings to exhibitions, conferences, and stage performances taking place on demountable platforms. The seating capacity of the undivided hall is over 4000 people. The performing platforms can be located at the side wall, called the "side-stage" location, and in the center of the hall, called the "center-stage" location.

If no stage is used, such as during conferences, the sound coverage is provided by means of distributed, bi-amplified, two-way loudspeaker systems. Each system consists of a large-format, high-frequency horn with compression driver and one vented, low-frequency enclosure. The nominal coverage angle of the high-frequency horn is 40° × 60°. The directivity index (DI) is 14 dB at 800 Hz. There are 34 such loudspeaker systems in the distributed system. When the hall is divided, the distributed loudspeakers are also divided into groups, each group forming an independent system with its own control, signal processing, and amplification equipment. Halls A, B, and C have their own sound control rooms located on the upper floor. The control rooms have operable control windows providing unobstructed lines of sight to the floor. When the hall is not divided, the sound is controlled from the centrally located control room of Hall B. In the undivided, or "no-stage", mode of operation of the distributed system, the signal to the loudspeakers is not delayed.

With the performing platforms in the side-stage or center-stage locations, the distributed loudspeakers are used as supplemental systems and the primary loudspeakers are the side-stage loudspeaker cluster, or four center-stage loudspeaker clusters, respectively. In the side-stage or center-stage mode of operation, the distributed loudspeakers operate with delayed signals. The use of loudspeakers in the different modes of operation in the undivided Plenary Hall is shown in Figure 5.24 and in the divided hall in Figure 5.25.

The overall acoustical frequency response of the distributed no-stage loudspeaker system, measured with a $\frac{1}{3}$-octave-band filtered pink noise signal, is plotted in Figure 5.26. A similar curve for the side-stage loudspeaker system is plotted in Figure 5.27. The undistorted sound levels, measured during the reproduction of wide-band recorded music from compact discs, is 103 dB for the no-stage loudspeaker system and 105 dB for the side-stage loudspeaker system. The crossover frequency of the two way loudspeakers is 500 Hz.

Halls A, B, and C are equipped with microphone receptacles at the programmed locations of head tables for conferences and the locations of performing stages. Each control room has a professional-quality control console for live sound reinforcement and for recording. Each hall is equipped with an automatic mixer to pick up microphone signals during conferences and to provide signal outputs for simultaneous interpretation systems.

Each head table and performing stage location is equipped with outlets for connecting floor monitor (foldback) loudspeakers, which are powered from the control room equipment. In addition, there is an independent floor monitoring system for use during large stage performances. This system uses an active microphone splitter, a special control console for floor monitoring, equalization and amplification equipment, and floor monitor loudspeakers. The 1-into-3 active microphone splitter provides independent, fully isolated microphone lines for use during house sound reinforcement, stage monitoring, and broadcasting or recording. A simplified block diagram of the Plenary Hall B sound system is shown in Figure 5.28. Sound systems for Halls A and C are similar.

Meeting Rooms

The convention center has two meeting rooms with flexible seating arrangements. Each room can be divided by means of operable partitions into two smaller, independent meeting rooms. The meeting rooms have carpeted floors and sound-absorbing treatment on the walls. The ceilings are hard and sound-reflecting. The reverberation time is 0.8 sec in the undivided rooms when they are empty. The operable partitions tested in the closed position have noise isolation class NIC-38. The ambient noise, with the air-handling system operating, meets the criterion curve NC-20.

The meeting rooms are equipped with duplex-type, distributed, ceiling-mounted loudspeakers, connected in four groups, one group serving each divided room. The control and amplification equipment is mounted on equipment carriers with casters. This allows the systems to be operated with the control equipment located in the meeting room or in the respective control/projection room. Each meeting room is provided with an automatic mixer, which can accept signals from microphones on stands or from a conference amplification system. The conference system consists of delegate stations with microphones and loudspeakers for use on the tables. Each room has a wall panel for connecting the control and amplification equipment. The panels include connectors for input to loudspeaker lines, for output from the mixer used with the simultaneous translation system, and for line-level input to the mixer. Push-button switches, installed in the wall panels, allow the respective loudspeaker groups to be connected for combined or divided room operation.

Simultaneous Interpretation System (SIS) and Conference Systems

The SIS system, serving the Plenary Hall, transmits in eight channels including six translated language channels, the floor language channel, and one spare channel. The SIS systems

218 • *Sound Reinforcement Systems*

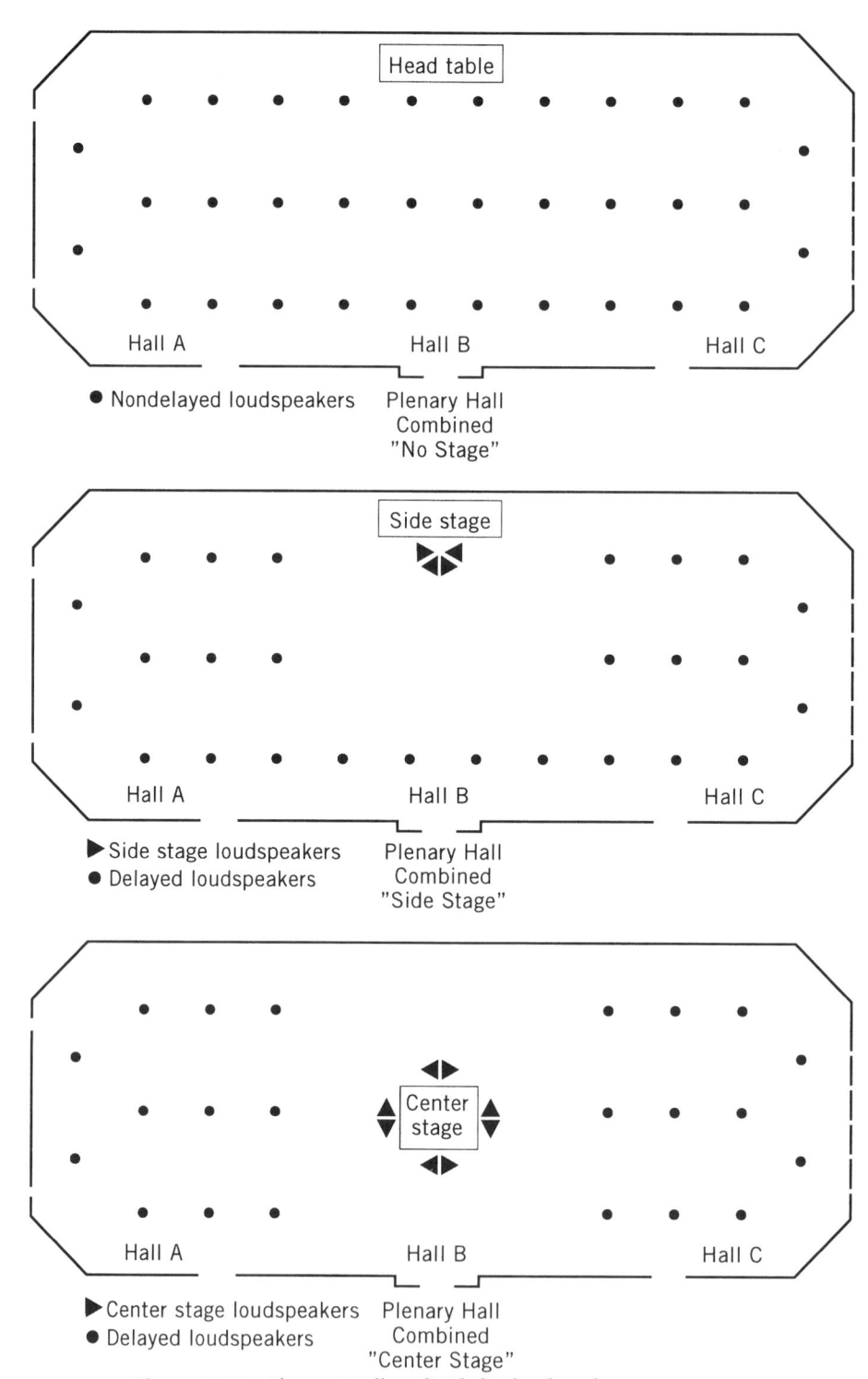

Figure 5.24 Plenary Hall undivided—loudspeaker arrangements.

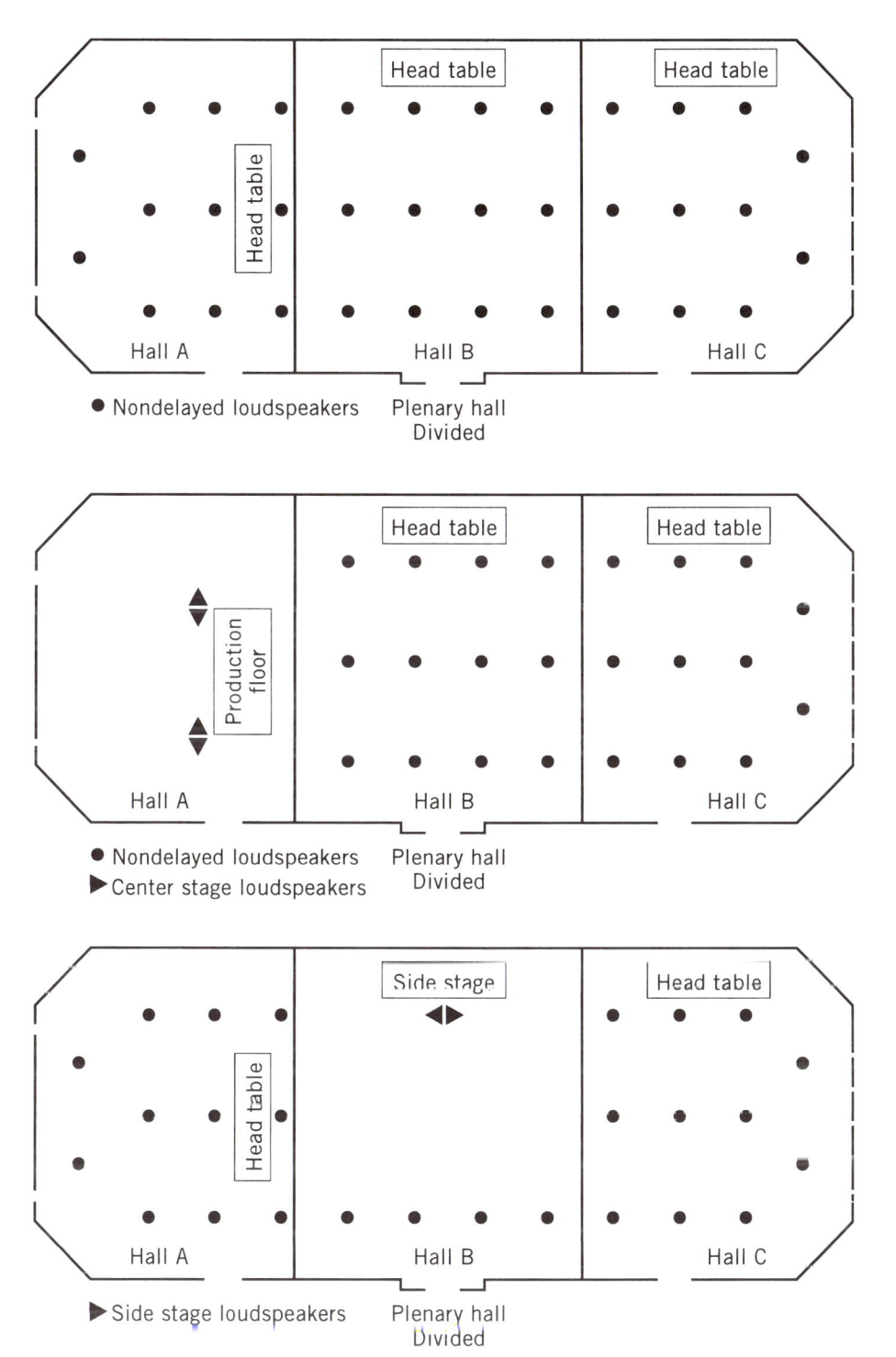

Figure 5.25 Plenary Hall divided—loudspeaker arrangements.

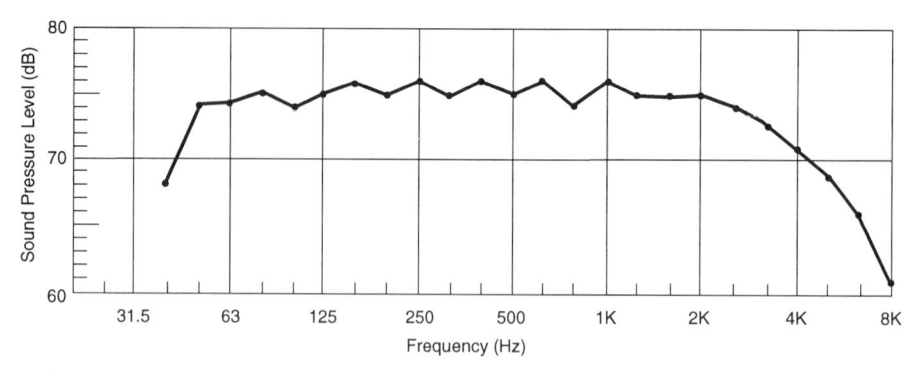

Figure 5.26 Frequency response of distributed loudspeaker system at Plenary Hall.

for the meeting rooms have seven channels, consisting of five translated language channels, the floor language channel, and one spare channel. The translation systems use wireless infrared transmission with frequency modulation. Each set of two meeting rooms has five booths for interpreters. Halls A, B, and C in the Plenary Hall each have six interpreters' booths. It is possible to divide the SIS into three independent systems or to operate it as one system for the combined Plenary Hall. The interpreters' booths are dimensioned to accommodate three interpreters each.

The infrared radiators in the Plenary Hall are installed at the bottom of the space frame. In the meeting rooms, they are installed under the central portion of the ceiling. In the Plenary Hall, the floor language signal is derived from the sound reinforcement system. In the meeting rooms, the signal is derived from the sound system or from the conference reinforcement system. The automatic mixers used with microphones for speech pickup have "live channel" indicators. The status signals from these indicators are used to activate light-emitting diode (LED) indicators used for speaker identification in each translation booth. The transmission via infrared radiation does not carry through walls and allows the SIS systems to be operated without interference using the same channel frequencies in adjacent rooms. The delegates' receivers are miniature, battery-operated units, with channel selectors, volume controls, and stethoscope headsets.

Conference reinforcement systems are provided for use in the meeting rooms when the quantity of microphones required is larger than the number of inputs available in the automatic mixers. The microphones used with the automatic mixers and the microphones in the conference system can be used together. The conference systems are portable and self-contained. They include units for the chairperson and delegates that are interconnected using single-cable technology and a common power supply. The delegates' units have a microphone, a loudspeaker, and key-type on-off switch. The station for the chairperson has a priority override switch. The conference systems have a capacity of up to 100 delegate stations.

Program Monitoring Systems

The program monitoring systems allow events taking place in the Plenary Hall to be heard through loudspeakers provided in the control rooms, projection rooms, stage manager's room,

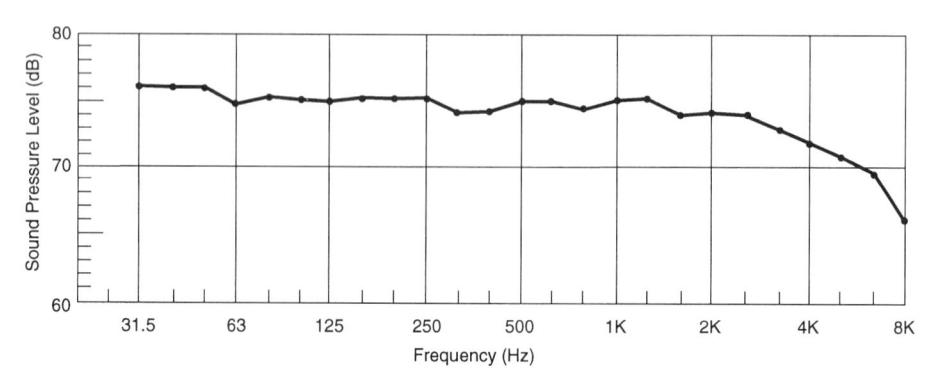

Figure 5.27 Frequency response of side-stage loudspeaker system at Plenary Hall.

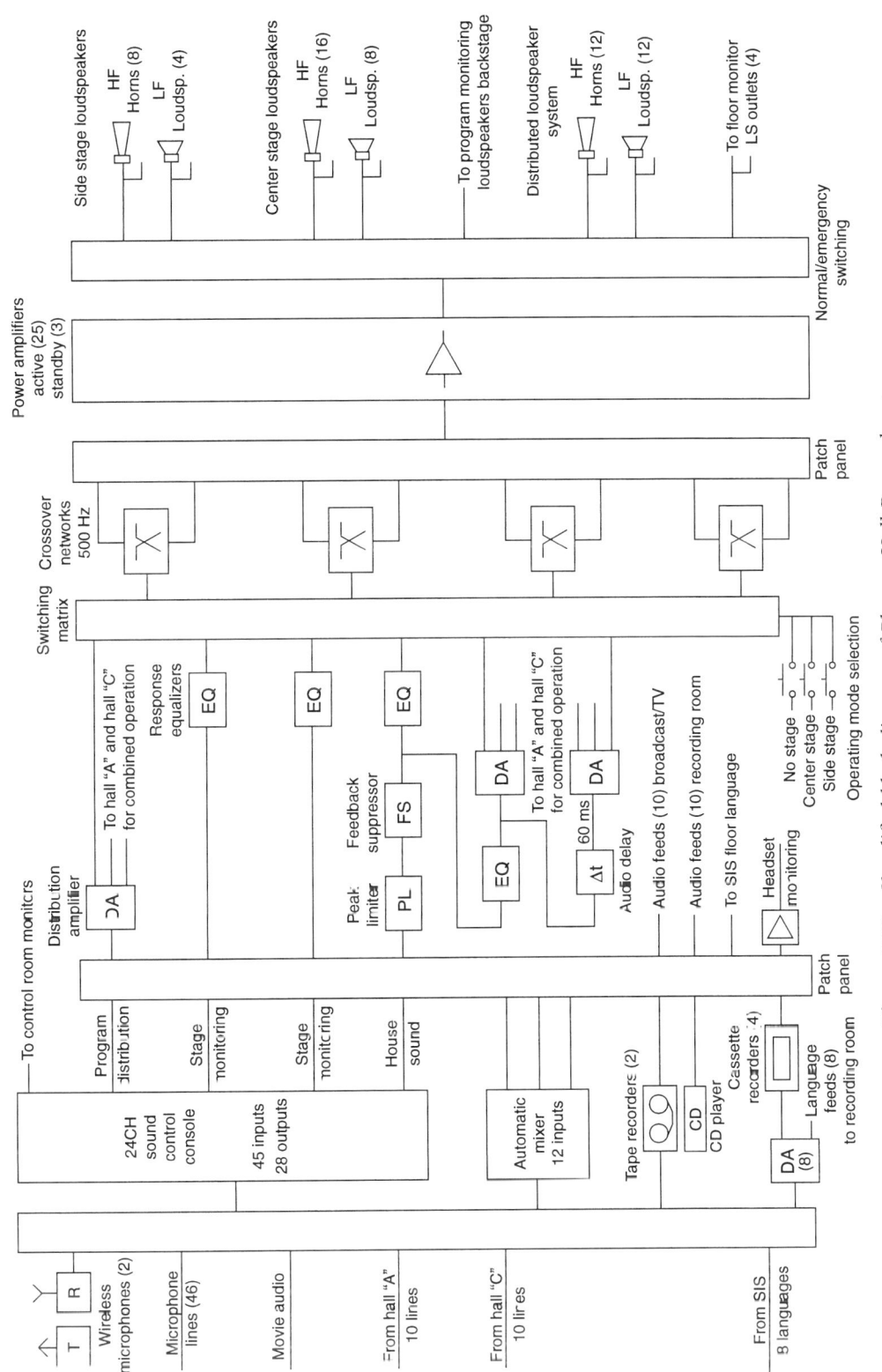

Figure 5.28 *Simplified block diagram of Plenary Hall B sound system.*

221

interpreters' booths, and in the associated corridors. A part of the monitoring system also allows paging calls to be transmitted from the stage manager to the sound control and lighting control rooms. The loudspeakers in the individual rooms served by the program monitoring system are provided with volume controls that have priority override for announcements. The system allows talk-back between the stages in Plenary Halls A, B, and C, and their respective lighting control positions.

Production Communications Systems

The production communication systems in the Plenary Hall support voice communications between the control rooms, the stage manager, and the staff positions on the floor and at the lighting bridges. It is possible to operate the system independently for divided Plenary Halls A, B, and C or for the combined hall. In the meeting rooms, the production communication systems support voice communication between the control rooms and the input panel locations in the meeting rooms.

The Plenary Hall production communication system consists of two components, one that is wired and one that is wireless. The wired system has two channels, with one channel assigned for communications by the lighting staff and the second channel assigned for sound system operations. The wireless system allows hands-free, two-way communication with one base station and six wireless beltpack-type remote stations. It is possible to operate the wireless and wired components of the system combined. The remote stations in the control rooms use loudspeakers or headsets. In the Plenary Hall, the production communication outlets are designed to work with beltpack-type remote stations and headsets.

Tape Recording and Tie Lines for Signal Distribution, Broadcast, and TV

Each control room is equipped with two reel-to-reel tape decks and four cassette decks for local archival recording of events and playback of prerecorded tapes. In addition, the complex has a radio/recording room for central recording and for the backup of the recording facilities available in the individual control rooms. The radio/recording room has eight cassette decks and two reel-to-reel tape machines. All audio signal distribution is at line level, using distribution amplifiers and 600-Ω lines.

The signals from all systems, as well as the signals from the microphone splitter, are available at an input connector panel, located at the outside docking point for broadcast/TV vans. The locations of the TV camera positions were coordinated with the Thai National Television. The transmission of video signals is via coaxial 75-Ω lines, using cables and connectors selected to meet the required standards.

Paging and Announcement System

The paging and announcement system provides for distribution of live or prerecorded announcements originating at the announcing position located in the communication center, with coverage to all public areas zoned as follows: entrance hall, foyer (Figure 5.29), cafeteria, restaurant and all-call (emergency). The system allows playback of background music.

The restaurant and the connected private dining rooms have a separate sound system, to support local announcements and playback of background music. Emergency announcements override the local program.

Figure 5.29 Foyer at Plenary Hall and meeting rooms. (Courtesy of Design 103, Architects).

Depending on the geometry and the acoustical conditions of the spaces to be covered, the paging and announcement system uses different types of loudspeakers. In principle, this is a distributed loudspeaker system. In spaces with large ceiling heights, such as the entrance hall and parts of the foyers at the Plenary Hall, the system uses bi-amplified, two-way loudspeakers, with large-format, directional, high-frequency horn loudspeakers and vented, low-frequency enclosures, installed within the space frame and aiming down. In the areas of the cafeteria and of the foyers covered with a sloping glass roof, the system uses short, line-source loudspeakers. In the areas with intermediate-height ceilings, which are acoustically treated, the system uses distributed, coaxial loudspeakers, 20 cm in diameter. Areas with low soffits are covered with distributed 10 cm, cone-type loudspeakers. The signal to the various loudspeaker systems is adjusted to obtain uniform sound levels throughout the covered areas.

Case Study

HONG KONG CULTURAL CENTER SOUND SYSTEMS, TSIM SHA TSUI, KOWLOON, HONG KONG

During the 1980s, the Hong Kong Government and the Hong Kong Urban Council implemented an extensive program of construction of regional cultural centers in Hong Kong and in the towns of the New Territories. The Hong Kong Cultural Center is the largest of these projects (Figure 5.30). It is located at the tip of the Kowloon peninsula, in the district of Tsim Sha Tsui. It is built on reclaimed land and on the grounds of the former Kowloon-Canton Railway Station. The Hong Kong Cultural Center is a multipurpose complex, which consists of the Museum of Art, the Museum of Space Science, and auditoria for the performing arts, which include the 2100-seat Concert Hall, a 1700-seat Grand Theater for opera and a 300- to 500-seat Studio Theater. The Cultural Center is the home of the Hong Kong Philharmonic Orchestra and a performing place for the Hong Kong Repertory Theater, the Hong Kong Chinese Orchestra, and the Hong Kong Ballet Company. The Cultural Center is frequently visited by performing companies from overseas. The following is a review of the features of the sound systems serving the performing spaces in the Cultural Center.

Concert Hall

The Concert Hall was designed to have the best possible conditions for listening to music. This requires long reverberation and envelopment with sound. Such conditions are not favorable for speech, which requires an acoustically dry environment. The purpose of the sound system in the Concert Hall is to assist in making announcements to the audience, which should sound clear and be intelligible. The sound system is used to reinforce narration during concerts and, increasingly, to provide reinforcement for vocalists, enhancement of the sound or musical instruments and reproduction of sound effects, or reproduction of the sound from electronic instruments. All this calls for a provision in the Concert Hall of the highest quality sound system, capable of a wide range or undistorted and uncolored reproduction of sound.

In a space with a large volume, such as the Concert Hall, these requirements are best met by a sound system that uses a single cluster of directional loudspeakers located so that the directional realism of sound is preserved. This requires the placement of the loudspeaker system in such a way that

Figure 5.30 Hong Kong Cultural Center. (Courtesy of Provisional Urban Council of Hong Kong.)

Figure 5.31 *Hong Kong Cultural Center, Concert Hall. (Courtesy of Provisional Urban Council of Hong Kong.)*

the reinforced sound is received by the listeners with a slight delay following the sound from the natural source. In a concert hall, the optimum location for such a loudspeaker system is the center of the space in front of the orchestral platform, where the loudspeakers are suspended from the ceiling. In practice, few large halls can be adequately covered from only one central loudspeaker cluster. In halls that have deep balconies, the audience under the balconies is shielded from sound arriving from above, and additional coverage must be supplied from soffit-mounted overhead loudspeakers. In halls that are long, where the ratio of the distance from the central loudspeaker system to the most distant seats and the distance from the system to the nearest seats is large, it is necessary to add supplemental loudspeakers to increase the sound delivered to the distant seats. Also, covering the front rows of seats, particularly the extreme left and right front seats, is difficult from a single central loudspeaker system. The addition of loudspeakers installed in the stage apron helps improve coverage in the front rows. All of these supplemental loudspeakers must be fed with delayed signals to make their sound "disappear" with respect to natural sound and with respect to the sound from the central loudspeaker cluster. The use of such supplemental loudspeakers became easy with the invention by Dr. Francis Lee of MIT of digital audio delays, which use no moving parts and require no maintenance. The delay mechanisms used before had loops of magnetic tape or rotating magnetic drums to delay the signal. All of these machines were cumbersome and unreliable.

There is a relatively narrow range of usable suspension heights for central loudspeaker clusters. If a loudspeaker cluster is placed less than approximately 9 m above the stage floor, the sound on the stage tends to be too loud and the danger of feedback is great. If a loudspeaker cluster is located more than approximately 10.5 m above the stage, the delay between the natural sound and the sound from the loudspeakers becomes excessive. This makes the sound unpleasant, both for the performers and for the audience in the front of the hall.

The operating position of the central loudspeaker cluster in the Hong Kong Concert Hall is 9.5 m above the level of the orchestral platform. The height of the cluster above the platform is adjustable by means of a telemetered winch. The loudspeaker cluster can be lowered to the platform for servicing. When not in use, the cluster can be hoisted up into a storage position under the ceiling. The Concert Hall has an adjustable orchestral canopy above the platform, which is provided to improve on-stage communication between the musicians. The suspension height of the canopy varies depending on the type of music being performed. When the organ is played, the canopy is hoisted into its uppermost position. Since most of the time the optimum suspension heights of the orchestral canopy and of the loudspeaker cluster are different, the loudspeaker cluster cannot be integrated with the canopy and it has to be winched independently.

Considerable effort was made to keep the loudspeaker cluster for the Concert Hall small and to make it relate visually to the canopy. The interior of the Concert Hall, looking toward the organ, is shown in Figure 5.31 and the view of the central loudspeaker cluster is shown in Figure 5.32. The central loudspeaker cluster is a bi-amplified, two-way system, using small-format, high-frequency horns and vented low-frequency enclosures. The crossover frequency is 500 Hz. The frequency response of the central loudspeaker system, following equalization in $\frac{1}{3}$-octave bands, is plotted in Figure 5.33. The maximum sound levels, measured during reproduction of wide-band music signals recorded on compact discs, are over 103 dB.

Figure 5.32 Central loudspeaker cluster at the Hong Kong Cultural Center Concert Hall. (Photo by author.)

The Concert Hall uses supplemental distributed loudspeakers installed in the stage front and in the underbalcony soffits. The Concert Hall is a surround hall, with seating for choir or audience extending behind the orchestral platform. Based on the designs of sound systems in churches, which have similar seating arrangements, the coverage of the choir areas is provided by means of loudspeakers installed in the backs of the seats and operated with delayed sound. This solution gives much better results than attempts to cover seating behind the orchestral platform from the central loudspeaker cluster. The chair-back loudspeakers are small, high-quality coaxial loudspeakers, used in automobile installations.

In addition to permanently installed loudspeakers, the Concert Hall sound system uses movable roll-on loudspeakers, intended for use during performances of popular music. These loudspeakers, located on the left- and right-hand sides of the orchestral platform, help maintain balance between the sound radiated from the stage and the sound delivered from the central loudspeaker system. The system includes portable stage monitor loudspeakers for reinforcement of sound back to the performers.

Microphone receptacles are provided in the floor of the orchestral platform and to the left and right of the performing area. Receptacles above the ceiling are provided for suspended microphones for recording.

The permanent sound control position is located in a sound control room in the rear of the Concert Hall, above the balcony. The control room has a large operable control window. Due to the elevated location of the sound control

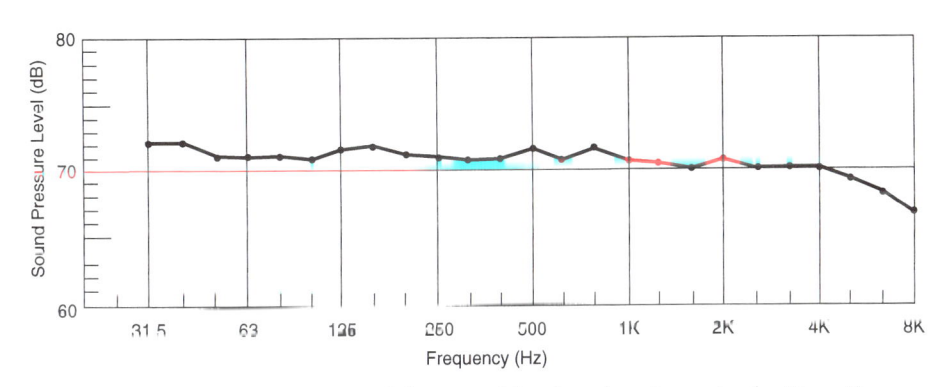

Figure 5.33 Frequency response of the central loudspeaker cluster in the Hong Kong Cultural Center Concert Hall.

room, the operator's position is not in the direct coverage of the main loudspeaker system. Instead, the control room is equipped with a delayed, dual-channel, stereophonic system for acoustic monitoring. The acoustic monitoring system uses microphones installed at the balcony in the Concert Hall and a pair of loudspeakers installed in the front of the sound control position. An auxiliary sound control position is provided in the main floor of the Concert Hall, for use with portable sound control consoles. The simplified diagram of the Concert Hall sound system is shown in Figure 5.34.

Grand Theater

The Grand Theater is a multipurpose hall, designed for staging a variety of performances and events including Western-style opera, Chinese opera, ballet, small orchestras, chamber music and solo performances, popular music shows, variety shows, speech plays, lectures, and conventions (Figure 5.35).

The main loudspeaker system is located in front of the proscenium, approximately 10 m above the level of the stage floor. The proscenium loudspeaker system uses large-format, high-frequency horns and horn-loaded, low-frequency enclosures. This system is designed as a three-channel system, with separate groups of high-frequency loudspeakers for each channel. This allows for multichannel sound reinforcement and for playback in stereophonic mode. Since low-frequency information does not contribute much to the localization of sound, or to the stereophonic effect, the low-frequency signals from the three channels are combined into one low-frequency channel. If multi-channel reinforcement or playback are not required, the main loudspeaker systems are combined for monophonic, single-channel operation.

Since there is a large difference in distances from the proscenium loudspeaker system to the nearest seats in the orchestra and to the most distant seats in the balconies, the proscenium loudspeaker system alone cannot provide uniform sound levels for all seating areas. The sound coverage for the balcony is provided by supplemental, two-way loudspeakers, mounted at the rear of the lighting bridge. The rearmost seats under the balcony overhangs, in the balconies, and on the main floor use supplemental coverage by means of overhead loudspeakers recessed in the underbalcony soffits. The purpose of these supplemental systems is to create conditions of exposure to the sound field that are similar to those that exist for the audience seated in the main volume of the hall.

In addition to the permanently installed loudspeakers, the Grand Theater sound system includes movable loudspeaker systems for use during highly amplified performances of popular music. These roll-on systems, intended to be placed to the left and to the right of the proscenium opening, are shared in their application with the Concert Hall. The Grand Theater is equipped for wide-screen motion picture projection. Three full-range movie sound loudspeakers are provided for placement behind the projection screen. The motion picture sound system is separate from the house sound system.

The Grand Theater sound system uses a variety of portable stage monitor loudspeakers for reinforcement of sound back to the performers. Portable loudspeakers for reproduction of sound effects are also provided. The system allows reinforcement of orchestra pit sound for dancers on the stage, during ballet performances.

The microphone complement includes cardioid microphones for general, on-stand use, hand-held microphones for close microphone pickup of pop singers, microphones for instrumental pickup, omnidirectional microphones for recording, and miniature condenser microphones for recording, for use at footlight locations and for theatrical reinforcement. Ultradirectional "shotgun" microphones for intermediate and long-range sound pickup are also provided. The system includes wireless microphones, operating in the 150- to 216-MHz frequency bands, with crystal-controlled transmitters and receivers with diversity reception. The simplified block diagram of the Grand Theater sound system is shown in Figure 5.36.

The sound control room is located in the rear of the auditorium and has an operable control window. Because of the presence of the balcony, there are no direct lines of sight from the operator's position to the proscenium loudspeakers. The listening conditions for the operator are improved by means of a dual-channel, stereophonic, acoustic monitoring system. The monitoring system consists of two microphones installed in front of the balcony rail, dual-channel amplifiers and delays, and two monitor loudspeakers, mounted above the sound control window, facing the operator. The gain of the monitoring system is adjusted to provide the same sound pressure levels at the operator's position as measured in the audience seats in the auditorium. The theater has an auxiliary sound control position in a forward location at the balcony. This position is equipped for use with a portable mixing console provided in the system, or with consoles used by touring groups.

The sound control console, installed permanently in the control room, is a broadcast-quality, modular mixing desk, with 24 inputs and 4 matrix outputs, which allow for independent control of the sound reinforcement functions, the foldback and monitoring functions, and the reproduction of sound effects and for the provision of signal mix for recording or distribution to backstage areas, broadcast, TV feed, and so forth.

The system is equalized in $\frac{1}{3}$-octave bands for the best possible overall frequency response, lack of sound coloration, and maximum gain before feedback. During reinforcement of plays, with distant microphone pickup, the stability of the system can be further increased through frequency shifting. The sound output from the main loudspeaker systems and supplemental systems is balanced to within ±1 dB for uniformity of coverage. The level adjustments are made by means of calibrated pads, not by using power amplifier volume controls.

The Grand Theater auditorium is equipped for hosting international conferences. It has five permanent booths for

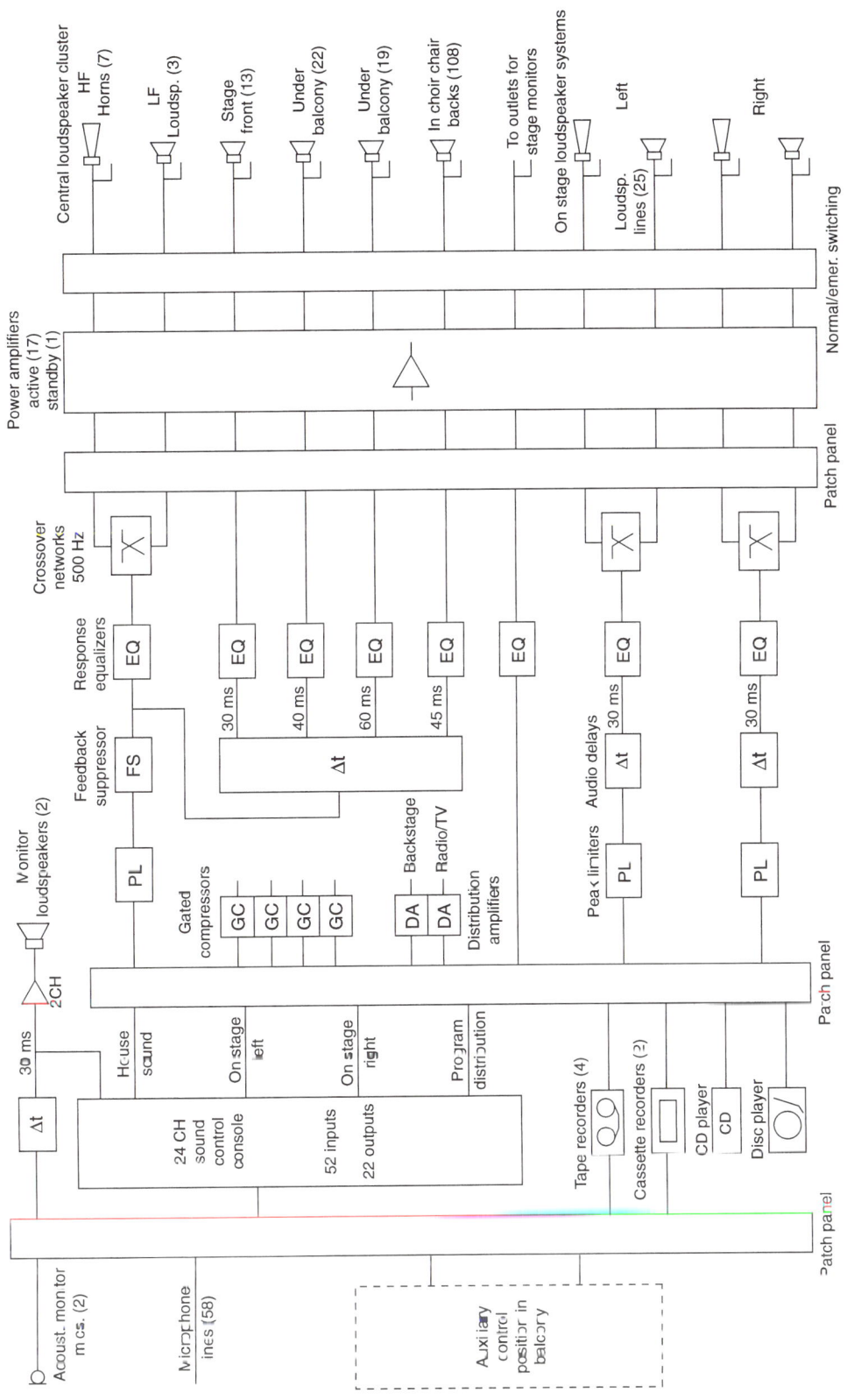

Figure 5.34 *Simplified block diagram of the Hong Kong Cultural Center Concert Hall sound system.*

Figure 5.35 *Hong Kong Cultural Center Grand Theater. (Courtesy of Provisional Urban Council of Hong Kong.)*

interpreters and a wireless infrared SIS system. Outlets for audience participation microphones are floor-mounted in the orchestra and in the balcony.

Studio Theater

The Studio Theater is equipped to stage speech plays, small orchestral and ballet performances, and lectures or other speech presentations. The seating is flexible and it can be arranged to form surround, partial surround, or proscenium seating. The stage grid extends over the entire ceiling area and numerous lighting bridges span the hall. The Studio Theater has a gallery surrounding the seating/acting area. The view of the Studio Theater is shown in Figure 5.37.

The Studio Theater sound system is intended primarily for reinforcement of speech and for playback of prerecorded sound including sound effects. It permits movement of the apparent locations of the sound source.

The loudspeaker systems consist of distributed loudspeakers, mounted overhead, at the bottom of the lighting bridges, and of surround loudspeakers on the walls at the gallery level. There are numerous outlets for portable stage monitoring and sound effects loudspeakers, recessed in the floor near the acting areas and mounted on the walls.

The reinforcement of live events is done primarily by means of overhead loudspeakers. Due to the steeply raked seating, the distance from the audience to some of the overhead loudspeakers may be shorter than the distance to the performers. In order to preserve the precedence of natural sound over the amplified sound, a slight delay is introduced for the signal to the loudspeakers. Longer delays can be used to create special sound effects, such as an echo or the sensation of a larger space.

The permanent sound control position is at the gallery level. There is an auxiliary control position at the gallery railing, and the sound can be also controlled from several locations at the acting floor level. The sound effects can be reproduced in four channels. The system has both dual- and four-channel tape machines. A simplified block diagram of the system is shown in Figure 5.38.

Program Monitoring and Backstage Paging Systems

The Concert Hall and both theaters have extensive systems for performance monitoring and backstage paging. Performance monitoring is available during amplified events, with the signal derived from the mixing consoles or, during unamplified events, with the signal picked up by microphones suspended in the upper volumes of the halls. The program monitoring and backstage paging extend to all the dressing rooms, backstage circulation areas, rehearsal and practice rooms, elevators, and all the points covered by the production communication systems.

The backstage performance monitoring and paging systems use recessed, ceiling-mounted loudspeakers in all spaces that have suspended ceilings, and surface-mounted loudspeakers elsewhere. The sound volume in the dressing rooms can be adjusted using wall-mounted controls. The volume controls are provided with priority relays to override the attenuation during paging calls.

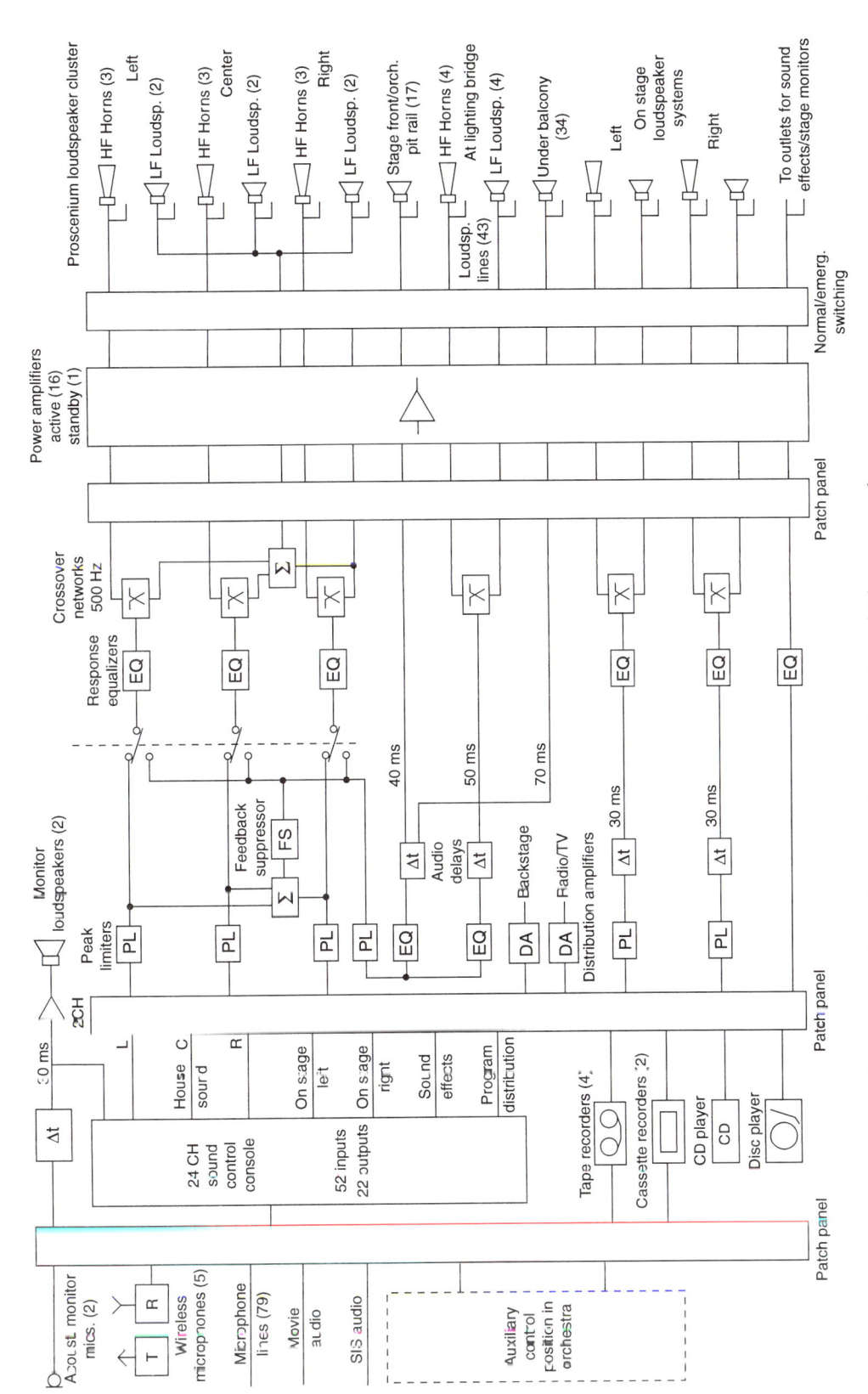

Figure 5.36 Simplified block diagram of the Grand Theater sound system.

Figure 5.37 Hong Kong Cultural Center Studio Theater. (Courtesy of Provisional Urban Council of Hong Kong.)

Production Communication Systems

The production communication systems provide voice communication between the stage manger, the production desk, the staff positions, such as sound and lighting control, and the other technical stations within the stage, stage house, galleries, catwalks, and the follow-spot positions. The production communication systems use remote stations with loudspeakers or beltpack stations with headsets where applicable. The Grand Theater is equipped with a four-channel intercom system, with channel assignment for different functions. The Concert Hall and the Studio Theater have two-channel systems, with one channel assigned to lighting and one to sound operations. The production communication systems have redundant power supplies for fail safe operation.

Cue Light Systems

The cue light systems provide a visual backup for the production communication systems and are controlled from cue light control panels, provided in the stage managers' console in each theater. Cue lights are provided at all key staff positions. Depending on the location, the cue lights use permanent fixtures or portable fixtures on extension cords. The red light indicates "stand by" and the green light is a "go" signal.

Stage Managers' Consoles

The Concert Hall and two theaters each have a stage manager's console, with equipment for backstage paging, stage announcements, foyer paging, lobby paging, production communications, cue light control, control of cassette decks with prerecorded announcements, and for audience recall. The stage manager's consoles move on casters and connect with the respective systems via flexible cables and multicontact connectors.

Sound Distribution and Paging Systems in Foyers, Lobbies, and Other Public Circulation Areas

These systems use ceiling-mounted recessed loudspeakers, loudspeakers recessed in the balcony rail, large two-way loudspeakers recessed in the walls, or short-line source loudspeakers, depending on the characteristics of the space where they are installed. Announcements originate at a paging position in the central foyer or at the stage managers' consoles in the respective theaters. The system allows announcements to be zoned for selected areas.

Management Intercom System

Because the Cultural Center complex is large, it was necessary to provide a voice communication system for management functions, independent from the theater operations. The management intercom system provides means for direct voice communication between stations, or groups of stations, installed in selected locations in the Concert Hall, the Grand Theater, Studio Theater, and the foyers. It uses components of a standard microprocessor-controlled intercom system, modified to meet the requirements of this specific application. The stations are recessed in the walls, concealed behind lock-

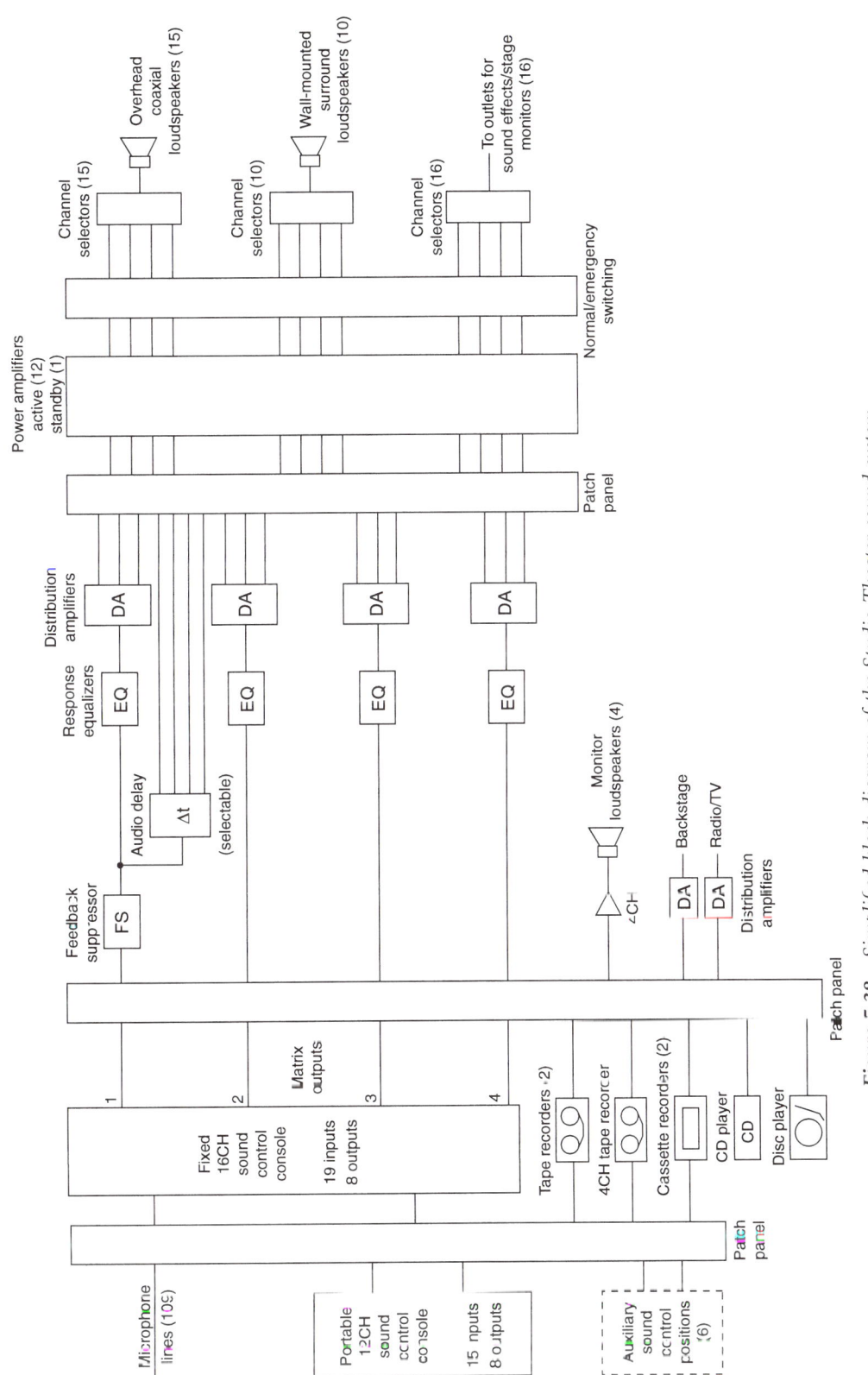

Figure 5.38 Simplified block diagram of the Studio Theater sound system.

able doors. The stations are provided with handsets to allow for private conversations. Hands-free communication is also possible.

Each sound system described in this chapter has its own set of unique requirements. Each system is different, ranging from small and simple to large and very complex. They all share one thing in common—they meet the needs and requirements of the job they were designed to do. Work on each of these projects started by listening to the clients and discussing their needs. It ended with the knowledge that those needs had been satisfied.

6

Recent Innovations in Acoustical Design and Research

Gary W. Siebein
and Bertram Y. Kinzey, Jr.

6.1 INTRODUCTION

The purpose of much of the current research in the field of architectural acoustics is to develop methods to evaluate, model, predict, and aurally simulate acoustical qualities of buildings. This research will allow qualities of sound to be consciously designed as important elements contributing to the multisensory experience of architecture. In a few years an integrated system will likely be developed to evaluate, design, predict, and simulate the subtleties of the sonic environment of rooms. This system will be designed to reflect the perceptions of the sonic qualities people who use the rooms find important. It should be able to be used as an integral part of the design process of all buildings.

The research activities are interdisciplinary by necessity. Architects, researchers, and consultants from architecture, neuroscience, psychology, music, theater, engineering, speech, and other areas have all been actively engaged in this work. An outline of the major areas of research is presented in three major categories described below.

Understanding and Measuring Room Acoustic Qualities

- The perceived aural qualities of existing buildings are being studied to define the components of acoustical quality. This occurs both by qualitative evaluations by researchers, consultants, and musicians listening to many performances in many rooms and by laboratory studies of sound quality by human listeners using techniques borrowed from psychology and neuroscience.
- Acoustical measurement systems for studies in full-size rooms and in scale models of rooms are being developed and tested. These systems are based on impulse response testing and digital signal processing from electrical engineering.

- The physical acoustical qualities of existing buildings are being studied to determine how sound interacts with the complex architectural features encountered in most buildings.
- Studies that describe how the architectural features of rooms effect acoustical measures and acoustical qualities in rooms are resulting in both qualitative and quantitative models to predict the acoustics of rooms. This involves using advanced quantitative tests to determine primary perceptual factors, preference spaces, and correlated architectural features of rooms using techniques from statistics.

Acoustical Modeling and Aural Simulation

- Physical and computer modeling techniques are being developed to duplicate the complex sound fields found in actual buildings using techniques from computer science and physics. The modeling techniques are being refined to predict the acoustical qualities of buildings that have not yet been built.
- Technology to aurally simulate the acoustical qualities of buildings as part of the design process is under development relying on and in some cases stimulating advancements in virtual acoustics, audio recording, playback, and control from digital signal processing and virtual reality.
- Expert design systems are under development to assist in evaluating acoustical design decisions.

Other Directions in Architectural Acoustics Research

- New materials are being developed to diffuse sound in predictable ways.
- Systems for active noise control are being developed and used. Active noise control systems measure a disturbing sound and propagate a new sound wave with opposite phase to effectively cancel the original sound.
- Extensive tests with modern instrumentation are being conducted to measure the transmission loss of individual building materials as well as construction details. This information will assist architects and contractors in selecting appropriate materials and building systems to provide sound isolation between adjacent activities.
- Noise from plumbing systems has also been studied in detail. Effective methods to reduce noise from plumbing systems have been identified.
- Extensive testing has also been underway for impact noise on flooring systems. Several new testing methods have been developed in this area.
- The health effects of duct linings are being investigated to ensure quiet air-conditioning systems also result in high-quality indoor air.
- The energy effects of passive acoustical silencers are being investigated to optimize the combined acoustical and energy performance of buildings.
- The acoustical qualities of air-conditioning noise are being investigated to refine design criteria for acceptable sounds in buildings from equipment. (Please refer to coverage of this topic in Chapter 3.)
- Advances in sound system components and design have greatly improved the quality of sound reinforcement systems in buildings. (Please refer to coverage of this topic in Chapter 5.)

This active and multifaceted research agenda is fostered by a vital acoustical consulting community. It is merging the forefront of issues from practice with the leading edge of research from laboratories and universities around the world in a new hybrid research and design method. Interestingly enough, this work has been integrated and applied in a series of actual buildings over the years where results from the laboratory have been tested in actual building projects. The buildings have stimulated further research. This interactive cycle of research stimulating practice, which raises questions to drive research, has contributed to a dramatic growth of knowledge in recent years in "architectural acoustics or the science of sound as it pertains to buildings" (Sabine, 1964, p. 3). This new knowledge represents the most dramatic period in the history of the field since Sabine began his work almost 100 years ago.

For example, there has been a revolution in the design of concert halls in recent years that has had several major components (Siebein and Gold, 1997). First has been a rigorous examination of the renowned "shoebox" concert halls of the late nineteenth and early twentieth centuries such as the Grosser Musikvereinsaal in Vienna, the Concertgebouw in Amsterdam, and Symphony Hall in Boston (Johnson, 1990). A more complete understanding of the acoustical qualities of these halls has been undertaken using a variety of techniques developed in allied disciplines. These studies have identified the interactions among the architectural design features and acoustical qualities of the rooms (Siebein, et al., 1992; Chiang, 1994; Beranek, 1996). The techniques used in the studies are included in the first and second topics in the research issues described above.

The combining of these techniques in a hybrid design and research method has resulted in the development of new theories of concert hall design that have been centered on the development of impulse response test techniques (Siebein and Gold, 1997) (Please refer to the section on The Impulse Response for a discussion of this topic). These techniques assimilate principles of shape, proportion, material, configuration, and program from the historic precedents of the late nineteenth century. The acoustical and architectural qualities of the halls of the future have been extended beyond those of the precedent halls through an extensive research effort that has been integrated and explored in practice in an exemplary manner. The remarkable acoustical qualities of these halls have been obtained in a rich and varied contemporary architectural vocabulary. The acoustical attributes of the precedent halls have been identified through impulse response tests conducted in the rooms. Aural simulations have been used in laboratory studies to evaluate the subtle qualities of sounds heard in the precedent rooms. These techniques allow specific acoustic or architectural variables to be isolated and examined in detail. This hybrid method that combines architectural analysis of historic halls, qualitative evaluations of performances in existing rooms and simulated sound fields in the laboratory, impulse response measurements in buildings and models, and a series of actual buildings constructed over a 30-year period has produced the first series of successful halls since the end of the last century (Johnson, 1990; Siebein et al., 1992). These rooms comprise a research effort characterized by the deliberate experimentation of consultants and architects to test emerging theories in a series of actual buildings and to evaluate the consequences of this work in holistic architectural settings. These and other major results of recent acoustics research work will be presented in sections describing each of the topics listed above.

6.2 UNDERSTANDING AND MEASURING ROOM ACOUSTIC QUALITIES

Acoustical Qualities of Rooms

> *Because familiarity with the phenomena of sound has so far outstripped the adequate study of the problems involved, many of them have been popularly shrouded in a wholly unnecessary mystery. Of none, perhaps, is this more true than architectural acoustics. The conditions surrounding the transmission of speech in an enclosed auditorium are complicated, it is true, but are only such as will yield an exact solution in the light of adequate data* (Sabine, 1964, p. 219).

Contrary to popular opinion, good architectural acoustics is no accident. . . . The acoustical character of a proposed building can be rather accurately predicted . . . (Kinzey and Howard, 1963, p. 317).

Sound is one of the subtlest pieces of nature (Sir Francis Bacon quoted in Hunt, 1992, p. 1).

The ultimate evaluation of acoustical quality lies with the people who listen, speak, play, work, create, live, sleep, and otherwise use the rooms designed and built by those in the construction industry. This evaluation will differ from person to person, from activity to activity, from culture to culture, from one period in history to another, and even from one day to the next with the same person!

One of the most important developments in architectural acoustics in recent years has been the increasing number of studies investigating how people hear and what they like to hear when they listen to sounds, particularly speech and musical performances. Collectively, these studies begin to answer the question "What do we mean when we say that a room has good acoustics?" (Cremer and Muller, 1982, p. 451). These are often called studies of subjective qualities of sound. People are used as test subjects to evaluate live or recorded sounds. The general findings of this body of work has been to identify the qualities most people associate with "good" acoustics, to confirm that people can hear differences in the acoustical qualities at different seats within a room, to confirm that there are perceived differences of acoustical quality in different rooms, to isolate and identify some of the factors that contribute to acoustical quality, and to confirm that listening is a complex, multidimensional experience with many significant interactions among variables and many difficulties in describing phenomena precisely. Many doubt that it is possible to complete this extremely difficult task to "quantify sensations" (Cremer and Muller, 1982, p. 456). There are two basic formats for these studies: (1) analysis of live sounds heard in actual rooms and (2) analysis of sounds heard through loudspeakers or headphones in a laboratory. The results of the recent research are summarized below. A discussion of the research methods follows.

Summary of Recent Studies

There have been many studies reported in the literature since 1950. The following is a brief summary of several of the major results that have shaped current theories in the field. It is generally thought that there are only a few important qualities that affect overall acoustical impression at this time. These include the items listed in Table 6.1. The body of work discussed below represents a long-term research effort to identify and quantify the components of acoustic quality.

1. *People seem to prefer different reverberation times for different types of musical performances and other acoustical activities.* Preferred reverberation times varied from 1.3 sec for classical chamber music to 2.1 sec or higher for romantic music (Sabine, 1927; Cremer and Muller, 1982; Ando, 1985). Kuhl was able to draw some general conclusions from his study. Fifty percent of the subjects preferred a reverberation time of 1.5 sec for both Mozart's *Jupiter Symphony* and Stravinsky's *Sacre du Printemps*. A reverberation time of 2.1 sec was preferred by 50% of the subjects for the Brahms symphony. Kuhl (in Cremer and Muller, 1982) concluded that these pieces were representative of classical, modern and romantic music. Therefore, the reverberation times could be used as general guidelines for room design where these types of music would be performed.

2. *The early portion of the reverberant decay is important in people's perception of reverberation* (Schroeder et al., 1974, Jordan, 1974). It was generally well accepted that if the reverberation time was too long, subsequent notes in a musical piece or syllables in a train of speech would be masked by the increased level and duration of reverberation. Studies indicated that in actual speech and music only the first or initial portions of the reverberation process in rooms was actually heard by people. The acoustical measure early decay time was developed from this research. Early decay time (EDT) has been more highly correlated with the perceived quality of reverberance than reverberation time (RT60) in several studies (see Figure 6.1).

3. *It is possible to provide double sloped decays in rooms so that clarity may be provided by early reflections arriving shortly after the direct sound while a long reverberant "tail" provides a sense of fullness and reverberance* (Johnson, 1990). In much of the early literature on perceived qualities of sounds and the corresponding physical attributes of rooms that are responsible for these perceptions, reverberation and clarity are often presented as opposing qualities. This is an interest-

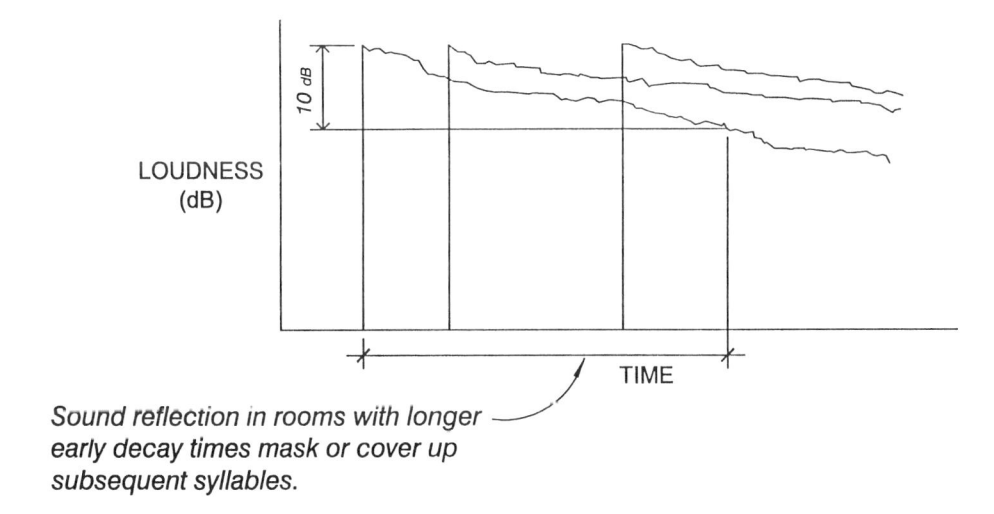

Sound reflection in rooms with longer early decay times mask or cover up subsequent syllables.

Short early decay time. Allows each syllable to be heard clearly.

Figure 6.1 Concept sketch of masking of sounds by reverberation in a room measured by early decay time.

ing point. Until the 1970s reverberation and clarity were both thought to be related to reverberation time (RT60). It was generally thought that the reverberation time (RT60) must be relatively short to achieve a sense of clarity. A longer reverberation time would provide sensations of reverberance and fullness or tone. The studies begun by Haas (1972) in the 1950s have shown that early sound reflections that closely follow the direct sound contribute to the clarity and loudness of sounds. These reflections can exist in the presence of lower levels of reverberation that persists for extended periods of time creating simultaneously a clear and reverberant sound (see Figure 6.2). The application of this concept in concert hall design is discussed in detail in the case studies section where several halls have explored ways to achieve simultaneous qualities of clarity and reverberance.

4. *The reverberation time should increase somewhat in the low frequencies to provide a sense of acoustic warmth* (Beranek, 1962). Barron comments that the preferred amount of increase in the bass sounds is a matter of preference. Many have established that the reverberation time at 125 Hz should be approximately 10 to 50% longer than the mid-frequency reverberation time (Barron, 1993, p. 44). Recent studies have indicated that the loudness or strength of the bass sounds compared to the middle-frequency sounds is also an important contrib-

THREE PARTS OF AN IMPULSE RESPONSE

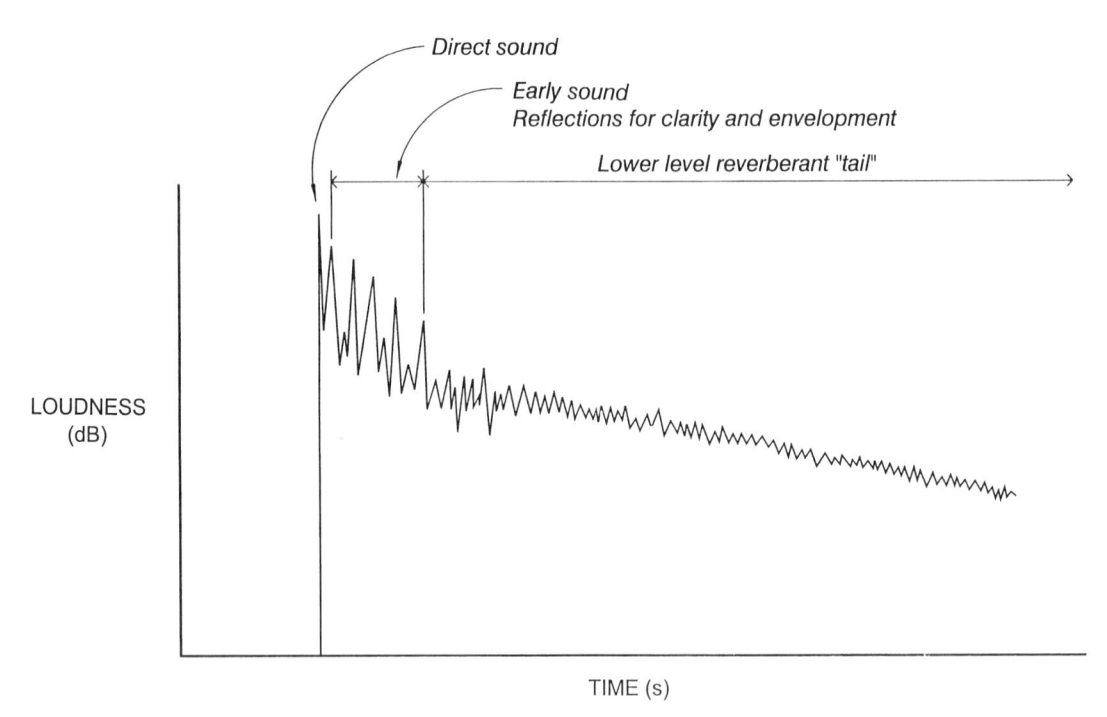

Figure 6.2 *Concept sketch of a double-sloped decay in a room. Notice the high levels of early reflected sound that increase loudness and clarity immediately after the direct sound. The longer reverberant "tail" persists for a much longer time, but at a lower level. Therefore, it provides fullness to the sounds without interfering with clarity.*

uting factor to the perception of warmth in a room, as shown in Figure 6.3 (Beranek, 1996).

5. *Several reflections from the ceiling are needed to arrive shortly after the direct sound (within the first 50 to 80 msec), especially in the rear portions of a room, to increase the overall level of the sound and to provide a sense of clarity* (Veneklasen, 1979, Thiele and Meyer, 1977; Cremer and Muller, 1982). A landmark paper published by Haas (1972) clearly showed that sounds arriving within 20 to 50 msec of the direct sound were integrated in the ear as a single sound with an increased level (see Figure 6.4). It was not until sounds were delayed by 70 msec or more that over 50% of his subjects detected them as a separate acoustical event or an echo. Thiele and Meyer (1977) and Reichardt et al. (1975) among others extended this concept with studies of how reflections affected the qualities of speech and music, respectively. The acoustical measures they originally proposed as a result of their work became early versions of the early-to-late sound index described later in this section. The importance of early sound reflections in increasing the loudness, clarity, and fullness of sounds has been one of the major contributions of recent research.

6. *Thresholds for the time delay, amplitude, and direction of reflections that are useful and provide support for the direct sound and those that are detrimental and decrease clarity, intelligibility, and spaciousness have been found.* There have been many studies in architectural acoustics and general psychoacoustics that have investigated how loud sound reflections can be at specific time delays after the direct sound to be perceived as reinforcing the direct sound, when they become detrimental and contribute to a loss of clarity or intelligibility, and when they become acoustical defects such as echoes. Generally sounds that are somewhat lower in amplitude than the direct sound (such as reflections from room surfaces) that arrive within 50 to 80 msec after the direct sound will increase the level of the sound. They will be heard as reinforcing the sound.

The perception of direction is also important. This is usually attributed to the law of the first wavefront. This states that the sound will usually be localized in the direction from which the first sound wave

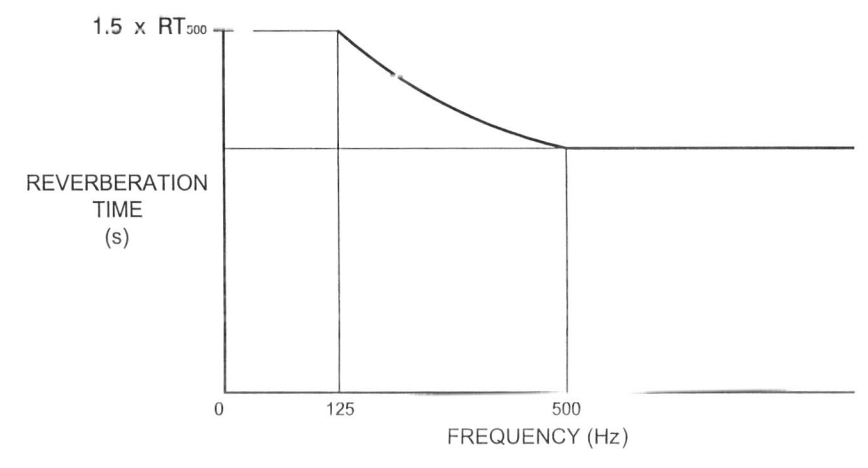

Figure 6.3 Reverberation time vs. frequency curve showing increase in reverberation time in the lower frequencies thought to contribute to a sense of acoustic warmth.

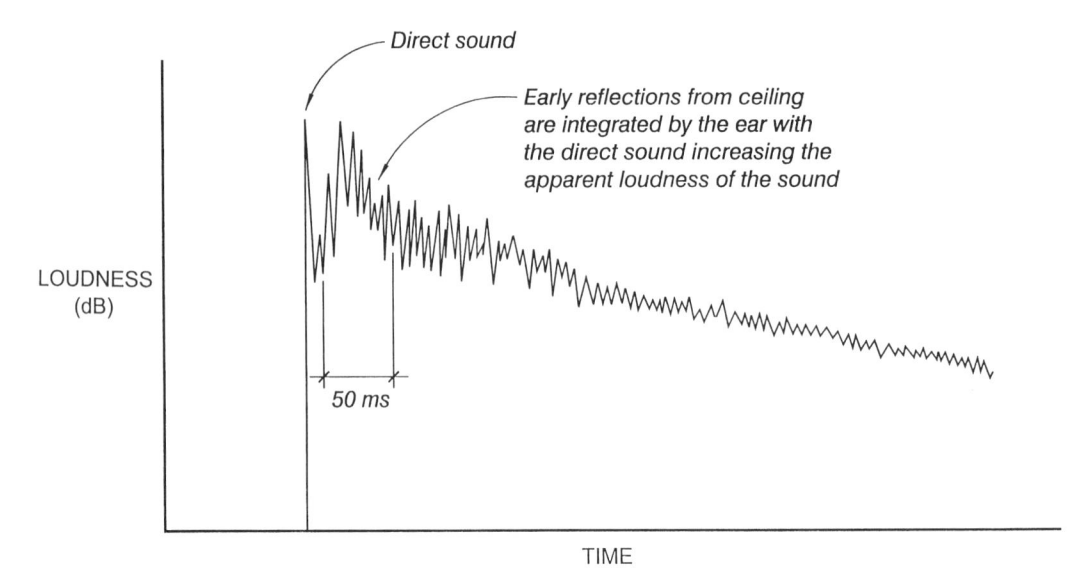

240 • *Recent Innovations in Acoustical Design and Research*

Figure 6.4 Sketch of impulse response of a room showing early sound reflections arriving shortly after the direct sound that are thought to increase the perceived loudness of the sound.

that arrives at the listeners location comes (Haas, 1972; Cremer and Muller, 1982; Barron and Marshall, 1981).

Lochner and Burger (1961) and Bradley (1986) have developed useful-to-detrimental energy ratios that compare the "useful" or early energy to the "detrimental" or later (reverberant) energy in an impulse response. The useful energy has been proposed by several researchers to occur somewhere between 35 msec after the direct sound up to 95 msec after the direct sound. The reverberant or late energy is that which comes after the useful energy up to the decay of the impulse. These measures have been used primarily to assess speech intelligibility in rooms.

The concept of a sound being divided into several components defined by perceived acoustic qualities is important. All of this work is based on the idea that an impulse response can be subdivided into several components: the direct sound, the early sound reflections, the later or reverberant sounds, and the ambient or background noise (see Figures 6.5 and 6.6).

A. The direct sound is the sound wave that travels directly from the source to the listener without striking any of the surfaces of the room. It is the first sound wave that arrives at a listener's location. It contributes to sensations of loudness, clarity, and localization. It will generally decrease due to geometric spreading or divergence as it moves farther from the sound source.

B. The early sound reflections are sound waves that strike one of the room surfaces and are reflected to the listener's location. Reflections that arrive within short time intervals after the direct sound (less than 80 msec for music) are usually combined with the direct sound by the ear. These reflections add to the direct sound increasing its apparent loudness. If the reflections arrive within 40 msec or less after the direct sound, they will also contribute to a sense of acoustic

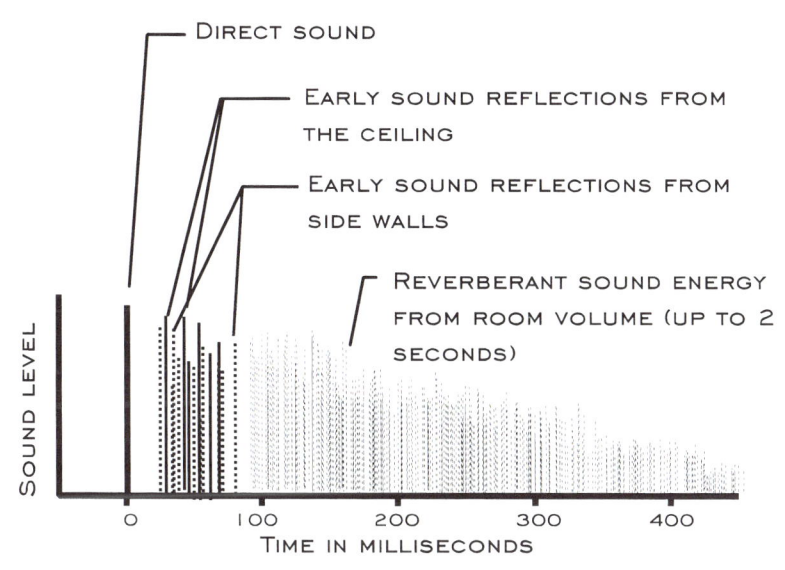

Figure 6.5 Sketch of impulse response in a room showing the direct sound, the early sound reflections arriving shortly after the direct sound, the later or reverberant sounds, and the ambient or background noise.

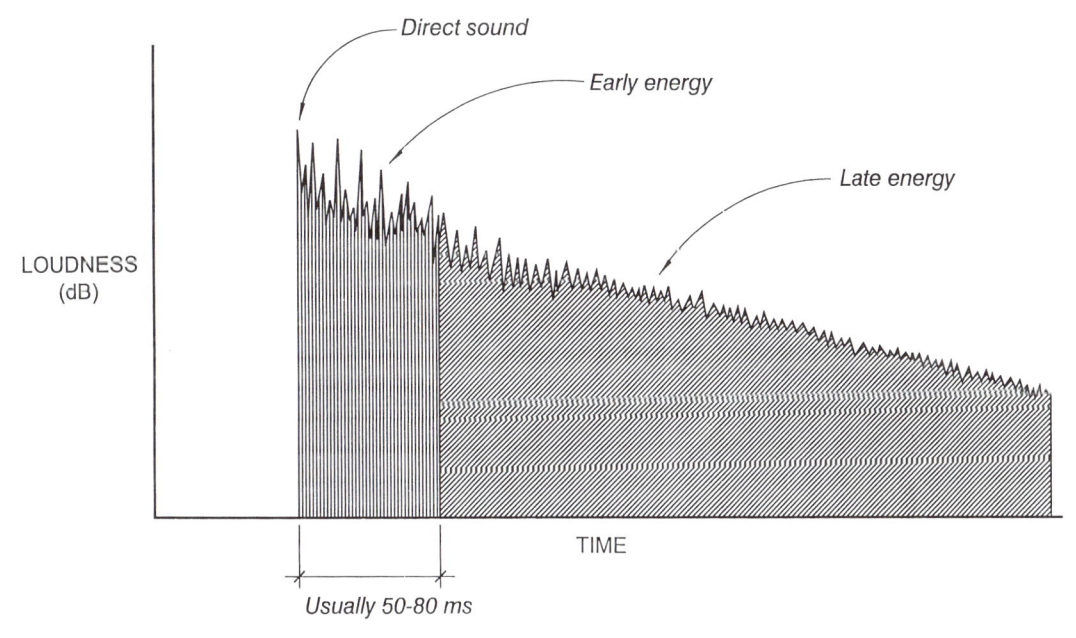

Figure 6.6 Sketch of impulse response of a room showing how an early-to-late temporal energy ratio is computed. The time period at which the early and late sounds are divided is still being debated, but the use of distinguishing the relative contributions to the total sound of the early reflections or sounds that reinforce the direct sound and the later or reverberant sound is fundamental to much of current thinking about room acoustics.

intimacy. Early reflections that arrive from the sides of the listener's head also contribute to sensations of envelopment and widening of the acoustic image of the sound source. The combination of these early reflections with the direct sound is what makes it possible to have similar levels of loudness at seats located throughout a large room.

C. The reverberant sound field consists of sound waves that have been reflected from multiple surfaces before they arrive at the listener's ears. They travel long distances between reflections and therefore are progressively reduced in loudness from the direct sound and early reflections. The reverberant sound field may persist for 2 sec or longer in concert halls. It contributes to sensations of reverberance. If the reverberant sounds arrive from many directions and are not exactly the same at the two ears of people listening, it will also increase the sensation of acoustic spaciousness in the room. If the reverberant sound field has strong low-frequency or bass components, it will increase the sense of warmth in the room. If it has strong higher frequency or treble components, it will contribute to the perception of brilliance.

7. *The arrival of the first reflection (initial time delay gap) should follow the direct sound very closely as in Figure 6.7.* Beranek (1962) postulated that initial time delay gaps of 20 msec or less were a distinguishing feature of the most preferred halls in a major survey of 54 concert halls reported in *Music, Acoustics and Architecture*. This work was one of the first studies to link qualities of sounds rated by listeners in rooms (listed in Table 6.1) with physical acoustic measures and architectural features of the rooms that determined the acoustic measures. Beranek postulated that initial time delay gap was related to the acoustical quality of intimacy. Intimacy is the sense of the relative size of the room in which sounds are heard. One generally prefers to listen in a small room where one is close to the sound source. Reflected sounds follow very closely after the direct sound in a small room because of

TABLE 6.1 Acoustical Qualities That Contribute to Overall Impression

Primary Qualities
  Clarity
  Loudness
  Reverberance
  Spaciousness and/or envelopment
  Localization
  Timbre, tonal balance, and relative balance [bass to
    mild (warmth), treble to mid (brilliance)]
  Ensemble, hearing on stage, and support
Secondary Qualities
  Source with
  Intimacy
  Diffusion
  Balance of soloist vs. orchestra
  Balance of chorus vs. orchestra
  Dynamic range—pianissimo to fortissimo
  Background noise or echoes audible

the proximity of room surfaces to the listeners. In a larger room one can provide acoustic sensations of intimacy by locating reflecting surfaces close to listeners so the sound reflections will follow closely after the direct sound. Beranek suggested that the initial time delay gap could be measured from ray diagrams on plan and section drawings of rooms (Beranek, 1962).

8. *Reflections arriving from the sides (lateral reflections) within 80 ms. after the arrival of the direct sound provide a sense of spaciousness, spatial impression, envelopment and increase the apparent width of the sound source.* Barron (1993), Marshall (1967), and Johnson (1990) have shown that early lateral reflections have been an important feature of many traditional shoebox-shaped concert halls. The relatively narrow room width, parallel walls, and several tiers of narrow balconies on the sides and rear of the rooms combine to produce many early lateral reflections. Subsequent laboratory studies have identified the contribution of these early lateral reflections to several important acoustical qualities.

   a. The apparent size of the sound source appears to broaden somewhat. This attribute is known as the apparent source width (ASW). It can become plainly evident to anyone who takes a pair of stereo loudspeakers in their living room and begins to move them about. One can start with both speakers together in the center of the room and listen to music. After a brief period of time, gradually move the speakers farther apart and listen. Then continue moving them farther apart listening at several intervals until they are at the corners of the room.

   b. There is also an enhanced spatial quality to the sound. In other words, the sound begins to become more three dimensional. It is not just coming from a point in the room, but from all of the room surfaces. If several speakers are added to the system on the sides

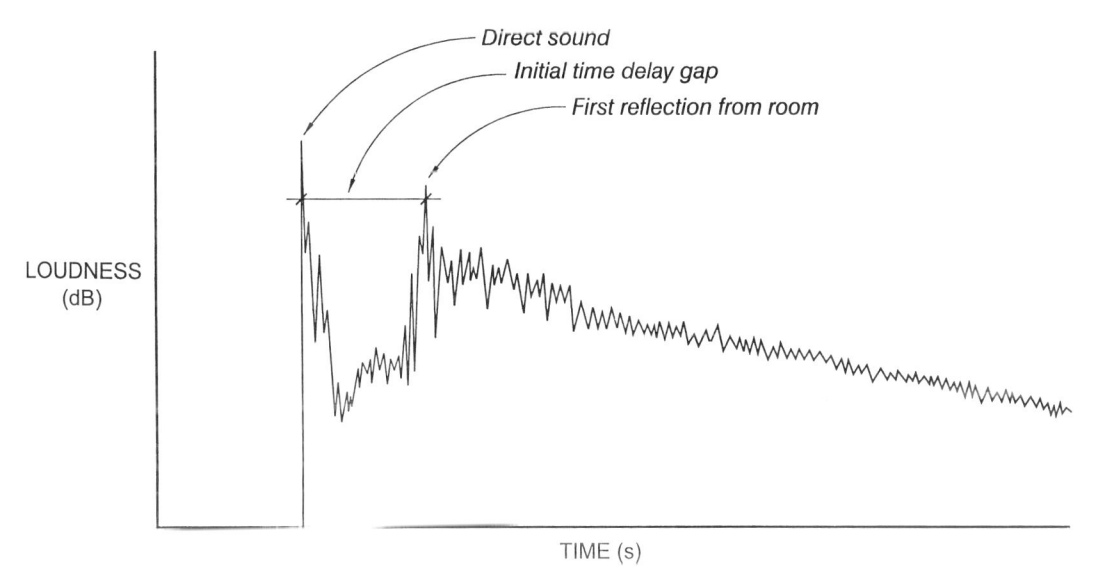

Figure 6.7 Concept sketch showing the initial time delay gap as the difference in arrival times between the direct sound and the first major reflection from the room surfaces.

of the room, the sensation of acoustic spaciousness will continue to increase. The early lateral reflections in concert halls create a similar effect on sound. This sensation of spaciousness can be increased by raising the volume of the sound as well.

c. If several additional loudspeakers are placed toward the rear of the room and reverberant "fill" is added to the sound, people will begin to feel enveloped in the sound field. It will appear as if they are immersed in a rich, full sound that surrounds them. They are submerged in the sound, not just "looking at it" in a static sense. This is similar to what occurs in a concert hall when relatively diffuse reflections occur throughout the reverberant process. The envelopmental sound will be different at both ears of the listeners or binaurally dissimilar. People will feel enveloped in the sound.

9. *Musical sounds are preferred to be configured somewhat differently at the left and right ears of listeners.* The technical term "binaural dissimilarity" occurs when the sounds that arrive at the left and right ears of a person are somewhat different. This results from a diffuse sound field that is comprised of reflections from many directions. This quality was found to be a major contributing factor to overall acoustic impression in many studies. Interaural cross correlation (IACC) is the usual measure of this quality (Gottlob in Cremer and Muller, 1982; Schroeder et al., 1974; Ando, 1985; Barron and Marshall, 1981). Gold (1994) and Soulodre et al., (1993) have shown that binaural dissimilarity in the reverberant sound field contributes to the acoustic quality of envelopment or spaciousness discussed above.

10. *The sound level should be adequate at all seats in the house* (Sabine, 1927). Loudness is probably the most important acoustic quality in a room. If the sound is not heard loudly enough, many of the more subtle qualities of the sound may go unnoticed because they are masked or are not strong enough to be heard above the ambient noise in the room. Loudness is a difficult quality to assess because it actually has many attributes. For example, sometimes in a music or theatrical performance, the sound is made deliberately quiet or soft for dramatic emphasis. At other times, the full orchestra or chorus may build to a dynamic crescendo. It is important also to balance the loudness of the orchestra with a soloist.

11. *Preferred listening levels have been identified.* These vary with the size of the room, the type of music, and so forth (Ando, 1985), although overall levels of about 79 dBA seen to be preferred (see Table 6.2).

12. *The timbre and frequency response of sounds should be preserved by the room* (Sabine, 1927). In general, sounds should be reflected from the room surfaces uniformly across the frequency spectrum. Selective absorption of narrow bands of frequencies by poorly chosen materials or mountings should be avoided. A recent study by Bech

TABLE 6.2 Preferred Listening Levels for Various Types of Music

STUDY	MUSIC	COMPOSER	PREFERRED LEVEL
Ando, 1985	Royal Pavanne	Gibbons	77–79 dBA
	Sinfonietta Opus 48; IV Movement Allegro con brio	Malcolm Arnold	79–80 dBA

(1995) has examined the effects of sound reflections on perceived timbre of sounds.

13. *The sound field should be reasonably diffuse.* There has been a general consensus among acoustical consultants that diffusion is an essential attribute in rooms. There has been little study regarding quantitative assessment of diffusing surfaces in rooms and the effects of diffusion on perceived qualities of sound by listeners (Thiele and Meyer, 1977; Beranek, 1962; Cremer and Muller, 1982). Chiang (1994) and Frick (in Beranek, 1996) have developed qualitative scales for rating the relative amount of sound diffusing material in a room. They conclude that large amounts of diffusing materials on many of the room surfaces is one of the characteristics of most of the successful concert halls in the world.

14. *A stage enclosure should allow musicians to hear each other, provide support, and ease of ensemble for musicians* (Gade, 1989). Musicians need reflections from an enclosure to be able to gauge the loudness at which they are playing. They also need to hear reflected sound from other musicians on stage so they can achieve a sense of ensemble and play in unison with each other. The sound from the musicians also needs a volume where it can blend and move into the room with a degree of cohesion. Gade (1989) has identified the concept of support and the acoustical measures ST100 and ST200 to evaluate the acoustics of the stage environment.

15. *The acoustical qualities perceived by many people vary within a given room and among different rooms at significant levels* (Hawkes and Douglas, 1971; Barron, 1988; Cervone, 1990). Hawkes and Douglas and Cervone found that there were distinct differences in the perceptions of people for seats in different parts of several major concert halls. In fact, in several large rooms, there were greater differences found among locations within the room than in sound qualities found in different rooms!

16. *The qualities that comprise overall acoustical impression vary among listeners and also among rooms.* Particular seating locations can be found within rooms that are preferred by listeners. The physical factors that contribute to these preferences can be identified. Strength, clarity, and timbre were found by Wilkens to explain 90% of the variance in a study of German concert halls. Barron found that reverberance, intimacy and envelopment were strongly related to overall impression in a study of British concert halls. Cervone et al. found that clarity, intimacy, and envelopment were related to overall impression in a study of American multiuse halls (Wilkens in Cremer and Muller, 1982; Barron, 1988; Cervone, 1990; Chiang, 1994).

17. *There are significant main effects and significant interaction affects among many of the qualities that comprise overall acoustic impression.* While the main effects of several important variables contributing to acoustical quality have been identified in the literature, there are also important interactions among the variables. For example, envelopment will increase as loudness and reverberance increase (Barron, 1988; Cervone, 1990; Chiang, 1994). The combined effects of multiple variables make isolating the particular effects of any one variable more difficult to determine. This finding has led to the necessity for conducting laboratory studies of isolated components of sound fields.

18. *The components of auditory spaciousness have been identified,* including the direction, amplitude, and time delay of reflections and the extent of reverberation required for this important quality (Blauert and Lindemann, 1986; Barron and Marshall, 1981).
19. *A proposal for estimating the relative preference of people for sound fields in rooms using four acoustical measures has been proposed by Ando.* The results of Ando's (1985) work suggested four factors determined subjective preference of sound fields: listening level, initial time delay gap (measured by the autocorrelation function), reverberation time, and wave incoherence at both ears (measured by the interaural cross correlation). Ando's model theoretically predicts the preference of people listening at seats in a large room based on acoustical measures made in the room or derived from computer models of rooms. There has been little application of this theory in the design of rooms to date, but it certainly offers promise for the future. In a similar study, distinctness, reverberation time, and the interaural cross correlation were found to relate to preferences of listeners in a study of music recorded in 25 halls (Schroeder et al., 1974).

There are a number of serious concerns about the ability of any of these methods to provide detailed insights into the perceived acoustical qualities of rooms. While providing important information to acoustical consultants and architects, the extrapolation of the results of this research must be done very carefully. There are important issues related to each of the methods described above to warn designers that the totality of the human perception of sounds in architectural spaces is not as well known as it should be at this time. However, there is an urgent need for additional research in this area sponsored by those who use, design, construct, and manage buildings to provide continuing insights into the complex relationships among people, sound, and buildings that create the sonic environment in which we will dwell in the future.

Other Issues Related to the Perception of Sounds

Speech intelligibility tests have been borrowed from other areas of acoustics and applied to rooms. Lists of words are read or played through a loudspeaker source. People are asked to write the words as they are heard or to circle the correct choice among several words. The percentage of words heard correctly is scored. There are several variations of this basic method that are typically used. It is important to note that the components of speech intelligibility provide only some of several important acoustical qualities of rooms for listening. The concept that early sound reflections are useful and increase the loudness of sounds thus increasing intelligibility and that later arriving sounds from late arriving reflections, reverberation, and background noise in the room decrease intelligibility has been well established. A series of useful to detrimental energy ratios have been proposed to measure these factors (Lochner and Burger, 1964; Bradley, 1986). Speech transmission index (STI) and rapid speech transmission index (RASTI) are measures based on the modulation transfer function that have been shown to account for these factors and are also strongly related to speech intelligibility test scores (Houtgast et al., 1980).

The STI is an important acoustical measurement that relates the levels of the direct sound and early reflections to the reverberant sounds and background noise for a simulated speech signal. The STI is thought to account for the relative degradation of speech by the combination of background noise, reverberation, and distance in a specific acoustic environment. The STI values range from 1.0 or ideal to 0.0. Rooms with STI values above 0.75 are thought to provide for good speech intelligibility. STI values of 0.45 and less are thought to provide poor to bad speech intelligibility.

Annoyance is a third aspect of listening in buildings. It is actually difficult to define noise-induced annoyance. It has been defined by Fidell and Green (Harris, 1991) as an adverse mental attitude toward noise exposure. It is not a behavior. There are acoustic and nonacoustic factors that contribute to noise-induced annoyance. Many of these factors are oftentimes in apparent conflict with each other. For example, as the loudness of a disturbing sound is increased, people generally become more annoyed. However, just the presence of an annoying sound at a fairly low level can sometimes cause severe annoyance, such as water dripping from a faucet. The subtleties of annoyance are a match for the subtleties of concert music in many respects. Annoyance is generally proportional to the duration of exposure. Fluctuations in sound level or duration or spectral content of sounds tend to be more annoying than steady sounds. The spectral content of sounds can contribute to annoyance. People will gradually become accustomed to some noises. When people are involved with the production of noise or they perceive a necessity for the noise, they will not usually complain about noise. Overall A-weighted sound level appears to be a reasonably well-accepted measure of sound level and the associated annoyance from broadband noise. There have been numerous measures proposed to account for tonality, impulsive noises, time variation, and rapid onset rates for sounds (Harris, 1991).

Methods to Assess Qualities of Sounds Heard in Rooms

Analysis of Sounds in Actual Rooms

Evaluation of Live Performances. People use a written form (a survey instrument) to evaluate a live music performance or other acoustical event during or shortly after listening. The survey, which is often a bipolar semantic rating scale, will allow people to respond on a scale between polar conditions such as *not reverberant* and *too reverberant*. It is important to use words that people understand. In concept, the words should represent opposite conditions of one quality (see Figure 6.8). The major advantage of this type of survey is that the responses of people to the real experience are obtained.

Interview Studies. A second way to evaluate live performances is an interview study. Sometime after the performance the researcher interviews the listeners or performers or conductors to obtain information about the acoustical qualities of the performance. Some of these studies have relied on the long-term memory of the subjects. They are asked about concerts attended weeks, months, and even years prior to the interview. In the case of historical research, this may entail reading the diaries of people who attended performances several hundred years ago or reading critical reviews in newspapers only several years old! "Nevertheless, many important facts have become known in the course of time just by interviewing many concert and opera goers" (Kuttruff, 1991, p. 173).

Listening to Recorded Music in a Hall. Some studies have played anechoically-recorded music through loudspeakers located on stage to listeners seated in an actual hall. *Anechoically recorded music* is music that has been recorded in a room that is totally sound absorbent. This type of room is called an anechoic room or a room without echoes. No sound reflections from the room surfaces are added to the direct sound produced by the musicians. The anechoic music is played through a loudspeaker or loudspeakers in a concert hall. Sound reflections and reverberation are added to the music as it moves from the loudspeaker to the microphones in the room just as they are added to live music played in the room. The reflections and reverberation of

ACOUSTICS EVALUATION SHEET   Sheet 1

NAME:_____   HALL: _____

POSITION:_____   ROOM MODE:_____

CLARITY	1 not clear enough	2	3	4	5	6	7 extremely clear	0 cannot tell
INTIMACY	1 not intimate enough	2	3	4	5	6	7 extremely intimate	0 cannot tell
ENVELOPMENT	1 not surrounding enough	2	3	4	5	6	7 extremely surrounding	0 cannot tell
BALANCE	1 too much bass	2	3	4 well balanced	5	6	7 too much treble	0 cannot tell
REVERBERANCE	1 not reverberant enough	2	3	4	5	6	7 too reverberant	0 cannot tell
LOUDNESS	1 not loud enough	2	3	4	5	6	7 too loud	0 cannot tell
OVERALL IMPRESSION	1 very bad	2	3	4	5	6	7 very good	0 cannot tell
BACKGROUND NOISE	1 not audible	2	3	4	5	6	7 too loud	0 cannot tell
ECHOES	1 none detected	2	3	4	5	6	7 clearly heard	0 cannot tell

COMMENTS: _____

Figure 6.8 Example of a questionnaire used to gather data in a live performance study (Cervone, 1990).

the room change the character of the music. This technique has the advantage of using a constant source of sound for comparisons of sound qualities in different rooms thereby eliminating some of the variables that are present in the evaluation of live performances. However, the music that is usually played through one or two loudspeakers is not the same sound source as a full orchestra performing on stage.

The evaluation of live performances is a very useful method to obtain general information regarding the acoustical qualities of an actual room. The primary factors involved in a judgment of acoustical quality for a room can be determined as well as the variation among these factors with different types of music at different locations within a room. Some general distinctions regarding the differences in acoustical qualities of different rooms can also be found. It is difficult to find significant results regarding fine details of the listening process using this method. A laboratory test using recorded or simulated sounds is often used to isolate and study individual variables.

Analysis of Sounds in the Laboratory

There are two basic approaches to conducting the analysis of sounds under laboratory conditions. One method requires the evaluation of actual music or speech with recorded or simulated room effects. The other method requires the evaluations of acoustic test signals such as "clicks." This latter method is used widely in psychoacoustic studies of thresholds, for example.

Evaluation of Recorded and Simulated Sound Fields. People are brought to a laboratory listening room. Recorded or simulated sounds are presented over a loudspeaker array or over headphones. A *recorded sound field* is one where sounds were recorded in an actual room. The sound fields are usually recorded with a binaural test manikin. This is a fiberglass or plastic dummy with a head that has ears and ear canals built into it (see Figure 6.9). Microphones are located either just inside the pinna or at the location of the eardrum. Extensive electronic filtering for the effects of the pinna, ear canal resonances, and so forth are built into many acoustic manikins. Sometimes the recordings are of an actual concert or performance in the room. Other studies have used music that was recorded in an anechoic chamber and played back in an actual room over loudspeakers (see Figure 6.10). The music recorded at locations in the room consists of the original anechoic music signal plus the acoustical effects added by the room and the playback system.

A *simulated sound field* is usually configured in one of two ways. Digital reverberation units and delay units are used in an audio system to add reflections and reverberation to sounds that were originally recorded in an anechoic environment (see Figure 6.11). A signal is played through the audio system and several reflections from different directions with different time delays as well as reverberation are added to the signal. The signals can also be filtered using an equalizer to represent the absorption of sounds by room surfaces and air. The relative loudness of each component may also be controlled individually. The result is a sound that has the acoustical effects of being in a room without ever leaving the audio system. It is easy to model extreme conditions that would be difficult to find in real buildings with this method.

A second way of making a simulated sound field consists of *convolving an impulse response* measured in a room or in a computer model of a room with sounds that were originally recorded in an anechoic environment. This method employs a computer to mathematically add each reflection with its corresponding time delay to the original sound theoretically resulting in all of the acoustical qualities of a specific location in a room (see Figure 6.12).

Once a sound signal is obtained by any of the methods noted above, it is presented to people in three basic ways: (1) in a large anechoic room with a loudspeaker array or (2) through two loudspeakers separated by a barrier and corrected for cross talk (also played in an anechoic environment) or (3) over headphones. The loudspeaker array must be in an anechoic environment with no distractions from the outside. Several research institutes have constructed these virtual acoustic environments where the response of human subjects to music can be studied. The loudspeakers must be several meters away from the subject so plane waves are received at the ear for frequencies of interest. Enough loudspeakers must be used so the acoustic location of each sound reflection is maintained and the spatial effects of reverberation are reproduced.

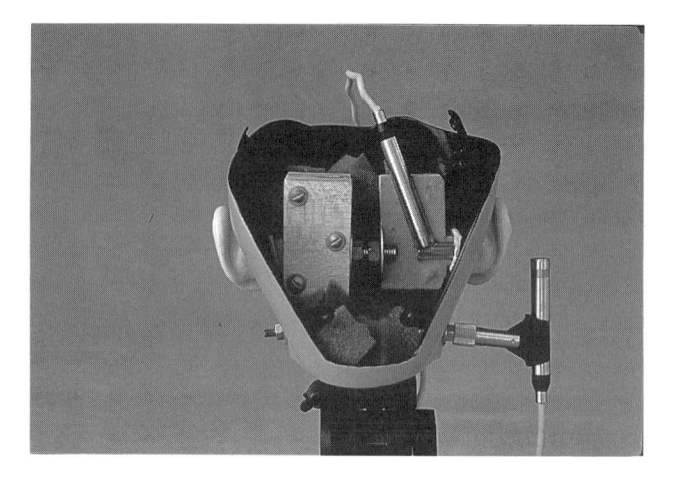

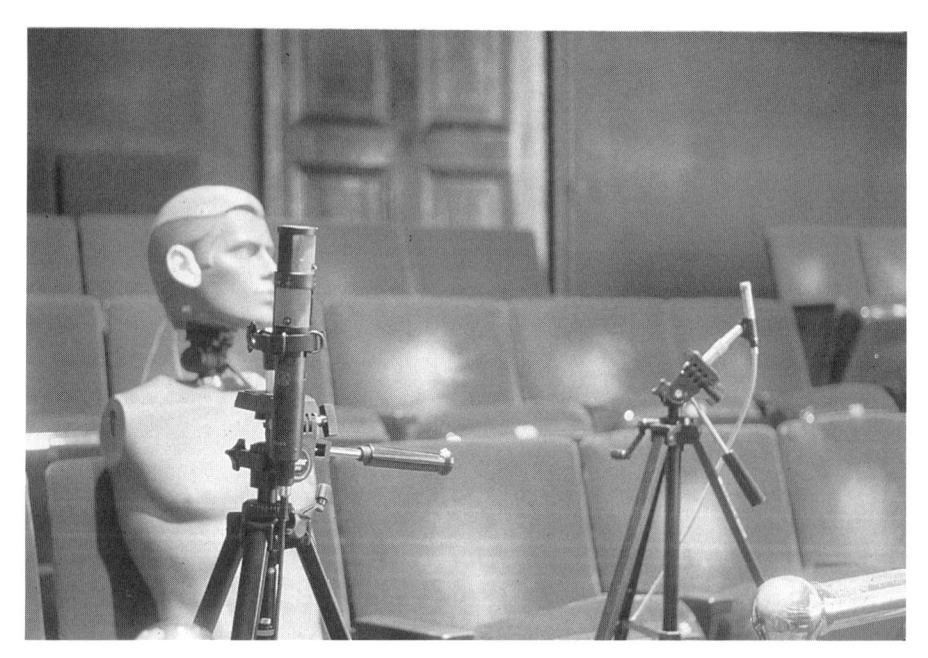

Figure 6.9 *Acoustical test manikin used to record high-quality binaural music and/or test signals in digital format in rooms. Commercially available manikins also come with sophisticated filtering to account for the effects of ear canal resonance, head and pinna effects, and other important details of human listening.*

The signal must have enough channels so reflections can be sent to different loudspeakers. A minimum of 6 loudspeakers have been used in some studies with more than 20 speakers used in more detailed simulations (see Figure 6.13).

Several studies (Schroeder et al., 1974; Soulodre et al., 1993) have used only two loudspeakers: one representing the acoustic signal that reaches the left ear and the other representing the acoustic signal that reaches the right ear, in listening tests. An electronic correction for "cross talk" was used to ensure that the signal intended to reach the left ear, for example, was not corrupted by the signal intended for the right ear of the subject. Soulodre et al. (1993) used a board placed between the two speakers that was pressed against the subjects nose to ensure separation of the left and right sounds (see Figure 6.14).

Headphones present an economical and easy alternative to the elaborate loudspeaker array, but they present several difficulties as well. The fidelity of low-frequency

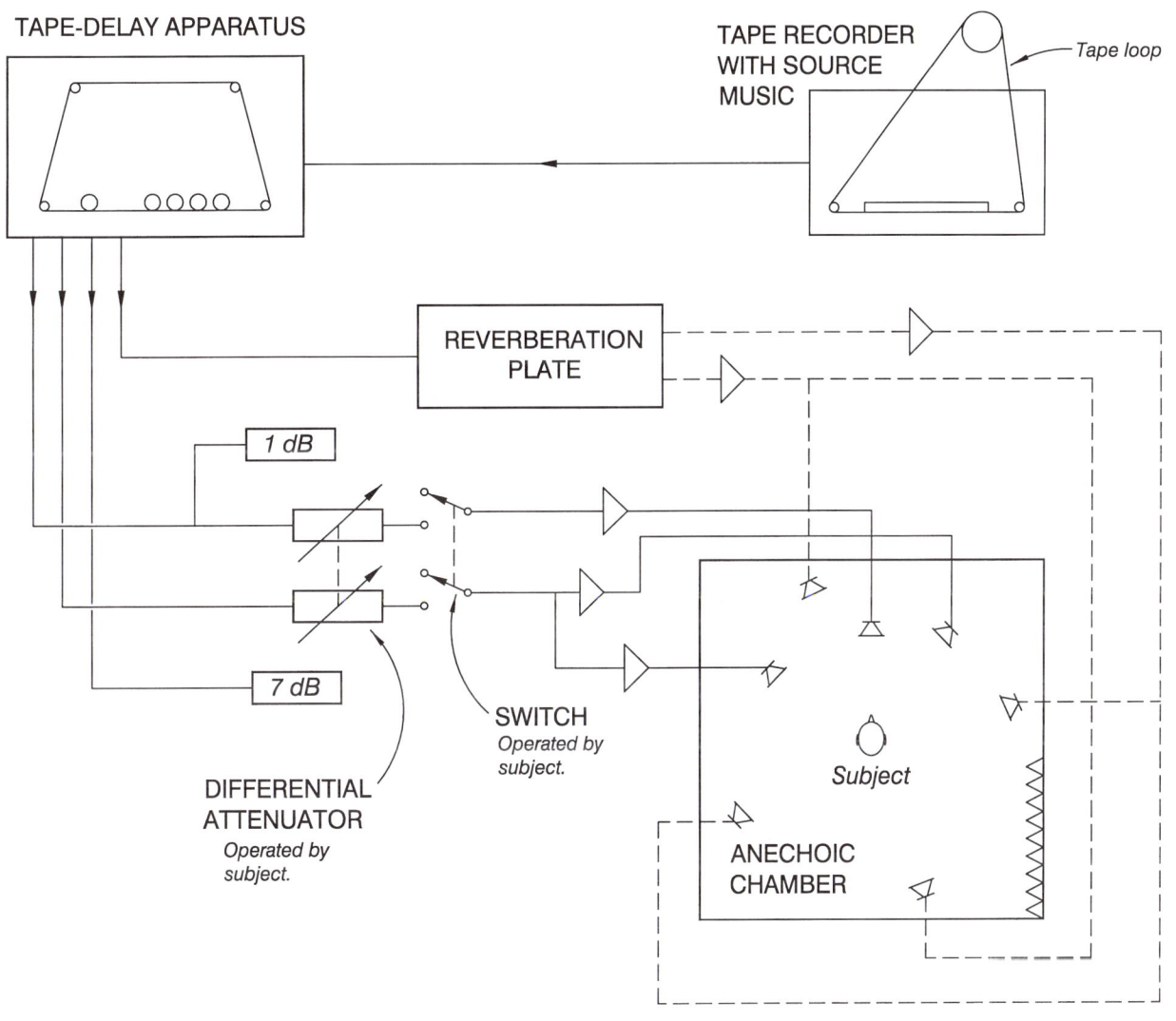

Figure 6.10 *Concept sketch of an early anechoic listening room for performing acoustical evaluations of music. (From Barron and Marshall, 1981.)*

sound presented over headphones is questioned, as is the ability of listeners to accurately localize sound sources.

Evaluation of Acoustic Test Signals. This type of study is conducted similarly to the evaluation of recorded sound fields noted above except the sounds used for evaluation are acoustic test signals with specific properties pertinent to the study. Reducing the number of variables present in a test situation helps to clarify the results of the study. It is often difficult to separate the acoustic and nonacoustic factors that contribute to the evaluation of actual rooms where lighting, other people, interior design, temperature, and the like all affect the total environment in which one is immersed. Similarly, the information content of an actual music or speech sound becomes a part of the evaluation of live or recorded sounds. In psychoacoustics, much of the testing of how people hear is done with acoustic test signals that carry no information content to isolate the acoustic factors involved. This has been particularly true regarding the response of people to the localization, loudness, and pitch of sounds.

Signals that are used consist of pure tones, random noise with varying frequency content, time-varying random noise, and signals with discrete arrival times, separations

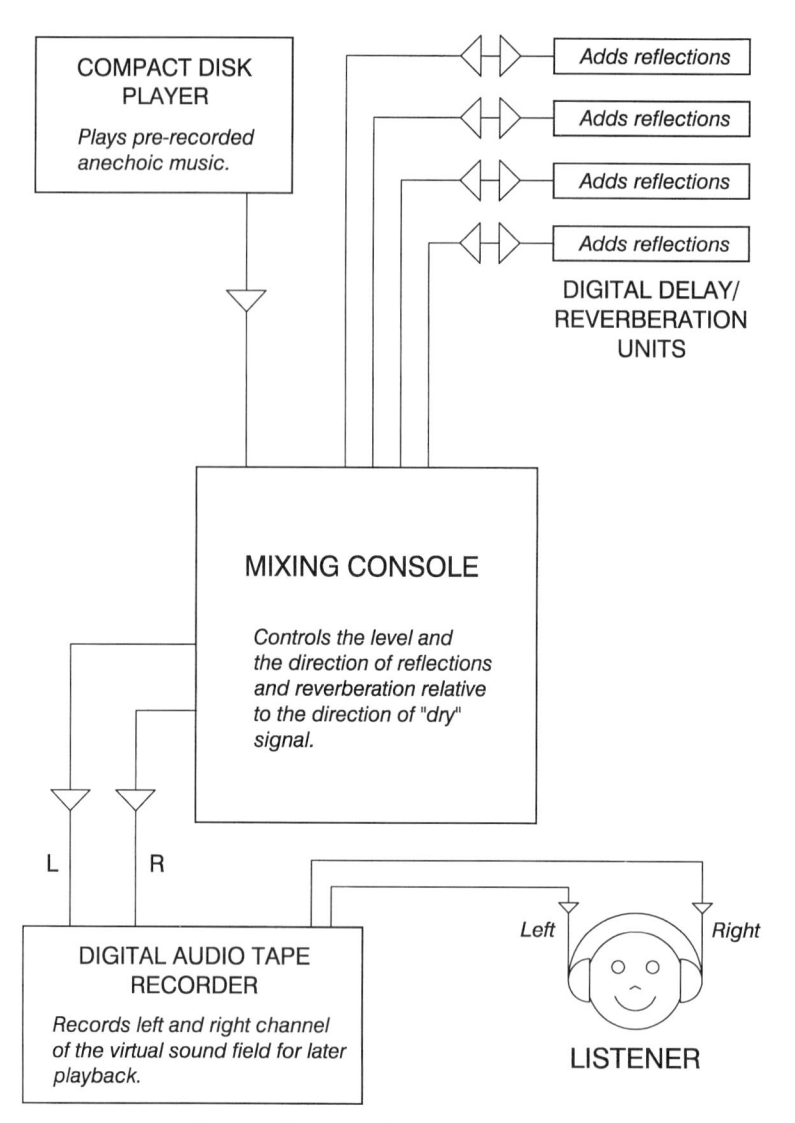

Figure 6.11 *Diagram showing the method of digitally simulating an impulse response in a room for laboratory listening studies. (From Gold, 1994.)*

in arrivals, and other controlled sounds depending on the phenomenon under investigation.

There are three basic procedures for evaluating sounds under laboratory conditions.
Semantic Differentials. People use a survey form as described above to evaluate the qualities of a sound presented to them.
Adjustment Tests. People are presented with a sound field. They have control over one of the variables. For example, they can adjust the amount of reverberation or direction of arrival of primary sound reflections in the signal until it is perceived as *spacious* as possible or as *spacious* as a reference signal.
Paired Comparison Tests. People are presented with two sound fields, one after the other. They are asked to judge which sound field is more spacious or more clear than the other. People are making relative rather than absolute judgments of sound quality in this type of test. The subjects relative judgments are often used to rank order the sounds. People do not have to use their own variable criteria to assess the sounds.

Note of anechoic musical passage recorded at 48,000 samples per second.

Music Note & Room Response

Room Response

Impulse response of room 48 kHz or 48,000 samples per second

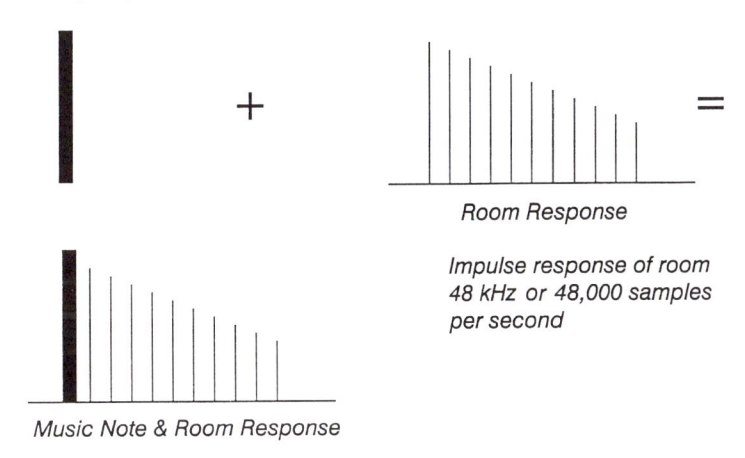

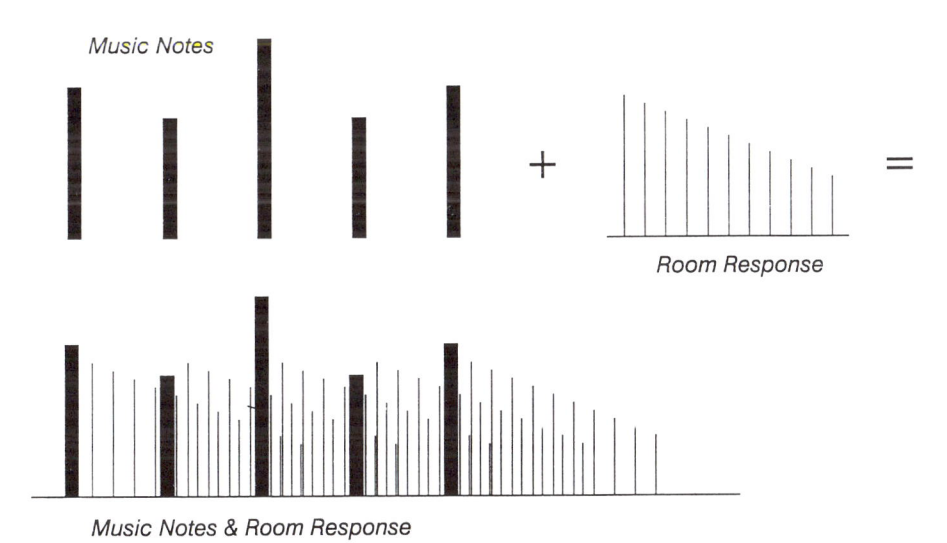

Music Notes

Room Response

Music Notes & Room Response

Figure 6.12 Schematic diagram of the process of convolving anechoic music recordings with the impulse response of a room to simulate music as it would be heard in the room.

Data obtained with paired comparison tests is often more consistent with smaller standard errors than data produced with other methods.

Preference tests are a variation of paired comparison tests. People are presented with two sounds and asked simply which they prefer for various reasons. Subtleties of vocabulary are not a factor in this type of experiment. The question is simply which of the two choices is better?

Acoustical Measurement Systems

Multichannel, digital, acoustical measurement systems are being designed to physically measure the sound field in rooms in ways that represent the sounds that people actually hear. The measurement systems are based on the concept of deriving the impulse

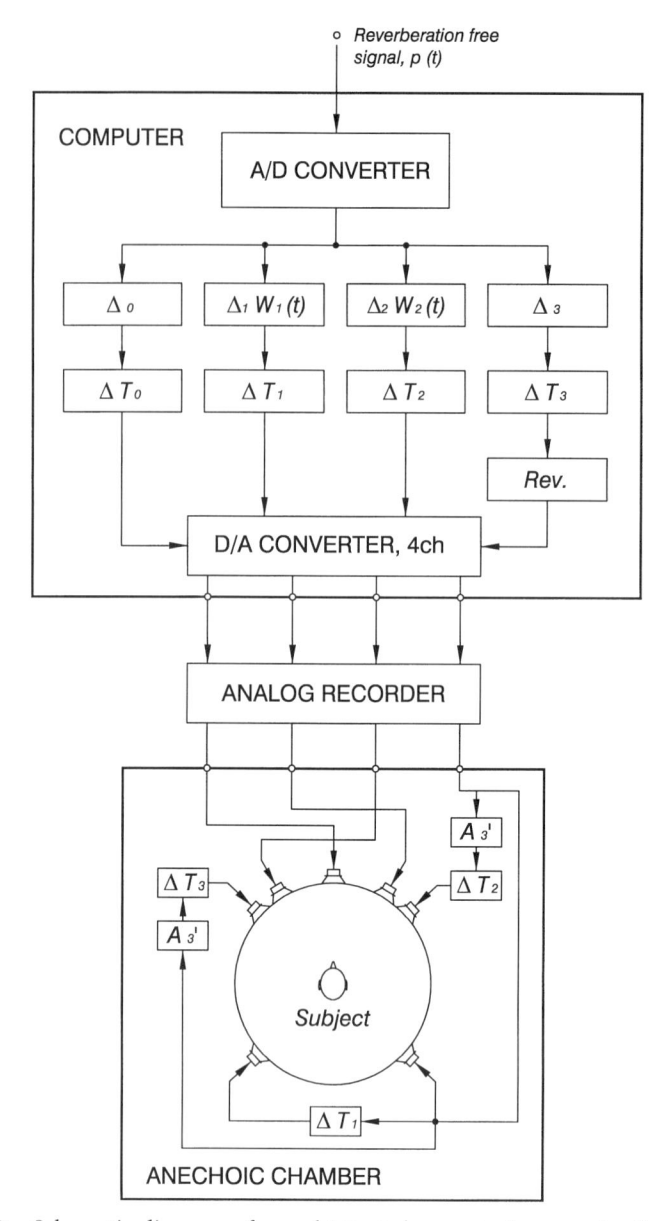

Figure 6.13 *Schematic diagram of a sophisticated acoustical system for listening to simulated sounds. (From Ando, 1985.)*

response of the room from an acoustic test signal. Impulse responses are recorded binaurally with microphones located in the ears of a manikin. This way the sound field that is measured is the same sound field that reaches the ears of people. A dodecahedral loudspeaker that is virtually omnidirectional was built to more accurately and repeatably make acoustical measurements in rooms. Similar loudspeakers have been constructed for use in 1:10 scale models.

Impulse Response of Rooms

The impulse response is a means of graphically depicting the amplitude, arrival time, frequency components, and direction of arrival of the direct sound and all subsequent reflections from the room enclosure for a specific source–receiver path. It is analogous

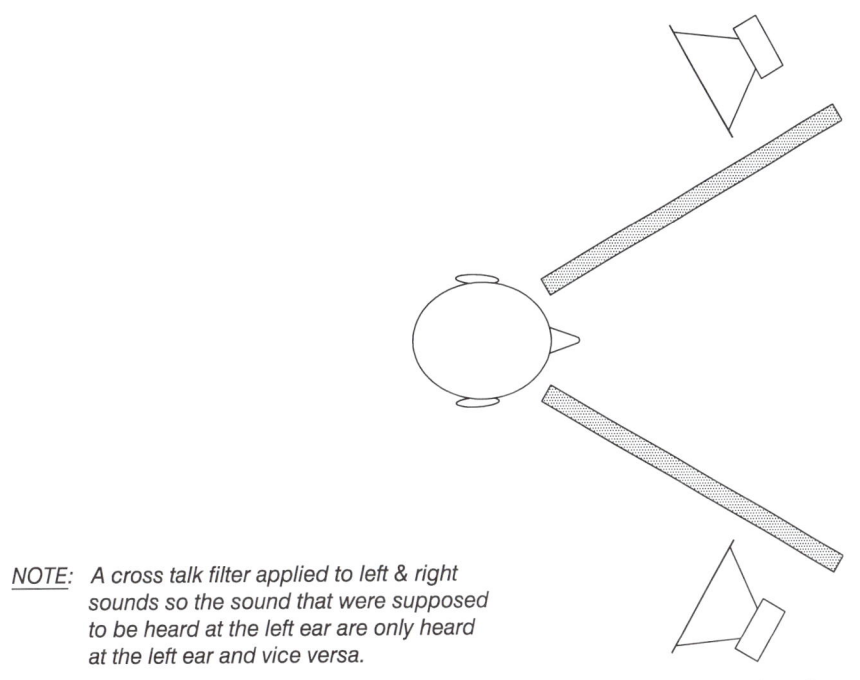

NOTE: A cross talk filter applied to left & right sounds so the sound that were supposed to be heard at the left ear are only heard at the left ear and vice versa.

Figure 6.14 Schematic diagram of Soulodre's setup for using two loudspeakers for simulated listening tests. (From Soulodre, 1993.)

to an electrocardiogram of a heart beat that a doctor might use to evaluate the health of a patient. By looking at the relative amplitude and timing of a sequence of pulses (the series of peaks and valleys one observes on a cardiogram), the doctor can make diagnoses and prescribe treatments for patients. An acoustical consultant can look at an impulse response and make diagnoses about the "acoustical health" of the room and recommend design changes. The impulse response can help provide a conceptual understanding of the links between perceived acoustical qualities and the architectural features of rooms. Additionally, it is used extensively to characterize sound fields in both computer models, scale models, and full-size rooms. Once the impulse response has been obtained, the information can then be translated into a virtual sound field. This makes it possible to actually hear how architectural changes to a computer model or scale model of a room that are indicated in the impulse response might be perceived by listeners in the room.

The impulse response represents the pulsing of a room by a single, loud sound. It is hypothesized that each syllable of spoken words and each note played by musical instruments excite the room in a similar way. The loudness, frequency content, time of arrival, the direction of the direct sound, and all of the sounds reflected from architectural surfaces are included in this measurement. The impulse response will vary at each seat in a room. It represents the unique signature of sound that arrives from a given source to a specific listener. The contribution of the direct sound and reflections from each of the walls, the ceiling, and other architectural elements is depicted in the impulse response.

In a large room, impulse responses will vary significantly from location to location just as the perceived acoustic qualities do. Additionally, as the architectural characteristics of a room are altered so is the impulse response. The impulse response can be used as a basis for suggesting possible architectural approaches to the design of a room or to help determine which modifications can be made to an existing room.

The impulse response of a free-field or open-air situation is shown in Figure 6.15. The single "spike" is shown as a sound level (dB), or a log (pressure)^2 format because this is the way people respond to sound. There is only a direct sound path from a

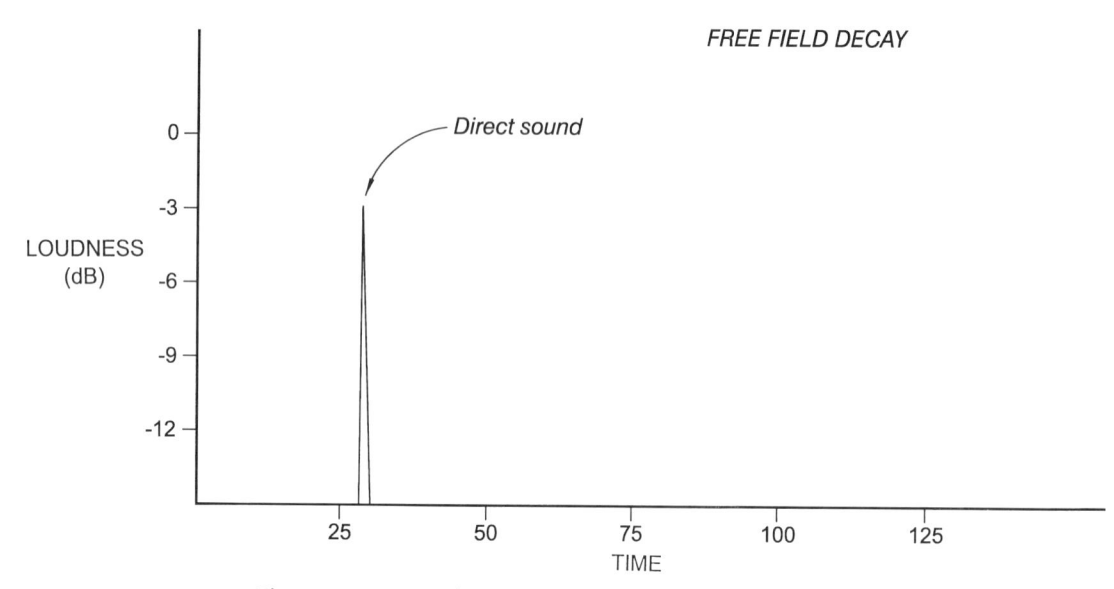

256 • *Recent Innovations in Acoustical Design and Research*

Figure 6.15 Impulse response of a free-field situation with a direct sound only.

person speaking to a listener. The loudness or amplitude of the sound decreases as the speaker and listener move farther away from each other. It also takes time for the sound to reach the listener. Therefore the arrival of the sound at the listener's ears occurs somewhat after it leaves the lips of the person speaking.

If one walks away from the person who is speaking, the sound level is reduced due to geometric spreading until it is 6 dB less than the original sound level by the time the distance from the sound source has been doubled to 20 m. If one continues to walk down the aisle, the sound level will be 12 dB less than the original signal and is heard as less than half as loud as the distance from the source is doubled again to 40 m. The sound level in the free field decreases by 6 dB as the distance from the source is doubled (see Figure 6.16).

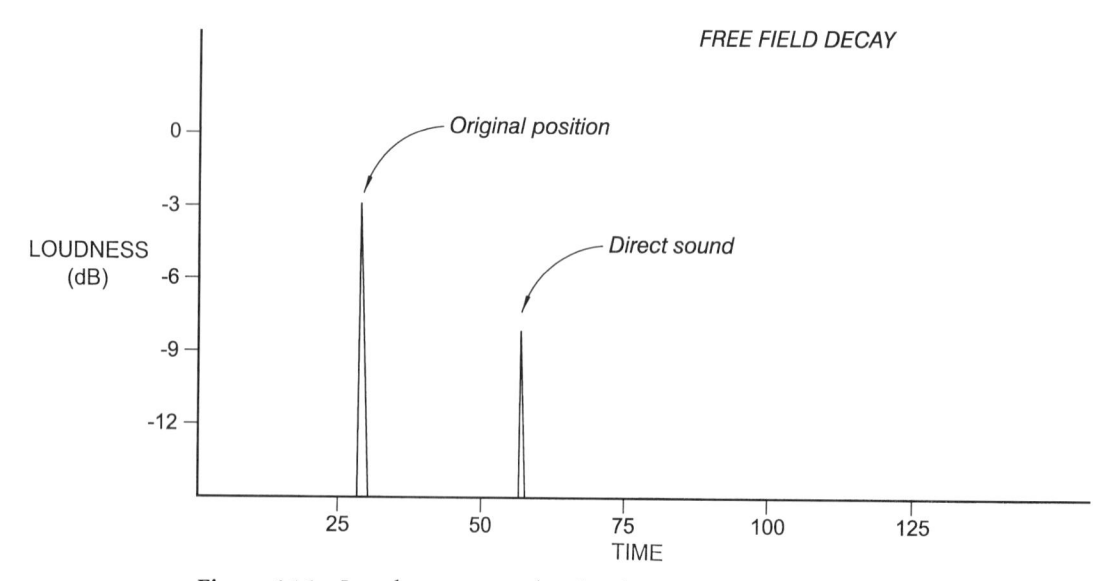

Figure 6.16 Impulse responses showing decrease in sound level as one doubles the distance from the source in a free-field situation.

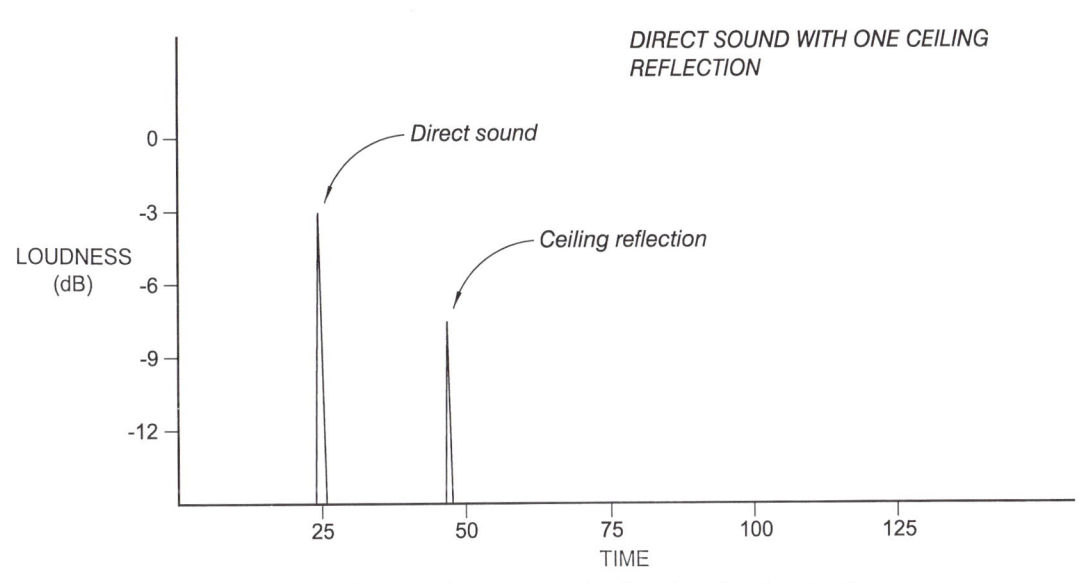

Figure 6.17 Impulse response of an amphitheater seating location showing the direct sound and reflections from the stage floor and an overhead ceiling canopy.

If a ceiling panel is added to create an amphitheater, a reflection will be heard by the listener arriving shortly after the direct sound from the overhead ceiling (see Figure 6.17). This will be perceived as an increase in the overall sound level at the listeners location. However, the relative decrease in sound level from the front of the amphitheater to the back will still be heard.

If a ceiling is designed to provide three or more reflections arriving shortly after the direct sound in the rear of a room, the sound level will be increased there as well, even though the direct sound has been decreased due to distance (see Figure 6.18). This helps maintain more even sound levels throughout the entire room.

If the rear wall of a large room is made of sound-reflective material, such as concrete block or gypsum board, the long delay time between the direct sound

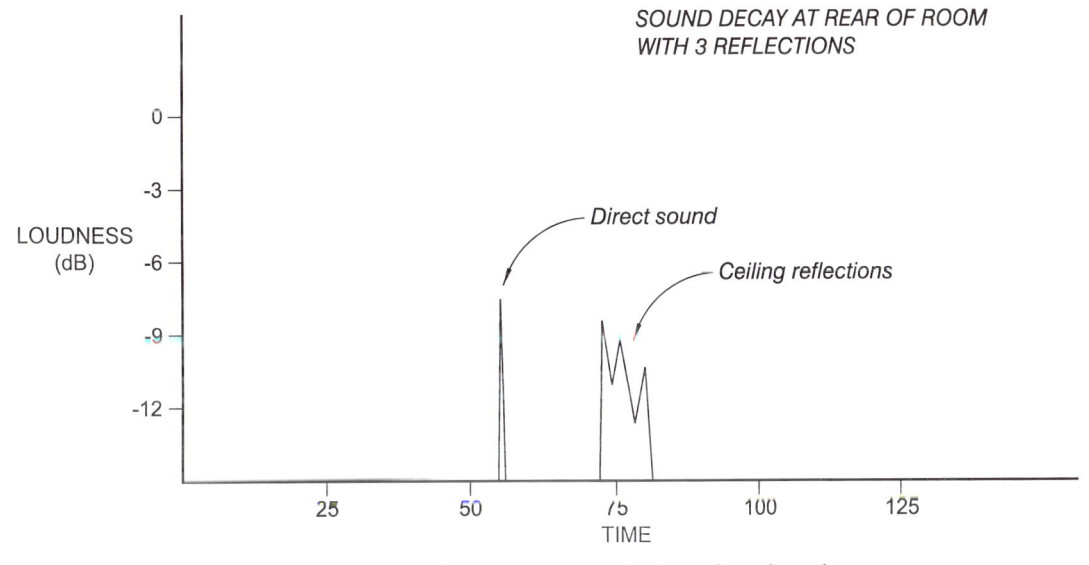

Figure 6.18 Impulse response for a small lecture room with the ceiling shaped to direct sound reflections to the audience to maintain relative loudness levels in the room.

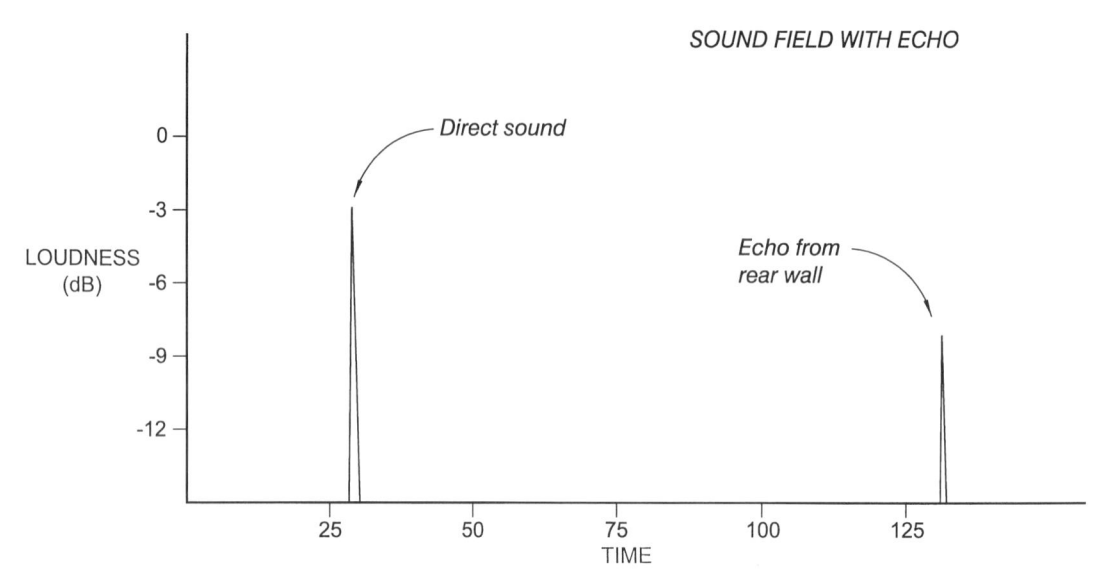

258 • *Recent Innovations in Acoustical Design and Research*

Figure 6.19 Impulse response showing the effect of a long-delayed reflection or echo off the untreated rear wall of a room.

(120 msec) and the reflected sound will result in a distinct acoustical event, or an echo to be perceived in the front of the room (see Figure 6.19).

If the room has hard parallel side walls, a "slapping" or "ringing" will occur as the sound bounces back and forth between them (see Figure 6.20). This is called a flutter echo.

If a wall is added on the left side of the room, a reflection is directed from that side to a listener. If another wall is added on the right side of the room, a reflection from that side is directed toward the listener, widening the image of the sound. These lateral reflections are thought to enhance the sense of acoustic spaciousness (see Figure

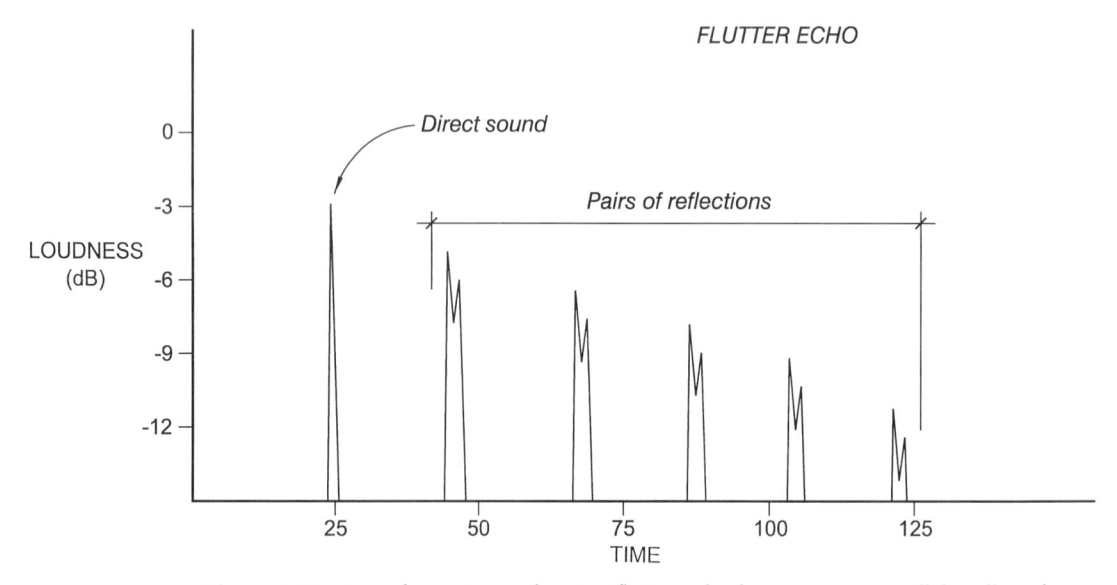

Figure 6.20 Impulse response showing flutter echo between two parallel wall surfaces.

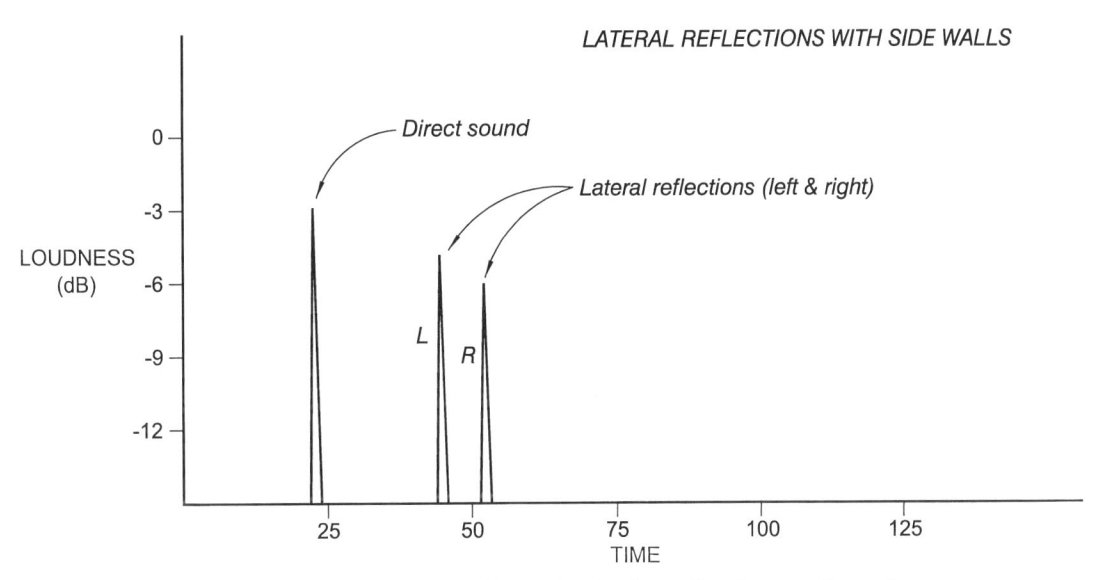

Figure 6.21 Impulse response showing the effects of early (lateral) reflections from the side walls.

6.21). Side walls that provide good reflections to the center of the room yet scatter the reflections to the degree that a flutter echo does not occur are considered beneficial lateral reflections.

In a small lecture hall, reflections from both the ceiling and the side walls can be designed to provide relatively uniform sound levels throughout the room. The reverberation time in the room will be a little less than one second. The sound is heard as both louder due to the early reflections and fuller due to the reverberation than the free-field condition (see Figure 6.22).

In a fine concert hall, there will be reflections from an overhead ceiling canopy, strong lateral reflections from specially designed side walls, and a reverberation time

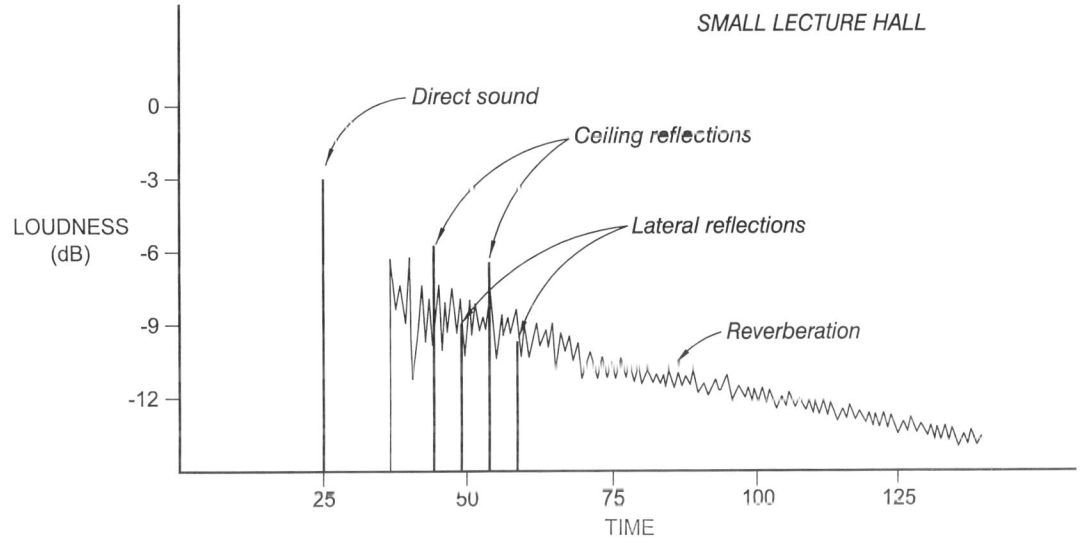

Figure 6.22 Impulse response of a small lecture hall. Notice the early reflections arriving shortly after the direct sound. The reverberant energy falls off relatively quickly due to the small volume of the room.

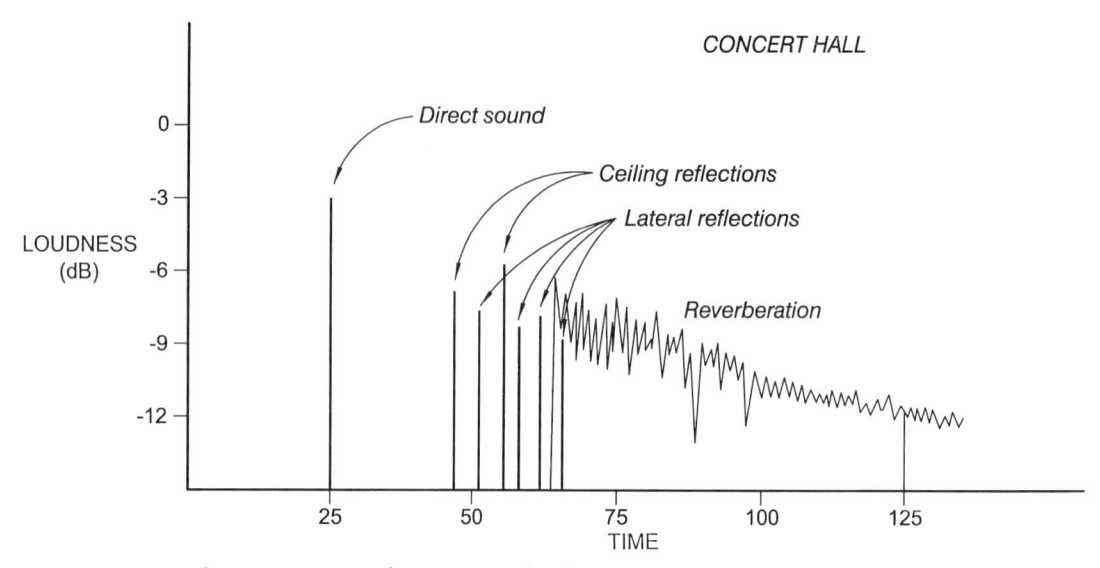

260 • Recent Innovations in Acoustical Design and Research

Figure 6.23 Impulse response of a fine concert hall. Notice the early reflections from overhead that will provide a clear sound, the reflections from the side walls that will provide a spacious and enveloping sound, and the extended reverberation time that will provide fullness to the sound.

of 2 sec to enhance the character of music performed in the room (see Figure 6.23). There is a delicate balance between the direct sound, the reflection sequence and the reverberant field. The sound is full and spacious. It also envelopes the listeners.

In a Gothic cathedral, with a large room volume, and all stone and glass surfaces, the sound will persist for over 4 sec. Music and speech become muddy as the sound is overpowered by the reverberation and clarity is lost (see Figure 6.24).

Figure 6.24 Impulse response of a large Gothic cathedral. The high level of the reverberant sound reflections and the prolonged time it takes for them to decay are the acoustical effects of a large room volume with very reflective materials.

Acoustical Measures Derived from The Impulse Response

A number of indices of acoustic qualities can be calculated from the impulse response. Several of the more "popular" measures are defined below. All of the acoustical measures were derived from $p(t)$, the sound pressure over time recorded by microphones at locations in a room. The notation t designates time with the origin ($t = 0$) chosen to be the arrival of the wide-band direct pulse.

Reverberation Time (RT60)

Reverberation time (RT60) is the amount of time it takes for a sound to decay to inaudibility after the source has stopped. Reverberation is caused by multiple reflections of sound from the room surfaces. It is found by backward integration of the impulse response. A least-squares line is fit to the resulting Schroeder curve. The slope of the line is extrapolated to an equivalent 60-dB decay to find the reverberation time. Reverberation time is usually expressed in seconds.

Early Decay Time (EDT10)

Early decay time is a modified measure of reverberation time. Reverberation time is the time required for a sound to decay 60 dB whereas the early decay time is the time required for the first 10 dB of decay multiplied by 6 to extrapolate the result to a 60-dB decay. Early decay time was proposed by Jordan based on previous research that suggested "the later part of a reverberant decay excited by a specific impulse in running speech or music is already masked by subsequent signals once it has dropped by about 10 dB" (Cremer and Muller, 1982). A study by Cervone showed a significant relation between early decay time and overall acoustic impression rated by listeners at live concert performances (Cervone et al., 1990).

Early-to-Late Energy Ratios (El$_t$)

The early-to-late energy ratios compare the early energy in an impulse response to the later or reverberant energy level. They are logarithmic ratios of the early sound energy integrated from $t = 0$ to $t = t$, relative to the late or reverberant sound energy integrated from $t = t$ to $t = \infty$. The most widely used temporal energy ratio uses $t = 80$ msec and is sometimes called clarity index. It is also called clarity or clearness index. This measure is also denoted as C or C_{80} in the literature. This measure was developed by Reichardt (Cremer and Muller, 1982) based on (1) the suggestion by Thiele that the distinctness of speech was dependent upon the levels of "useful" sound energy, the energy integrated from $t = 0$ msec to $t = 50$ msec, relative to the total sound energy and (2) the observation of music by Reichardt that led to a widely accepted value of 80 msec as the limit of perceptibility of music (Cremer and Muller, 1982). However, C_{80} has been found to be significantly correlated with EDT_{10} (Cervone et al., 1990).

$$C_t = 10 \log \frac{\text{early energy}}{\text{late energy}}$$

Relative Loudness (L) or Relative Strength (G)

Relative loudness level (also called overall strength) is the sound energy level at a seat in a room compared to the sound level at 10 m from the sound source in an anechoic environment. It effectively measures the contribution to loudness of the early reflections and reverberation in the room. This reference sound energy was first used by Gade and Rindel in 1985 (Gade, 1989). Relative strength was proposed to approximate the subjective sense of loudness. This measure is also denoted as L in the literature:

$$G = 10 \log \frac{\text{loudness of sound in room}}{\text{loudness of sound in anechoic room at 10 m}}$$

Bass Ratio Based on Early Decay Time, BR(EDT)

Bass ratio based on early decay time was proposed by Beranek (reverberation time was originally used) to evaluate timbre or tonal balance, especially warmth (Beranek, 1962). It is measured by the ratio of (a) the sum of the early decay times at 125 and 250 Hz to (b) the sum of the early decay times at 500 Hz and 1 kHz. This is the reverberation times in the lower frequencies divided by the reverberation times in the middle frequencies:

$$\text{BR(RT)} = \frac{\text{low frequency reverberation time}}{\text{middle frequency reverberation time}}$$

Treble Ratio Based on Early Decay Time, TR(EDT)

Treble ratio based on early decay time is proposed to evaluate timbre or tonal balance, especially brilliance (Chiang, 1994). It is measured by the ratio of the high frequency reverberation times to the middle-frequency reverberation times. Chiang has also expressed it as the ratio of (a) the sum of the early decay times at 2 and 4 kHz to (b) the sum of the early decay times at 500 Hz and 1 kHz. This measure was developed because treble and bass have been evaluated separately in the questionnaires used in several major live listening studies of acoustical qualities of rooms:

$$\text{TR(RT)} = \frac{\text{high frequency early decay time}}{\text{middle frequency early decay time}}$$

Early Inter-Aural Cross Correlation Coefficient, IACC$_{80}$

The interaural cross correlation, or IACC, measures the relative difference between the sounds that arrive at the left and right ears of a listener. It is the normalized maximum cross correlation of the sound pressures recorded at the left ear and right ear of a listener or a manikin with the listener facing the sound source. This measure is referred to as early interaural cross correlation, IACC$_E$ or IACC$_{80}$ if the integration time between $t = 0$ msec to $t = 80$ msec is used. Studies suggested that IACC$_{80}$ is correlated with the subjective attribute of spaciousness and with the overall preference of people for listening to music (Ando, 1985; Beranek, 1992).

Lateral Energy Fraction (LEF)

Lateral energy fraction is the ratio of the integrated sound energy in the first 80 msec. after the direct sound measured from the sides of the listener's head compared to the total integrated energy level of the early sound at the same location. The lateral energy fraction is supposed to relate to a sense of spatial impression with higher values of LEF providing a greater sense of spatial impression (Barron and Marshall, 1981):

$$LEF = \frac{\text{early lateral energy}}{\text{total early energy}}$$

Support, ST1

Support (Gade, 1989) is proposed to measure the "support" of the sound reflected from stage enclosures for the musicians. This is related to a sense of ensemble and balance among musicians on stage. Three source positions are used: the soloist position, the center of right-side strings, and in the second row of winds at far left. ST1 was measured one meter from each of the three source positions and one meter above the stage floor. Gade concluded from his study that the optimum value of ST1 was about -12 dB $\pm$ 1 dB:

$$ST1 = 10 \log \frac{\text{early sound energy reflected from the state enclosure}}{\text{direct sound energy}}$$

The use of these measures is slowly making its way into the design process of many rooms. However, additional research remains to be performed to develop strong links as to how the architectural features of rooms contribute to the impulse response at various locations and to how the impulse response contributes to the various acoustical qualities perceived by people in rooms.

Measurements in Existing Rooms

Acoustical measurements have been taken at many different locations in many existing rooms by teams of researchers to refine the measurement techniques and to use as a database for statistical, computer, and physical model development. The use of impulse response measurement systems allows one to accurately measure the unique signature of sound that arrives at each seat in a room. Therefore, the architectural factors that contribute to important variations in the acoustical qualities that people actually hear at different seats in a room can be studied. The local variations in acoustical quality that were found within many rooms were often greater than the differences found between different rooms! The magnitude of changes in the sound fields in rooms with variable acoustics features such as coupled room volumes, curtains, orchestra enclosures, and suspended ceiling canopies has also been documented.

Effects of Architectural Features of Rooms on Acoustical Measures

Methods are being developed to quantify many of the detailed architectural design features found in concert halls, theaters, and churches such as highly articulated sound diffusing surfaces, suspended ceiling canopies, angled stage enclosures, coupled room volumes, and zig-zag wall panels, such as those shown in Figure 6.25. These architectural details were found to be highly significant in describing the variance of acoustical measurements at different locations in rooms (see Table 6.3). This is the first evidence that validates many of the traditional design approaches used in concert halls and theaters in a systematic way. Regression models related the architectural dimensions and details of complex rooms to acoustical measurements made at specific locations within the rooms with model $R^2 > 0.90$ (Chiang, 1994).

TABLE 6.3 Summary of Acoustical Qualities in Rooms with Explanation of How They Are Achieved, the Architectural Features of Rooms Thought to Be Responsible for Them, and How They Are Measured According to Current Research

ACOUSTIC QUALITY	ARCHITECTURAL FEATURE	DESCRIPTION OF EVENT	ACOUSTIC MEASURE
Envelopment and source width	Narrow rooms from 70–80 ft across multiple tiers of narrow balconies	Early sound reflections arriving at the listener from the side (up to 80 msec after the direct sound)	Lateral energy fraction (LEF) LF < 0.40
Clarity	Sound-reflecting ceiling Ceiling canopy Parterre walls	Sound reflections that arrive shortly after the direct sound	Clarity index (C_{80}) early-to-late energy ratio
Reverberance	Large room volume, sound-reflecting materials, shoebox shape, acoustical banners and reverberation chambers	Prolonging of sound in the room	Reverberation time (RT) (2.0 sec)
Loudness	Room size (1000–2000 seats) proximity to source and sight lines	Sound reflections from the ceiling and walls shortly after the direct sound	Loudness (L) or relative strength (G)
Intimacy	Orchestra in same room volume as audience	Arrival of the first sound reflection from a building surface shortly after the direct sound	Initial time delay gap (ITD) (<20 msec)
Warmth	Heavy massive building materials	Persistence of sound at low frequencies or extended low-frequency reverberation	Bass ratio (>1.0)
Brillance	Heavy massive building materials	Persistence of sound at high frequencies or extended high-frequency reverberation	Treble ratio (>1.0)
Spaciousness	Surface texture and sound-diffusing materials; large room volume	Late sound energy arriving from the sides (After 80–100 msec)	Interaural cross correlation IACC (<0.50)
Localization of sound	Clear sight and sound line between listener and source	Strength of direct sound relative to subsequent reflections	Early loudness level
Ensemble	Overhead and side wall sound-reflecting surfaces at performance area	Sound reflections that allow the musicians across the stage to be heard	Support

Source: Siebein and Gold, 1997.

6.3 ACOUSTICAL MODELING AND AURAL SIMULATION

Acoustical Model Studies

Acoustical modeling techniques have been developed to accurately represent the sound fields at individual seating locations in architectural spaces. Recent research and design applications have focused on physical acoustical modeling techniques and computer modeling. Modeling techniques allow basic design approaches to be evaluated in small, easy to manipulate models during the design process. It is possible to quickly study the effects of architectural design changes on acoustical measures. This research allows architects and acoustical consultants to "sculpt" the acoustical qualities of rooms as part of their design process. The impulse response obtained in model studies can be

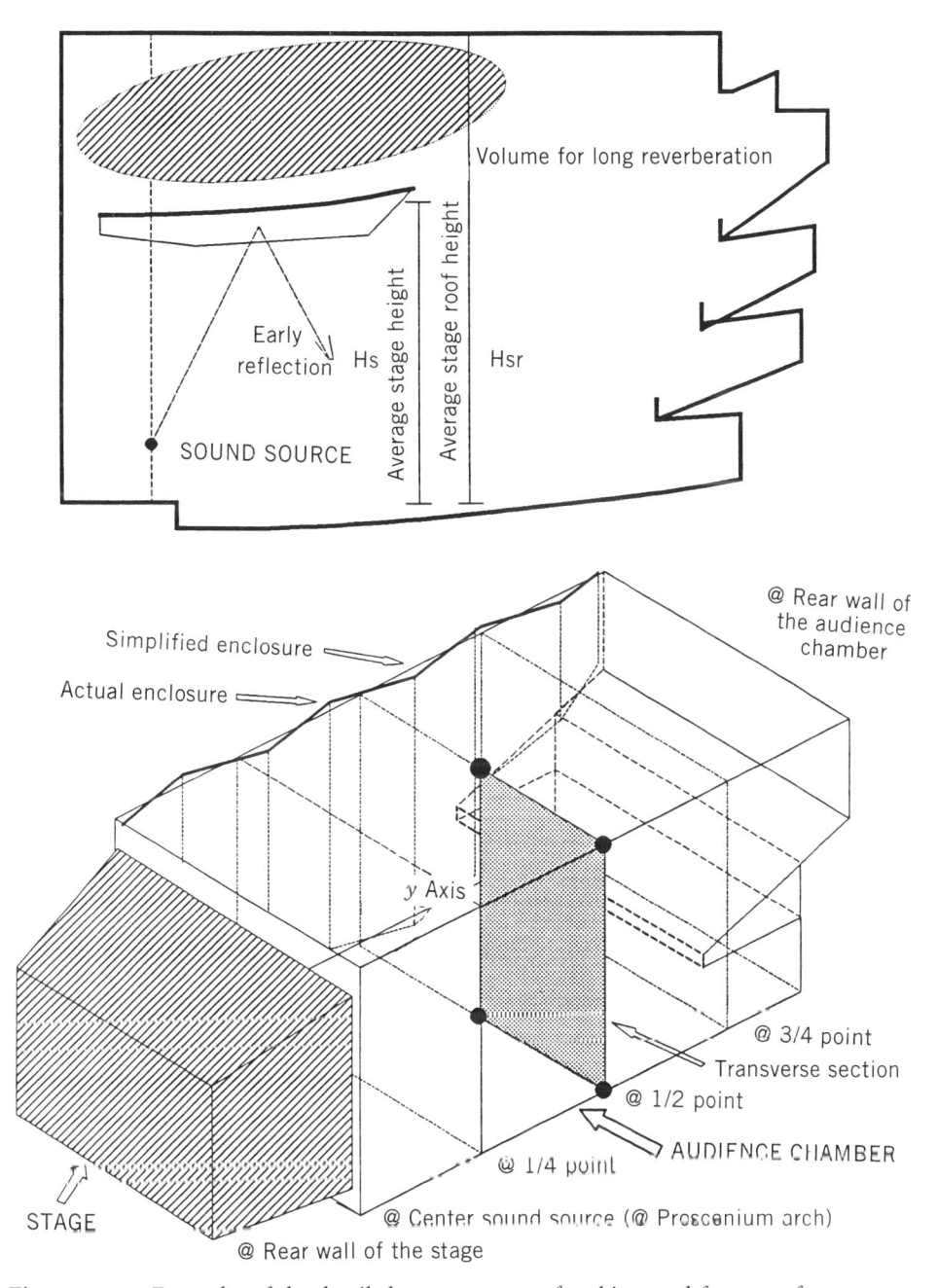

Figure 6.25 Examples of the detailed measurement of architectural features of rooms used in statistical models to estimate values of room acoustic measures (Chiang, 1994).

mixed with sound recordings to provide aural simulations of rooms while they are being designed.

> *The designer develops one or more design schemes for a given project. Then he critically evaluates the relative merits of each scheme; analyzing, evaluating, comparing and ultimately selecting or combining elements of preliminary solutions into a final design. This process requires a flexible yet economical method to explore the acoustical qualities of these solutions during the design process.* (Siebein, 1986, p. 1).

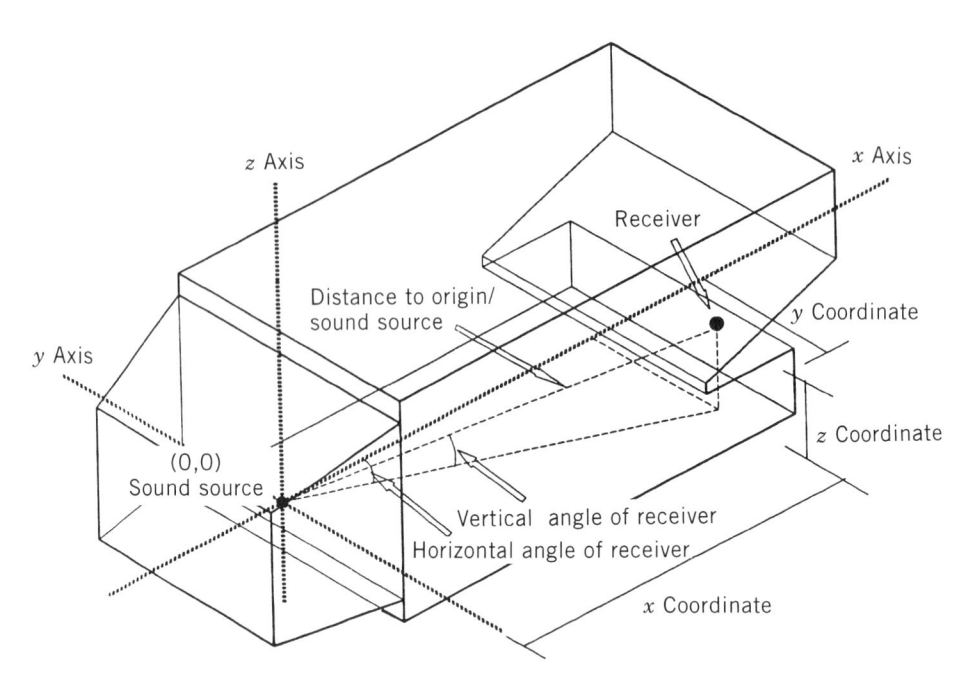

Figure 6.25 (*Continued*).

Physical Modeling Techniques

Physical modeling involves the construction of an actual scale model of a room being designed (see Figure 6.26). Small size models (at 1:48 or 1:96 scale) are often used to evaluate basic design decisions of the size and shape of rooms early in the design process. Larger size models (at scales of 1:24 or 1:10) are built later in the design

Figure 6.26 *Students working in an acoustical model at the University of Florida.*

process to evaluate the detailed acoustical design of a room. A miniature sound source is used to propagate sound into the room. Sparks are often used to create an actual loud impulsive noise. Electronic signals such as maximum length sequences are also propagated through loudspeakers in larger models. The direct, reflected, and reverberant sounds from the source are picked up at selected locations in the room by small microphones. Impulse responses are recorded on computer disks or on a storage oscilloscope. Several commercial systems are available for purchase that include the sound source, an analog-to-digital board to record the signals, and software to filter, calculate the impulse response, and derive acoustical measures.

In a 1:10 scale model, the linear dimensions of the model room are reduced to one-tenth their original size. The materials chosen to construct the model must have absorption coefficients that are similar to those in the full size room at frequencies that are 10 times higher. For example, concrete has an absorption coefficient of approximately 0.01 at 1000 Hz. A model material used to simulate concrete would have to have an absorption coefficient of 0.01 at 10,000 Hz. The frequencies produced by the sound source must be increased by the scale of the model. If sounds from 125 to 4000 Hz are of interest in the full size room, then the frequencies of interest in a 1:10 scale model would be 1250 to 40,000 Hz. As the frequencies are increased, the air in the scale model becomes much more absorbent than the corresponding air in the full size room primarily due to water vapor in the air. Therefore, in small size models, one must either use inert atmospheres in the models of nitrogen, use very dry air (<3% relative humidity), or correct mathematically for air absorption in the computer program. Current research in acoustical modeling has involved developing miniature sound sources, corrections for air absorption, and a 1:10 scale manikin for binaural recording of sounds.

Computer Model Studies

Computer models of room acoustics start with a three-dimensional model of a proposed room. Many available software programs include interfaces so that computed-aided design (CAD) models can be imported into the acoustical program. Sounds are propagated from simulated sources on stage into the room. The impulse response of the room can then be estimated at selected locations in the room. Many of the acoustical measures described in the preceding section can also be calculated at specific locations as well.

The directional and frequency characteristics of sound sources are listed in commercially available computer modeling programs. The sources can be located in the room and sound waves propagated into the room from the source. There are two basic methods used today in computer models to develop the impulse response of the room: a ray-tracing method and an image method. In the ray-tracing method, a selected number of sound rays are propagated from the source into the room. The rays will reflect off surfaces in the room as sound waves do. The amplitude and arrival time of the sound waves will be recorded at selected locations in the model. The room surfaces are assigned absorption coefficients in octave band frequencies. Many current programs also can approximate the diffusion and diffraction of sound that is so important in auditorium acoustics. The approximation of these effects is still under investigation by researchers. The recording of the reflections results in the estimation of the impulse response of the room. Some programs provide abbreviated estimation procedures where the reverberant energy is approximated by an envelope derived from the reverberation time of the room. This means that all of the multiple high-order reflections that occur in the reverberant field are not individually calculated. The image method works backward. The computer locates all of the surfaces from a given receiver location that can project reflections to it from the source. Rays are then traced from the receiver to the surface to the source on stage.

Computer model studies show the distribution of sound pressure levels in rooms, ray diagrams of specific sound paths between the source and receiver, estimate the impulse response at specific locations, and calculate many acoustical design parameters.

Aural Simulation

The acoustical responses recorded in scale models and computer models can be convolved with music or speech to aurally simulate sounds as they would be heard in the completed building while it is being designed. These techniques were described in the preceding section on laboratory sounds. The acoustical qualities of a room, including the early reflections and reverberation, are added to music or speech to enable one to experience the aural qualities of one room while actually being in a different room. Virtual sound fields are being used for demonstration, education, and acoustical research. As a research tool they are being employed to investigate the relationships between quantitative acoustical indices and subjective preferences. Additionally, work is being done to aurally model rooms based on impulse responses that have been generated from computer models. This would allow one to listen to speech or music performed in a room that only exists as a computer or scale model!

The source for the virtual sound field can vary greatly; a human speaker, an individual instrument, or an entire orchestra may be used as the sound source. The source is recorded in an anechoic environment. Ideally, the anechoic environment does not add any sound reflections or reverberation to the source material. This is called a "dry" recording.

Moving reflections directly overhead relative to the direct signal usually results in perceived increases of loudness with no change of direction. Reflections that arrive from the sides usually increase loudness as well; but more importantly, they tend to increase the sense of the apparent aural space. This sense of acoustical spaciousness has been a major area of interest. The subjective sensation has been described as "the difference between feeling 'inside' the music and looking at it as through a window" (Barron, 1981) by the manager of the Concertgebouw Orchestra of Amsterdam.

Expert System for Acoustical Design

An expert system for preliminary acoustical design of performance spaces has been developed utilizing the extensive database of acoustical and architectural measurements accumulated to date (Mahalingham, 1995). This system generates three-dimensional sketches of auditoria based on the relationships found among acoustical qualities, acoustical measures, and the architectural features of rooms found in the research described above. Much research remains to be done before there is enough scientific understanding of room acoustics for this to be a truly artificial intelligence system. The current prototype for the system allows one to input the seating capacity and a variety of acoustical measures as criteria to be met in the room. One then selects a basic shape for the plan. The computer optimizes the acoustic measures with the related architectural features of rooms from the statistical models described above to produce a three-dimensional model of a room that meets the combination of criteria defined. While the concept for this system is intriguing, it is important to note that the optimization process is extremely tentative at this time because of the uncertain relationships among acoustical measures and architectural features of rooms. Additional research in this area is essential for progress to continue.

6.4 OTHER DIRECTIONS IN ARCHITECTURAL ACOUSTICS RESEARCH

New Sound-Diffusing Materials

Work by Manfred Schroeder (1986) developed a quantitative understanding of the diffusion of sound. Schroeder produced equations based on number theory to design a sound-diffusing surface. One can determine the frequencies to be diffused based on the width and depth of a series of "wells" on the panel. Figure 6.27 shows some sound-diffusing panels. The panels with the parallel wells will diffuse sound in the front-to-back plane. The panel comprised of a series of rectangular wells will diffuse sound in all directions.

These quadratic residue diffusers, or QRDs, have been manufactured by several producers and are commercially available in a variety of configurations as prefabricated units. They are becoming widely used in stage enclosures to provide sound reflections so musicians can hear each other, in music rehearsal rooms that typically lack sound diffusion, in control rooms, concert halls, and studios among others. The certainty of the sound diffusion from these panels and the ability to predict the sound diffusion in advance are major advantages of panels designed using these methods compared to zig zag surfaces or other site-fabricated panels.

Active Noise Control Systems

Active noise control systems consist of a computer-controlled system that will propagate sound waves that are 180° out of phase with a disturbing sound. The disturbing sound

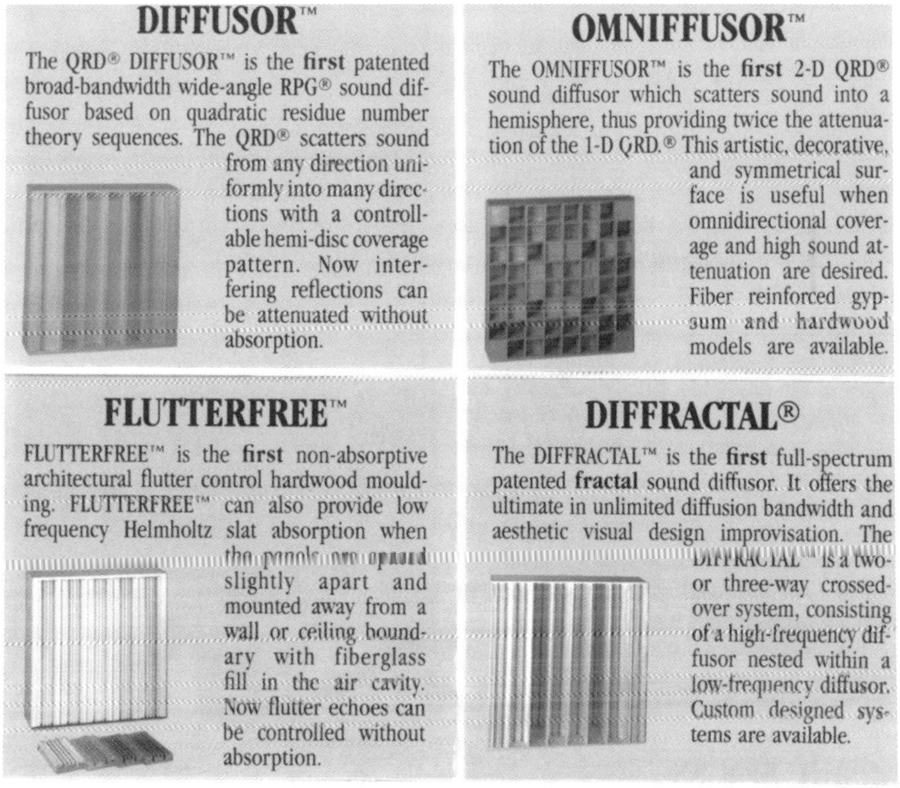

Figure 6.27 QRD sound diffusing panels. (Courtesy RPG Corporation.)

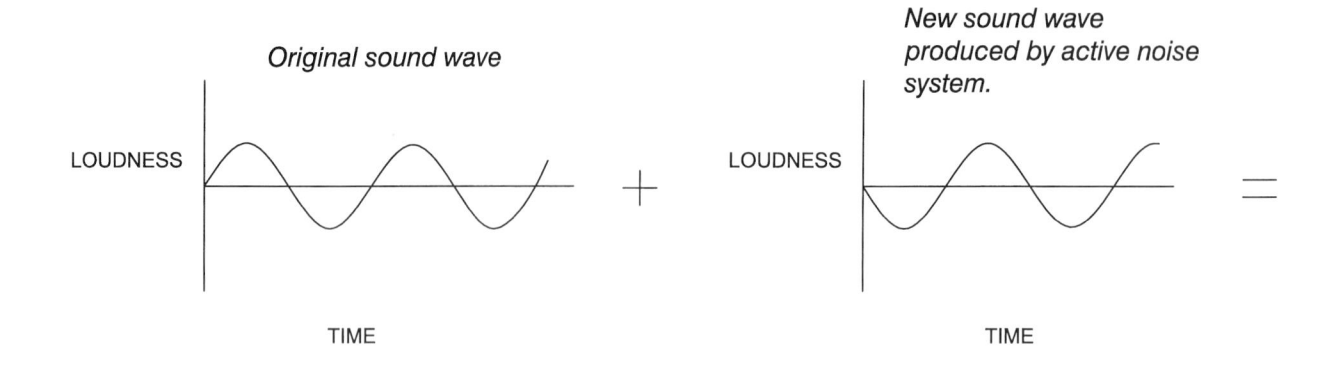

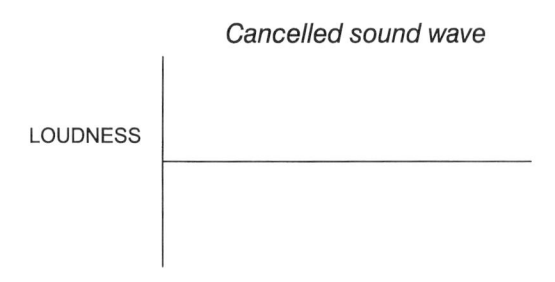

Figure 6.28 Schematic diagram illustrating destructive interference of sound waves. This is the principle used as the basis for designing active noise control system.

is effectively canceled by destructive interference with the new sound that is produced by the computer. In other words, the computer produces a sound that is exactly the opposite as the disturbing sound. When the two sounds interact, they sum to zero amplitude. This is illustrated in Figure 6.28.

The primary application for these systems in buildings to date has been in air-conditioning systems. When a fan produces high levels of low-frequency noise that are difficult to control with "passive" silencers, active noise control systems can be used effectively. A diagram of a typical active noise control system is shown in Figure 6.29. A microphone located in the duct is connected to a microprocessor. The processor

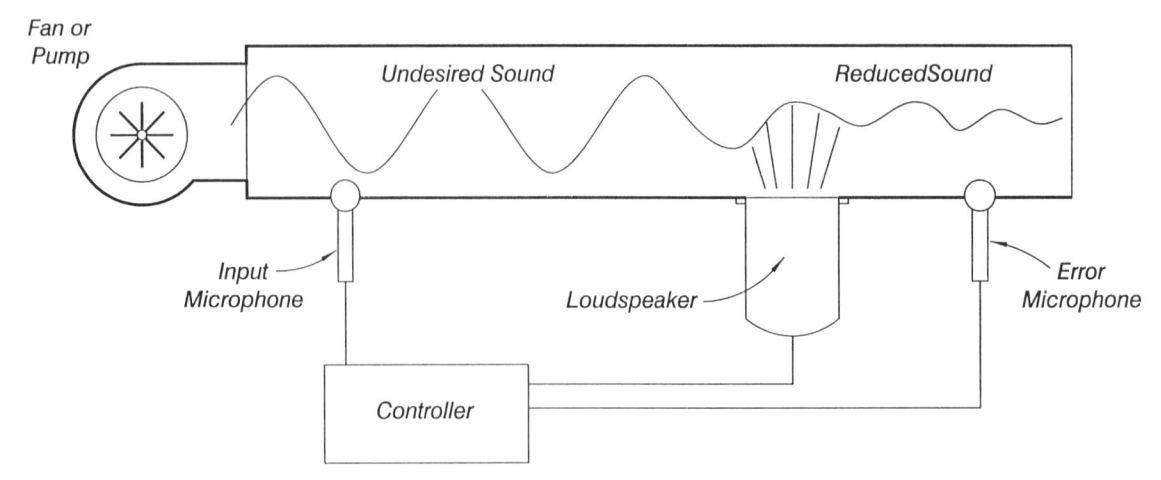

Figure 6.29 Schematic diagram of an active noise control system used in air-conditioning ducts in buildings.

calculates the spectrum of the disturbing sound waves in the duct and generates a set of waves that are 180° out of phase with the fan noise. This sound is amplified and propagated through the duct from a loudspeaker installed in the duct. The new sound from the loudspeaker combines with the fan noise to cancel or reduce it through destructive interference. An error microphone is often located downstream from the loudspeaker to provide feedback to adaptively change the signal to more effectively cancel the noise. These systems are becoming more widely used as the cost for the system decreases and as the number of successful installations increase.

Sound Isolation Testing

Tests for sound transmission loss and sound transmission class (STC) ratings have been conducted for many years. In recent years, however, a number of studies by national laboratories have significantly expanded the database available to architects and consultants on the sound transmission loss of wall and floor/ceiling assemblies. Warnock (1992) has measured many combinations of lightweight concrete masonry walls with various amounts of gypsum board on furring, plaster finishes, and the like. U.S. Gypsum has provided extensive test data on many wall assemblies using gypsum board.

Nightingale (1993) has measured the flanking sound transmitted through typical party wall assemblies due to construction details such as not cutting the flooring under a two-leaf partition using the sound intensity method. He has quantitatively found the amount of flanking noise in several directions from one unit to another under various conditions.

Plumbing System Noise Control

Plumbing noise has been a difficult problem in buildings for a long time. Plumbing system noise control arises due to vibration of piping from turbulent water flow. The vibrations are transmitted to the framing, flooring, and other building systems with which the piping comes in direct contact. Simple principles such as resilient mounting of pipes from the structure can reduce this noise by 10 to 20 dB. While these techniques are well-known and inexpensive, they are rarely used. It has only been in recent years that research conducted at the National Research Council of Canada has quantitatively examined plumbing noise in buildings. The attenuation of various pipe isolation strategies are summarized in Table 6.4.

Additional relief can be obtained by careful planning. In multifamily housing, align plumbing walls in units back to back. Do not locate plumbing lines from one unit in the ceilings of another. Do not locate the bathroom for the teenage children

TABLE 6.4 Attenuation of Plumbing Noise Achieved by Various Pipe Mounting Methods (MJM, 1992)

CONNECTOR	SOUND REDUCTION (dBA)
Rigid connection of pipe clamp to structure	0
Cork between pipe and oversided clamp or hanger	5–8
Closed cell elastomer sleeve (½" thick Armaflex) × 3" long between pipe and oversized clamp or hanger	15–19
Resilient pipe hanger	20

Source: MJM Associates (1992).

over the master bedroom in a single-family home. Avoid locating plumbing chases in walls adjacent to bedrooms and the like.

Impact Noise

Impact noise is often a problem in multifamily housing projects. Wood frame floors will transmit low-frequency sounds from footfalls that are below the range of frequencies considered in impact insulation class (IIC) ratings of floor/ceiling constructions. Continuous concrete floor slabs with expensive tile or wood floors will also transmit substantial impact sounds from footsteps in units above unless acoustical "underlayments," resilient ceilings, and floating floors are used.

One of the difficulties in selecting appropriate IIC ratings for construction assemblies is an apparent nonlinear response of people to transmitted impact sound levels. For example, a concrete floor slab with tile flooring has an IIC rating of 25 to 30. Most people would find sounds transmitted through this assembly objectionable. Carpet on the concrete slab would have an IIC rating of 70 to 80. Most people would find sounds transmitted through this assembly acceptable. Many common acoustical underlayment products for use with tile floors have IIC ratings of 45 to 55. Many building codes have IIC requirements in this vicinity as well. There are extremely mixed responses of people to transmitted sounds through construction assemblies with IIC ratings in this range.

Many of the disturbing impact sounds tend to be in the lower end of the frequency spectrum. Often times the disturbing sounds are below the 100-Hz minimum frequency considered in the IIC rating. The relatively continuous impacts of the standard tapping machine are thought not to represent the lower frequencies from a person's footfall either. A tire drop test is under development to better simulate the low-frequency sounds associated with footsteps particularly in wood frame floor systems. A notable piece of recent research by Blazier and Dupree (1994) found a quantitative relationship to estimate the degradation of IIC rating due to flanking at the perimeter of the floor. This situation occurs in many buildings where proper details for isolation of an acoustic underlayment material are not installed. Several inches of flanking path along the perimeter of a floor can significantly reduce the actual IIC rating of a floor system. Additional research is needed to develop qualitative rating scales for impact noises.

Noise Control in Mechanical Systems and Health and Energy Issues

Glass-fiber linings were a major noise control element in duct systems. "Lined" ducts significantly reduce sound propagating through ducts. They take up little or no extra room and are very inexpensive. Recently, concerns have been raised regarding indoor air quality in air-conditioning systems with internally lined air ducts. Moisture can condense on the lining from cool air moving through ducts. This dark, damp environment is thought to form a breeding ground for mold, spores, bacteria, and so forth that can be "blown" around the building causing sick building syndrome or building-related illness. When lined ducts are used, it is essential that the mechanical engineer provide adequate air movement and control the moisture content of air so this does not occur. It is worth noting that mold and bacteria can form on metal surfaces also if air moisture content is not properly limited.

There have been several studies concerning health risks associated with people breathing small glass fibers that become entrained in the airstream. These studies have classified glass fiber as a possible carcinogen. However, glass-fiber producers state that the only studies that showed cancer attributed to glass fiber were implantation studies where the fiber was implanted in an opening in the abdomen of a laboratory test animal. While there are many conflicting conclusions that have been reached from laboratory studies on this subject, this is an area where building designers and owners are moving very cautiously. Research will continue in this area for some time in the future.

Silencers are inserted in ducts primarily to control the transmission of fan noise to the duct system. These are necessary especially for the large ducts near air-handling units since a thin internal fiberglass duct lining would not be adequate to attenuate significant amounts of noise. It is often difficult to find adequate room to install silencers in the short runs of ducts that frequently occur in and near mechanical equipment rooms. Silencers also impose pressure drops in systems that must be overcome by additional fan power.

Active noise control systems discussed earlier are an alternative to internal fiberglass lining and silencers. There is a need to conduct research on the relationship between duct size, insulation, fan power, and acoustical consequences to determine the optimum energy conservation that can be realized when these factors are appropriately related to one another.

6.5 CONCLUSIONS

The rapid acceleration of our technological age has given us the ability to make increasingly detailed analyses of our architectural acoustic environment. In turn it is providing the tools for modeling building designs in order to predict with greater accuracy than in the recent past their acoustic performance in advance of any construction. There is every prospect that these tools will continue to be improved and, thereby, permit better and more economic processes for use in architectural acoustics design. The architectural profession should keep abreast of this increasing ability of the architectural acoustics profession to provide significant service in the process of building design and use it to contribute to an excellent creative effort.

FURTHER READINGS

Barron, M., *Auditorium Acoustics and Architectural Design*, E. & FN Spon, London: 1993.
Beranek, L. L. *Concert and Opera Halls: How They Sound*, Acoustical Society of America, Woodbury, NY, 1996.
Cremer, L., and Muller, H. A., *Principles and Applications of Room Acoustics*, Vols. 1 and 2, English translation with additions by T. J. Schultz, Applied Science, New York, 1982.

These three books carry the discussions of new acoustical measures and their applications into much greater detail than the current text.

REFERENCES

Ando, Y., *Concert Hall Acoustics*, Springer, Berlin: 1985.
Artec, Unpublished promotional material for firm. Artec, Inc., New York: 1990.
Banham, R., *The Architecture of the Well-Tempered Environment*, 2nd ed., University of Chicago, Chicago, 1984.
Barron, M., *Auditorium Acoustics and Architectural Design*, E & FN Spon, London, 1993.

Barron, M., "Subjective Study of British Symphony Concert Halls," *Acustica,* **66**(1), 1–14 (1988).
Barron, M., and Marshall, A. H., "Spatial Impression due to Early Lateral Reflections in Concert Halls: The Derivation of a Physical Measure," *J. Sound Vibration,* **77**(2), 211–232 (1981).
Bech, S., "Timbral Aspects of Reproduced Sound in Small Rooms," *JASA,* **97**(3), 1717–1726 (1995).
Beranek, L. L., *Concert and Opera Halls: How They Sound,* Acoustical Society of America, Woodbury, NY, 1996. Originally published as *Music, Acoustics, and Architecture,* Wiley, New York, 1962.
Beranek, L. L., "Concert Hall Acoustics—1992," *JASA,* **92**(1), 1–39, (1992).
Beranek, L. L., Ed. *Noise and Vibration Control,* McGraw-Hill, New York, 1971. Now published by INCE, Poughkeepsie, New York.
Beranek, L. L., *Music, Acoustics, and Architecture,* Wiley, New York, 1962.
Blauert, J., and Lindemann, W., "Auditory Spaciousness: Some Further Psychoacoustic Analyses," *JASA,* **80,** 533–542 (1986).
Blazier, W. E., and DuPree, R. B., "Investigation of Low Frequency Footfall Noise in Wood-frame, Multifamily Building Construction." *JASA,* **96**(3), 1521–1532 (1994).
Bradley, J. S., Gade, A. C., and Siebein, G. W., "Comparison of Auditorium Acoustics Measurements as a Function of Location in Halls," presented on the 125th Meeting of The Journal of Acoustical Society of America, Ottawa, May 1993.
Bradley, J. S., "Speech Intelligibility Studies in Classrooms," *JASA* **80**(3), 846–854 (1986).
Brebeck, D., "Sound and Ultrasound Absorption of Materials in Theory and Practice; Especially in View of Building Acoustic Scale Models in Measure 1:10," Doctoral Thesis, University of Munich, February, 1969.
Carvalho, A. P. O., "Objective Acoustical Analysis of Room Acoustic Measurements in Portuguese Catholic Churches," *Proceedings of 1994 National Conference on Noise Control Engineering, Fort Lauderdale, FL,* Institute of Noise Control Engineers, Poughkeepsie, NY: 1994.
Cervone, Richard P., "Subjective and Objective Methods for Evaluating the Acoustical Quality of Buildings for Music." Master's Thesis, University of Florida, 1990.
Cervone, R. P., Chiang, W., Siebein, G. W., Doddington, H. W., and Schwab, W., "The Subjective and Objective Evaluation of Three Rooms for Music Listening," *JASA,* Supplement 1, **88,** 120th JASA meeting, San Diego: (1990), session 6aa.
Chiang, W., "Effects of Architectural Parameters on Six Acoustical Measures in Auditoria," Ph.D. Dissertation, University of Florida, Gainesville, FL, 1994.
Chiang, W. H., "Comparisons of Acoustical Measures in Scale Models, Computer Models and Full-Size Rooms," Master's Thesis, University of Florida, 1992.
Cowan, H. J., An *Historical Outline of Architectural Science,* Elsevier, Amsterdam, 1966.
Cremer, L., and Muller, H. A., *Principles and Applications of Room Acoustics,* Vols. 1 and 2, English translation with additions by T. J. Schultz, Applied Science, New York, 1982.
Doelle, L., *Environmental Acoustics,* McGraw-Hill, New York, 1972.
Egan, M. D., *Architectural Acoustics,* McGraw-Hill, New York, 1987.
Eplee, D. F., "An Acoustical Comparison of the Stage Environments of the Vienna Grosser Musikvereinsaal and a Scale Model," Master's Thesis, University of Florida, 1986.
Fitch, J. M., *American Building: The Environmental Forces That Shaped It,* Schocken, New York, 1982.
Gade, A.C., "Acoustical Survey of Eleven European Concert Halls," Report No. 44, The Acoustics Laboratory, Technical University of Denmark, Copenhagen, 1989.
Gale, G., *A Theory of Science,* McGraw-Hill, New York, 1979.
Gold, M. A., "Subjective Evaluation of Spatial Impression: The Importance of Lateralization," *Proceedings of 1994 National Conference on Noise Control Engineering, Fort Lauderdale, FL,* Institute of Noise Control Engineers Poughkeepsie, NY: 1994.
Guericio, G. D., *The New Science of Sound.* World Monitor: 1989.
Haas, H., "The Influence of a Single Echo on the Audible Speech," *J. Audio Engr. Soc.* **20**(2), 146–159 (1972).
Harris, C. M., Ed., *Handbook of Acoustical Measurements and Noise Control,* McGraw-Hill, New York, 1991.

Hawkes, R. J., and Douglas, H., "Subjective Acoustic Experience in Concert Auditoria," *Acustica*, **24**(5), 243–247 (1971).

Heidigger, M., "The Question Concerning Technology," In Krell, D., *Basic Writings*, Harper & Row, New York, 1977.

Hook, J. L., "Acoustical Variation in the Foellinger Great Hall, Krannert Center for the Performing Arts," Master's Thesis, University of Illinois, 1989.

Houtgast, T., Steneken, H. J. M., and Plomp, R., "Predicting Speech Intelligibility in Rooms from the Modulation Transfer Function," *Acustica*, **11**, 195–200 (1980).

Hunt, Frederick V., *Origins in Acoustics*. Acoustical Society of America, Woodbury, New York: 1992.

Hyde, Jerald R., Segerstrom Hall: Acoustical Perspective. Unpublished promotional material by Segerstrom Hall, Costa Mesa, California: 1986.

Johnson, R., "Reflections on the Acoustical Design of Halls for Music Performance," V. O. Knudsen Lecture, presented on the 120th Meeting of the Acoustical Society of America, San Diego, November, 1990.

Jordan V. L., *Acoustical Design of Concert Halls and Theaters*, Applied Science, London, 1980.

Keet, W. de V., Ph.D Dissertation, Capetown University, 1969.

Kinzey, B. Y., Jr., and Howard S., *Environmental Technologies in Architecture*, Prentice-Hall, Englewood Cliffs, NJ, 1963.

Kuttruff, H., *Room Acoustics*, 3rd ed., Elsevier Applied Science, London, 1991.

Lochner, J. P. A., and Burger, J. F., *Acustica*, **11**, 195 (1961).

Lochner, J. P. A., and Burger, J. F., "The Influence of Reflections on Auditorium Acoustics." *J. Sound Vibration*, **1**, 426–454 (1964).

Lyon, R., and Cann, R. G., *Acoustical Modeling*, Grozier Technical Systems, Brookline, MA, 1975.

Mahalingham, Ganapathy. "Acoustic Sculpting: The Application of Object-Oriented Computing in the Development of Design Systems for Auditoria." Ph.D. Dissertation, University of Florida, 1995.

Marshall, A. H., "A Note on the Importance of Room Cross-Section in Concert Halls," *J. Sound Vibration*, **5**, 100–112 (1967).

McCleary, P., "An Interpretation of Technology." *JAE*, **37**(2), 2–4 (1983).

MJM Acoustical Consultants, *Research Report on Plumbing Noise in Multi-dwelling Buildings*. Canada Mortgage and Housing Corporation, Ottawa: 1992.

Nightingale, T. R. T., "Flanking Transmission in Wood-frame Constructions Caused by Simple Construction Faults," presented at the Acoustical Society of America meeting, May, 1993 in Ottawa, Canada.

Packard, R. T., Ed., *Architectural Graphic Standards*, Wiley, New York, 1990.

Polack, J. D., Marshall, A. H., and Dodd, G., "Digital Evaluation of Small Models: The MIDAS Package," *JASA*, **85**(1), 185–193 (1989).

Reichardt, W., Abdel Alim and Schmidt, W. Clarity., *Acustica*, **32**, 126 (1975).

Reynolds, G., "Sound Absorption and Sound Absorption Coefficients for Acoustical Modeling Materials," Master's Thesis, University of Florida, 1982.

Rozear, C. E., "An Acoustical Comparison of a 1:10 Scale Model to Its Prototype Room," Master's Thesis, University of Florida, 1985.

Sabine, W. C., *Collected Papers on Acoustics*, Harvard University Press, Cambridge, 1927; reprinted by Dover Publications, New York, 1964.

Schroeder, M. R., *Number Theory in Science and Communication with Applications in Cryptography, Physics, Digital Information, Computing and Self-Similarity*, 2nd ed., Springer, Berlin, 1986.

Schroeder, M. R., "New Method of Measuring Reverberation Time," *JASA*, **37**(4), 409–412 (1965).

Schroeder, M. R., Gottlob, D., and Siebrasse, K. F., "Comparative Study of European Concert Halls: Correlation of Subjective Preference with Geometric and Acoustic Parameters." *JASA*, **56**(4), 1195–1201 (1974).

Siebein, G. W., "A Method to Evaluate the Acoustical Consequences of Conceptual Decisions in the Studio Design Process," *Proceedings of the 1986 ACSA Technology Conference*, ACSA, Washington, DC, 1988.

Siebein, G. W. *Project Design Phase Analysis Techniques for Predicting the Acoustical Qualities of Buildings,* University of Florida, Gainsville, FL, 1986.

Siebein, G. W., and Gold, M. A., The Concert Hall of the 21st Century: Historic Precedent and Virtual Reality. Architecture: Material and Imagined, *Proceedings of the 85th ACSA Annual Meeting.* Association of Collegiate Schools of Architecture, Washington, D.C.: 1997, pp. 52–61.

Siebein, G. W., et al., *Listening to Buildings,* College of Architecture, University of Florida Research Report, 1992.

Souldore, G. A., Bradley, J., and Stammen, D. R., "Spaciousness Judgements of Binaurally Reproduced Sound Fields," presented on the 125th Meeting of the *Journal of Acoustical Society of America,* Ottawa, May 1993.

Thiele, R., and Meyer, E., "Acoustical Investigations in Numerous Concert Halls and Broadcast Studios Using New Measurement Techniques." English summary translated by T. D. Northwood in *Benchmark Papers in Acoustics,* Volume 10: Architectural Acoustics. Thomas D. Northwood, editor, Dowden, Hutchinson and Ross, Stroudsburg, Pennsylvania: 1977, pp. 181–208.

Veneklasen, P. S., "Design Considerations from the Viewpoint of the Professional Consultant," in *Auditorium Acoustics,* Robin Mackenzie, Ed., Elsevier, London, 1979.

Vitruvius, *The Ten Books on Architecture,* translated by Morris Hicky Morgan, Dover, New York, 1960.

Warnock, A. C. C., "Sound Transmission Through Two Kinds of Porous Concrete Blocks with Attached Drywall," *JASA,* **92**(3), 1452–1460, (1992).

CASE STUDIES OF RECENT HALLS FOR THE PERFORMING ARTS AND ACOUSTICAL MODEL STUDIES

Introduction

Recent findings in acoustical research have lead to theories as to which architectural features contribute to high-quality acoustics. These issues have significant implications in auditoria that influence the fundamental architectural characteristics of the room such as the volume, height, width, depth, stage type, seating configuration, and appropriate materials. Unfortunately, the incorporation of these elements is often in conflict with many social and economic factors. To be competitive in attracting audiences, larger numbers of people must be accommodated in auditoria to maintain affordability. Wide ranges of performance types including drama, opera, dance, and orchestral concerts usually occur in the same auditorium to economically support the facility. This has led to the uniquely American multipurpose hall. The push for more marketable auditoria has led to the need to be able to vary the acoustical qualities of these spaces relatively quickly depending on the type of performance. The challenge posed to the acoustical consultant is to creatively merge both recent and traditional acoustic theories within the context of the practical constraints of current building practices.

Three case studies of auditoria are presented. In addition, four acoustical model case studies are also described. Each of these halls incorporates strategies based on acoustical theories that are only recently beginning to be tested in the actual buildings. The emphasis here is how these theoretical strategies were actually implemented in the final design. Each of the rooms incorporates a different approach to achieve the final solution, yet they are each representative of the designers awareness of the latest theories developed through research in acoustics transformed into architectural space.

Case Study

SEGERSTROM HALL, ORANGE COUNTY PERFORMING ARTS CENTER

Completed: 1983
Architect: Charles Lawrence
Acoustical consultants: Jerald R. Hyde, Marshall Day Acoustics and Paoletti Associates, Inc.
Room type: Multipurpose
Seat count: 3000
Stage: Proscenium
Maximum room width: 176 ft (52 m)
Room volume (total): 1,082,810 ft^3 (27,800 m^3)
Variable features: Operable curtains at upper volume

Initial Design

Segerstrom Hall was proposed as a large, multipurpose auditorium with the ability to accommodate orchestral, solo, opera, drama, ballet, and musical theater performances. Excellent sight lines and close proximity to the performing area were also high priorities of the client. In order to accommodate the needs of the client, a fan-shaped auditorium plan was used. This project presented a significant challenge to the acoustic design team to achieve a level of acoustic quality in the room that would compare with that of smaller classical rectangular-shaped halls while incorporating variable acoustic features and a much larger audience area.

The design team was challenged to test state-of-the-art questions regarding acoustical research and design. Is it possible to build large multipurpose halls with satisfactory acoustical qualities? Can the acoustical qualities of a room be tested and previewed in advance of construction. "Segerstrom Hall provides a new set of answers to these questions. It demonstrates the reconciliation of research and Its built realization, acoustics and architecture, theater and symphony, art and science, and the process which brought it into being." (Hyde, 1986). The design was based on research by several of the acoustical consultants that found that lateral reflections (reflections that arrive from the sides of listeners' heads) that arrive shortly after the direct sound contribute to sensations of acoustic envelopment, intimacy, and spaciousness. The room evolved as an exploration of the lateral reflection concept. It also represented a radical departure from conventional concert hall design in both architectural and acoustical terms. It is a large, fan-shaped room with an asymmetrical seating plan and side walls in the middle of the large main floor area. It is difficult to tell where the ceiling starts and where the walls begin. This unique and striking solution was achieved through the open-minded collaboration of the acoustical and architectural design teams for the project.

Room Shape

A fan-shaped plan was used to achieve the desired seating capacity. This resulted in a room with a width of 160 ft at the rear of the hall. This width is far greater than most of the world's great concert halls. The wide configuration puts much of the audience area far from the side walls of the room. Lateral sound reflections, which were considered a key acoustic element by the consultants, would be lacking in the many areas away from the side walls in the fan-shaped plan. By subdividing the audience area and vertically staggering the smaller seating areas, side walls were effectively created within the overall seating area. The subdivision of the seating area also helped to maintain a sense of intimacy and provided good sight lines in the room. These side-wall surfaces in the middle of the room provided the lateral reflections that are typical in narrower concert halls. Second, large sound reflecting panels that 'float' within volume of the room were built. This strategy uses large reflecting panels with varying degrees of sound-diffusing surfaces to direct and distribute sound reflections to specific areas in the room during the critical early time periods. These panels are floated and carefully positioned within the large volume of the hall to achieve three acoustic goals. Floating the panels away from the structural envelope allows them to be closer to the audience. This provides sound reflections that arrive at listeners shortly after the direct sound, improving both loudness and clarity. The panels near the sides of the auditorium are angled to direct sound toward the center areas of the room, providing lateral sound reflections to the areas away from the side walls. This improves the sense of spatial impression in the center seating areas. Gaps between the panels are specifically designed to allow some of the sound energy to pass by the panels and travel into the ceiling space above the panels. This added room volume allows for the longer reverberation times (approximately 2.3 sec) needed for orchestral performances. There is a delicate balance between reverberance and clarity with this design approach. The floating panels enhance clarity by providing early sound reflections to listeners while the large room volume above and beyond the panels provides the volume needed to achieve appropriate reverberation times. The result of this approach provides both strong early reflections for clarity and a large, coupled room volume for reverberance to occur simultaneously. A similar result was achieved with entirely different architectural solutions in the two case studies that follow!

Many of the strategies above were incorporated in two previous halls by the consultants in New Zealand: Christchurch and the Michael Fowler Center. In developing these designs as well in Segerstrom Hall, the acoustical consultants constructed detailed acoustical scale models of their proposed designs. A 1:50 scale model was built early in the process to investigate the basic design approach and the sequence of early reflections in the room. Through studies of the impulse responses taken in the scale models, specific design decisions could be tested, evaluated and modified. The models were used to determine how the sound energy was distributed in the proposed design and to design the arrival time and amplitude of the sound reflection sequence arriving at each listener location. The acoustical consultants could make modifications relatively easily in the models and then retest the revised scheme. Through this process it was determined that 3000 seats would be the maximum capacity without hindering the acoustic quality and 200 seats were dropped from the program. A much larger model was built later in the process at a scale of 1:10. More detailed acoustical tests were conducted in the 1:10 model to confirm the results of the original studies. Music and speech that were recorded in an anechoic chamber were also played into the model at 10 times the normal speed to acquire the acoustical qualities of the room. Miniature microphones recorded the music and speech in the model. It was played back at one tenth the original speed so the designers could listen to simulations of how actual music would sound in the completed room.

Special Features

The sound-reflecting panels used in the room have very specific acoustical properties. Results of research in perceived spatial impression indicates that lateral reflections that are too strong will cause an image shift or confuse listeners as to the location of the sound source. For this reason, specific sound-diffusing surface characteristics were required for the large reflective panels. Research in this area has produced quadratic residue diffusers, which are designed to control the directional scattering of reflected sound. QRDs located on the surface of the floating panels allow designers to direct sound energy with controlled coverage to specific areas of the auditorium. This strategy was also tested and refined in the scale model studies of Segerstrom Hall.

Variable Acoustics

Segerstrom Hall was proposed as a multipurpose auditorium, therefore, a means of varying the reverberation was essential if high-quality acoustics was to be achieved for more than just one of the various performance types that were desired in the auditorium (see Figures 6.30, 6.31, and 6.32). Curtains located in the room above the suspended ceiling panels may be raised or lowered to control the duration of reverberation in the auditorium without affecting the sequence of early reflections. The curtains are used to close off upper regions of the space, subsequently reducing the overall volume. The curtains are highly sound-absorbent. By manipulating the room volume and amount of sound-absorbing materials the reverberation time may be varied from 1.7 to 2.0 sec.

CASE STUDY

McDERMOTT CONCERT HALL, MORTON H. MEYERSON SYMPHONY CENTER DALLAS, TEXAS

Completed: 1989
Architect: I. M. Pei
Acoustical consultant: Artec, Inc., Russell Johnson, Principal
Room type: Concert hall
Seat count: 2065
Stage: Open
Maximum room width: 84 ft (26 m)

(Continues on p. 286)

Figure 6.30 Interior view of Segerstrom Hall. *(Courtesy Jerald R. Hyde.)*

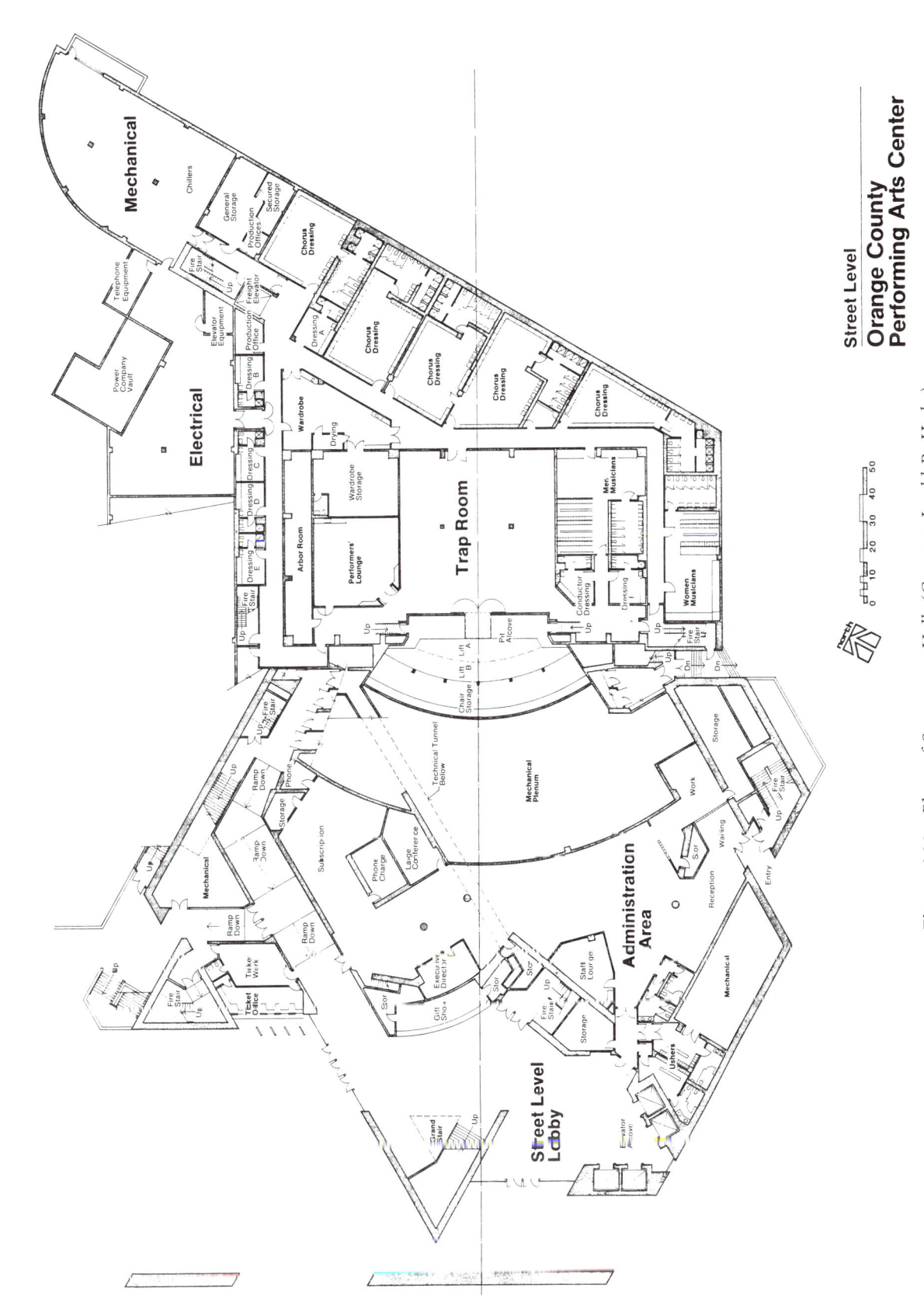

Figure 6.31 Plans of Segerstrom Hall. (Courtesy Jerald R. Hyde.)

279

Main Level
**Orange County
Performing Arts Center**

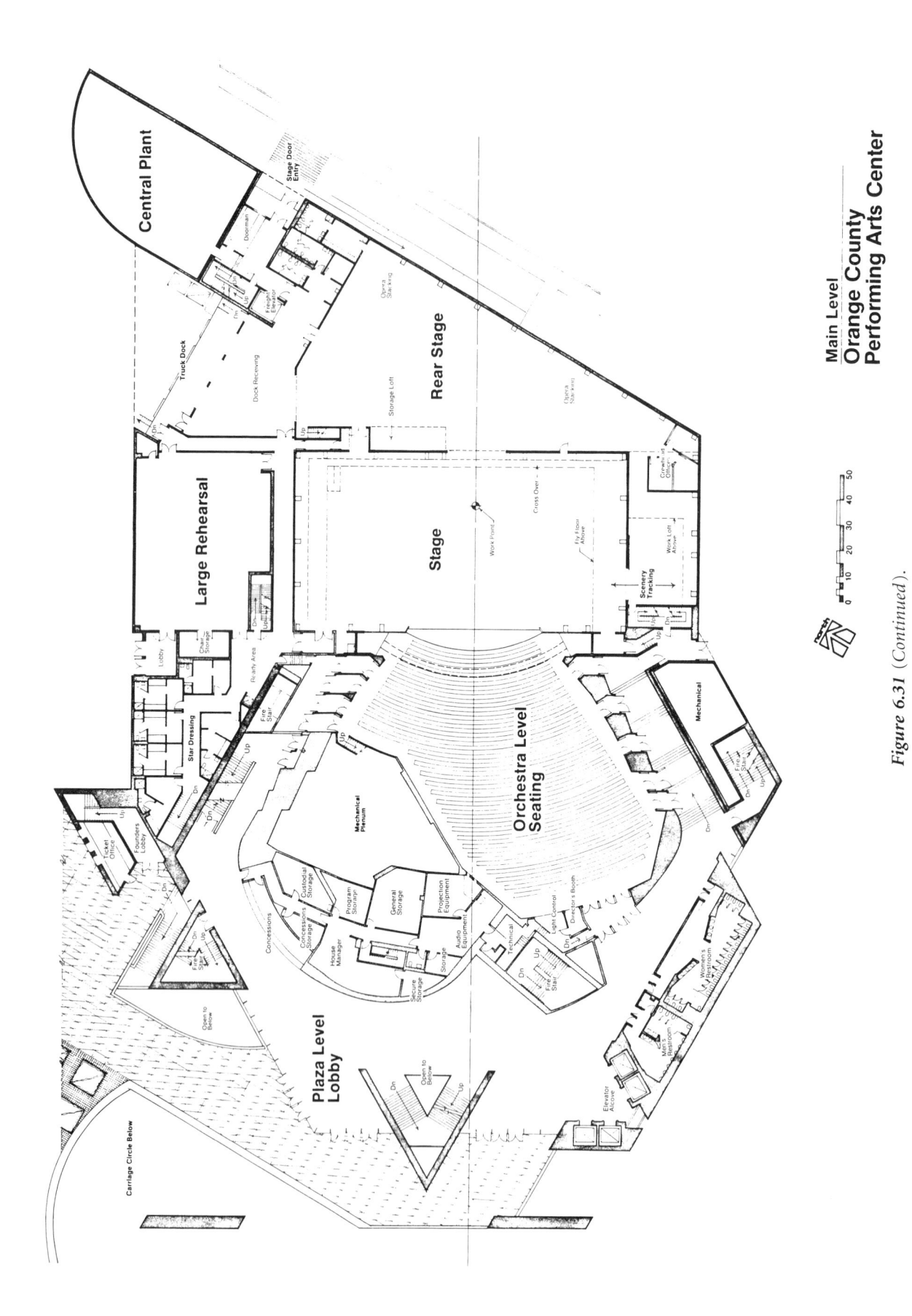

Figure 6.31 (*Continued*).

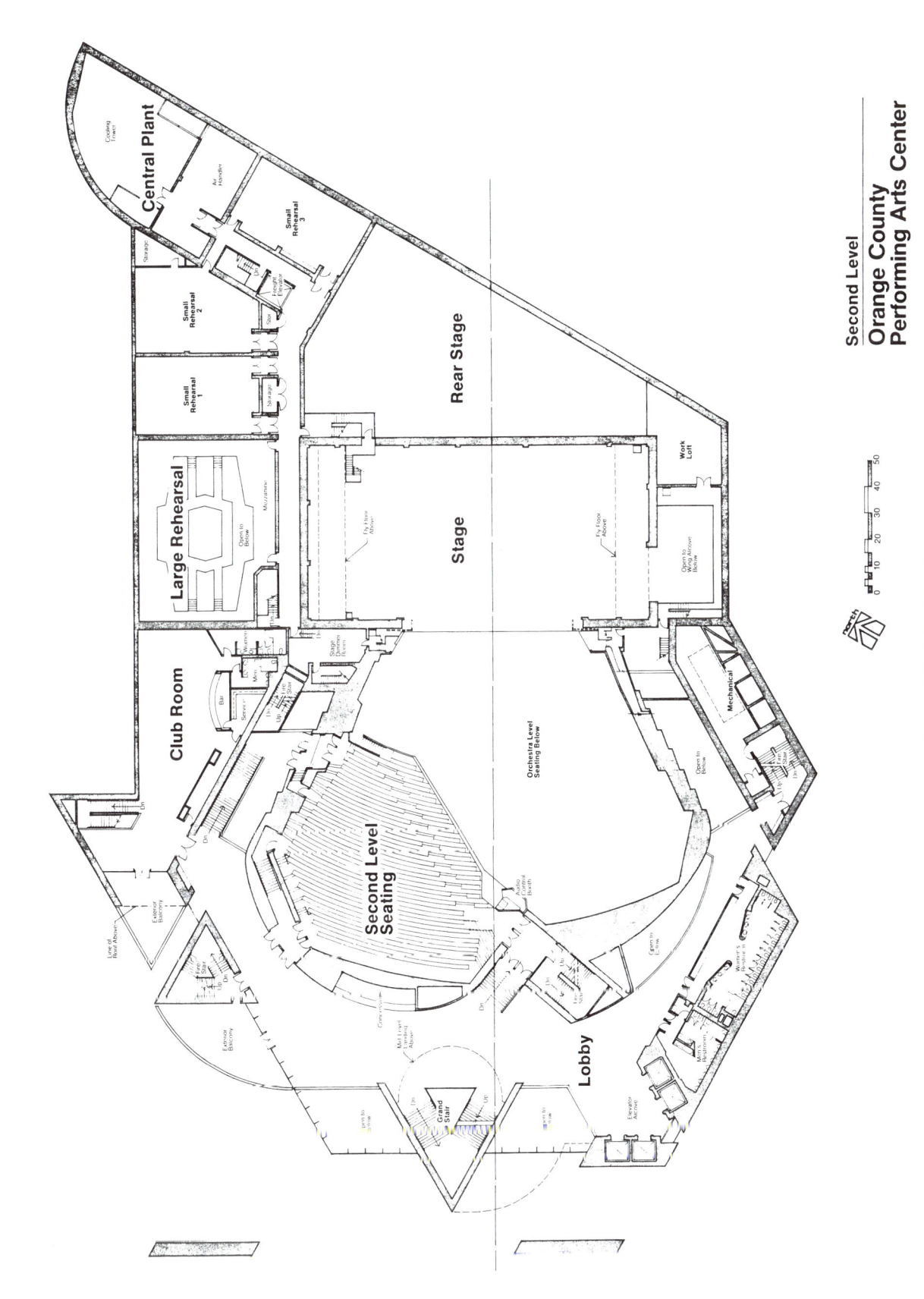

Figure 6.31 (*Continued*).

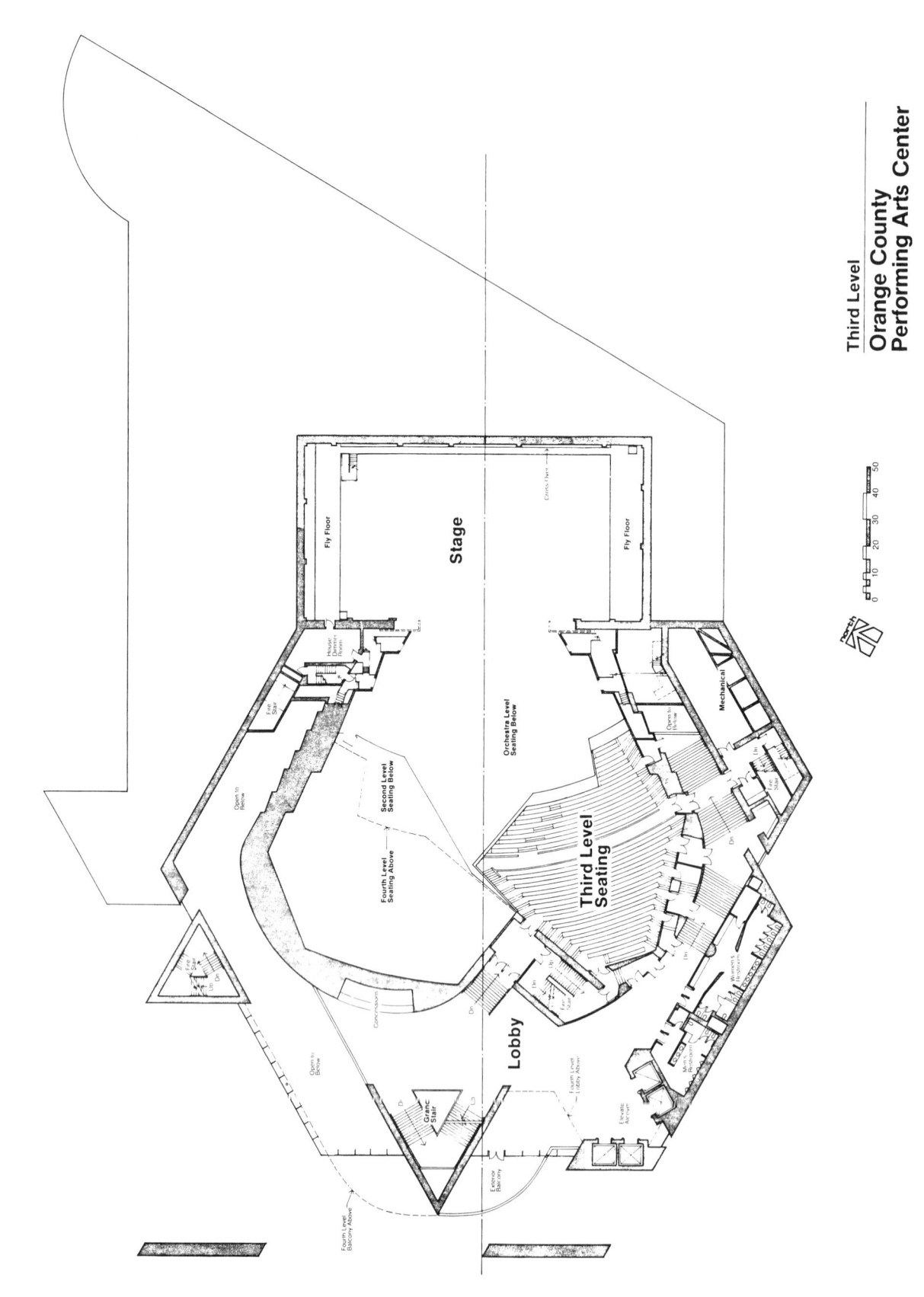

Figure 6.31 (*Continued*).

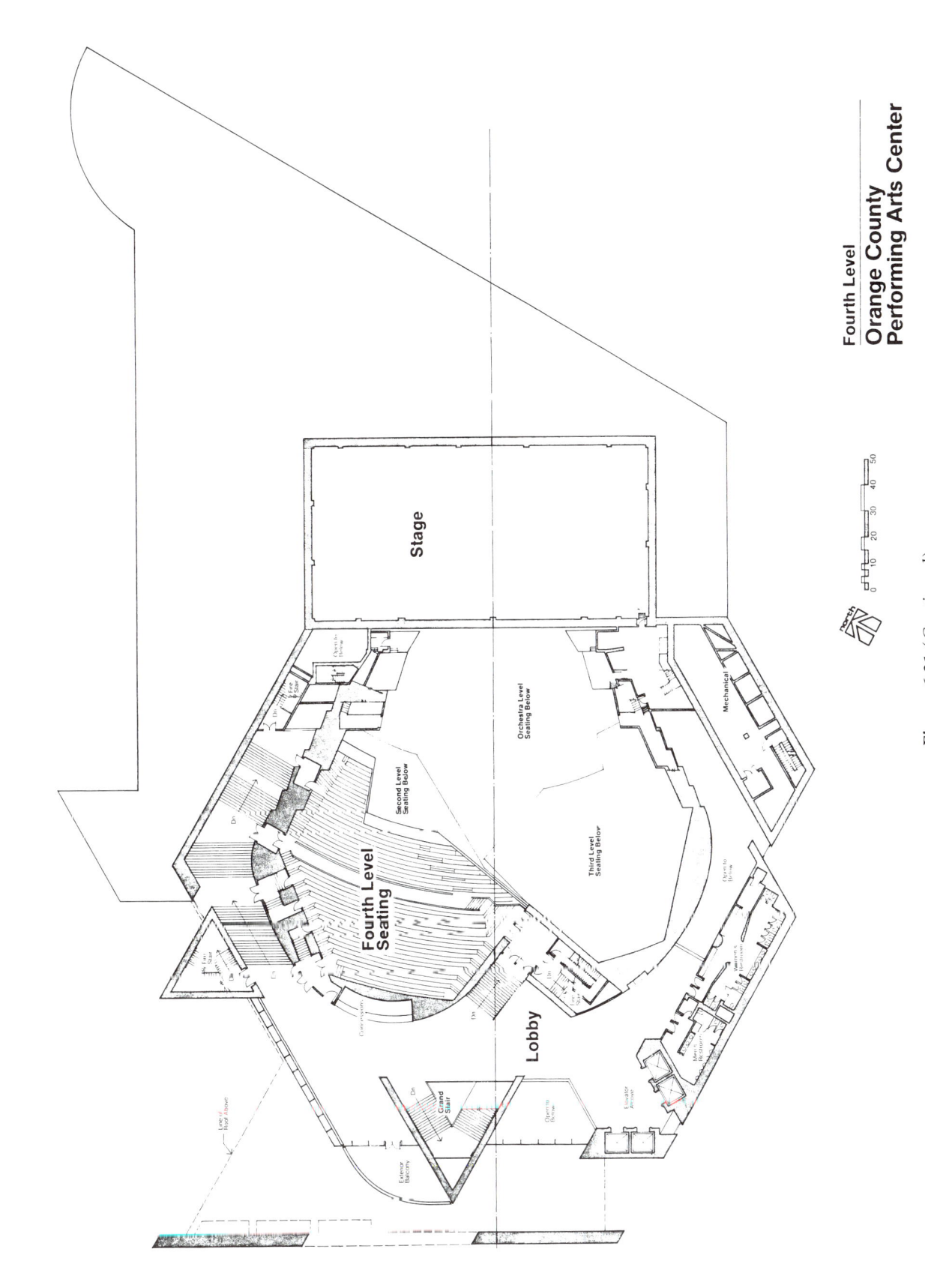

Figure 6.31 (Continued).

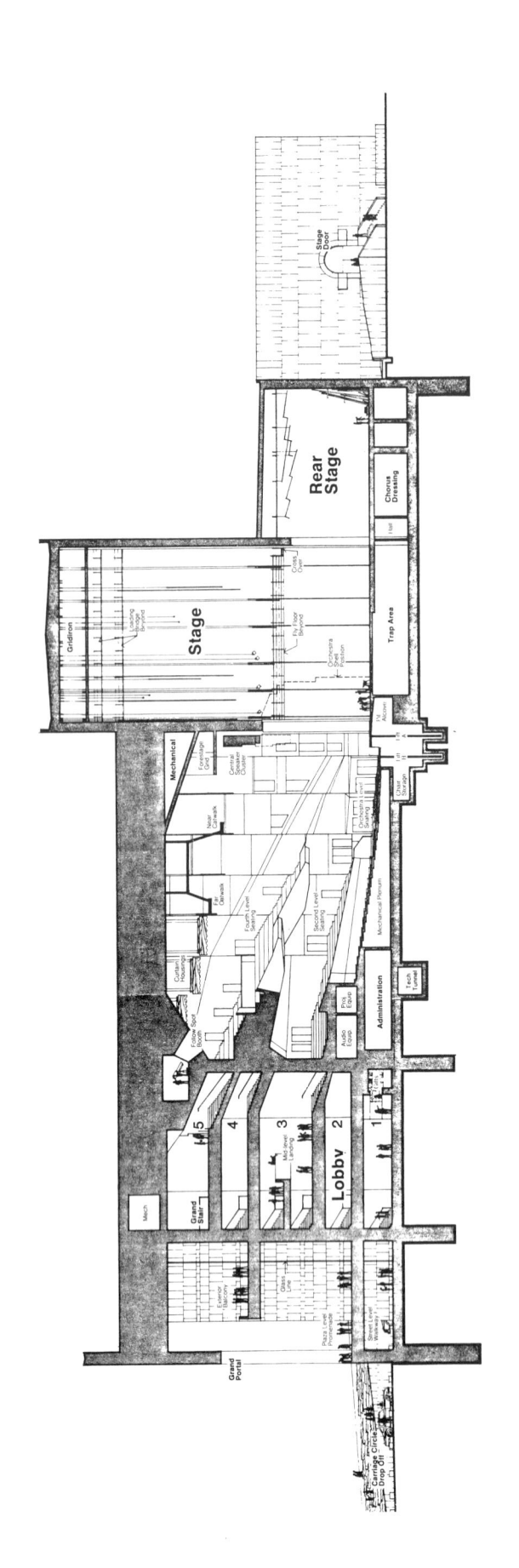

Figure 6.32 Sections of Segerstrom Hall. (Courtesy Jerald R. Hyde.)

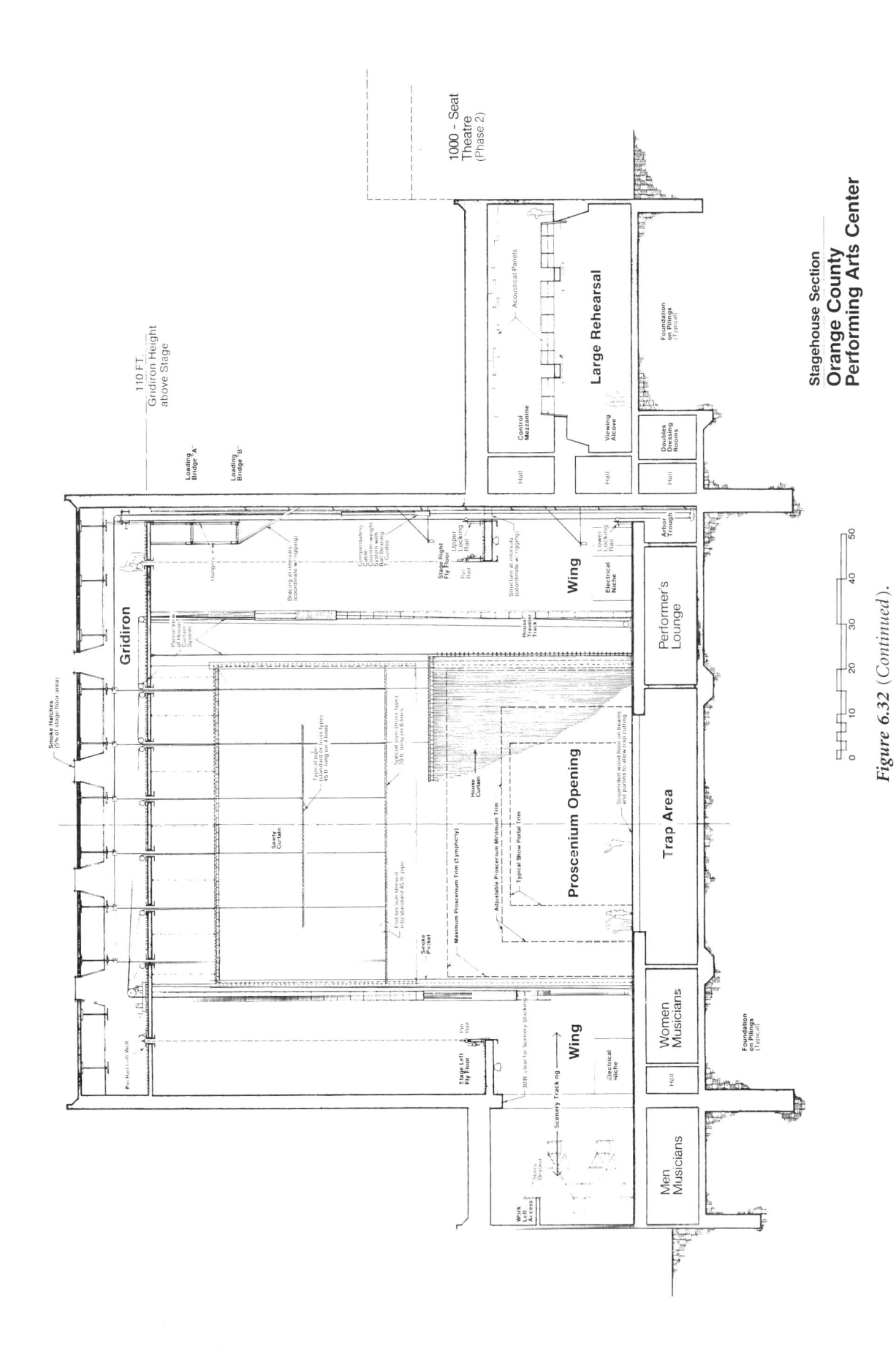

Figure 6.32 (*Continued*).

Room volume (total): 1,082,810 ft³ (27,800 m³)
Variable features: Operable curtains at upper volume; moveable ceiling canopy; reverberation chambers in upper portion of room

Initial Design

The Meyerson-McDermott Concert Hall was designed by I. M. Pei and Partners, Inc., architects, and Artec Consultants, Inc., acoustical consultants. Artec was hired by the Dallas Symphony Association to design a room with excellent acoustics: to allow the room "to speak truthfully and momentously; to have acoustical and architectural qualities of a worldwide distinction." The acoustical design of the room preceded the architectural design in what Artec calls a "revolutionary approach in the design of buildings for the performing arts." (Artec, 1990). Russell Johnson, the principal consultant and founder of Artec, has studied the architecture and acoustics of many historic concert halls and theaters. He listens avidly to musical performances in rooms around the world gaining an experiential basis for his critical evaluation of proposed designs. His firm is also well-known for using computer models and aural simulations in their design studies. Johnson takes an inherently architectural approach to acoustical design. He strives to integrate the properties of well-known, successful historic halls in buildings that reflect a contemporary attitude toward musical performance and architecture.

The basic acoustical design of the Meyerson-McDermott Concert Hall relied heavily on the use of historic precedent. Many of the design principles were adapted from opera houses, theaters and concert halls built in Europe between 1600 and 1910. These included the following features (illustrated in Figure 6.33).

1. The audience was stacked in several tiers of narrow balconies that surround the room. This results in a series of horizontal ledges along the perimeter that extend from the rear of the room to the vicinity of the stage. The bottoms of these surfaces reflect sound toward the sides of the heads of the audience on the main floor, increasing the sense of envelopment and spaciousness.
2. The width of the room was kept to 85 ft, which is relatively narrow compared to many recent halls.
3. The total area devoted to seating was limited to 2065 people. Seating capacities of concert halls were generally 600–1800 seats until the early 1900s. The performers and the orchestra are kept in a single, open room volume.
4. A generous amount of room volume per seat was provided (approximately 409 ft³ per person). This usually resulted in a room that was long and narrow with a high ceiling—a "shoebox hall." The ceiling in the Dallas hall is 84 ft high. The side walls of the historic halls were often parallel as are the side walls in the Dallas hall.
5. Heavy, massive finish materials that reflect all frequencies of sound were used. Traditional concert halls were often constructed of masonry walls with plaster coatings. Moreover, the walls and ceiling were not flat surfaces. Pilasters, deep coffers, moldings, statues, niches, and other surface elements of various sizes diffused the sound, spreading it throughout the room. A plaster skin 2½ in. thick covers the masonry walls in the Meyerson Hall. The exposed wood paneling is laminated in thick sections and adhered directly to the concrete wall without the use of furring strips. This technique eliminates the airspaces behind thin skin materials that absorb low-frequency sound.

The basic principles found in older rooms were interwoven with many innovative approaches that have evolved over a series of projects to transform the desired acoustical qualities of the pre-1910 concert rooms into a modern architectural vocabulary and to resolve the acoustical difficulties presented by many contemporary issues such as the large number of seats usually required in current halls (see Figures 6.34, 6.35, and 6.36). Artec has developed several innovative approaches to achieve the acoustical qualities of the precedent halls given the constraints posed by modern design practice. Newer features are combined with the features derived from historic halls in a synthetic manner that produces a classic sound with a contemporary expression. These features include the use of large reverberation chambers located along the upper walls of the room; retractable sound-absorbent banners on the side walls and large movable ceiling canopies over the stage. There are also four tiers of sound-reflecting soffits that surround the room rather than the two or three present in older rooms.

Reverberation Chambers. A large space made of cast-in-place concrete with a volume of over 300,000 ft³ surrounds the upper portion of the room in a "top hat" configuration. Large concrete panels operated by motors can open and close at the push of a button. This is a coupled space that allows sound to enter the chamber when the doors are opened, bounce off the walls and ceiling of the chamber, and then reenter the room as late reverberant energy. This sound has the effect of extending the reverberation time in the room while not interfering with the clarity of the music because the reverberant "tail" from the chamber is at a low enough level. This is an innovative acoustical attribute because it was previously thought that clarity and reverberance were mutually exclusive qualities of sound. Early reflections that arrive at locations in the room from the suspended ceiling canopy, the side walls, balcony soffits, and balcony faces in the narrow room provide clear strong sounds while the sound that enters the room from the reverberation chambers produces a longer, warm, tail to the sound. This results in a double slope during the sound decay period. The strong early energy dies off quickly while the lesser slope of the reverberant tail takes over. This concept is a noticeable shift from the accepted traditions of room acoustics that provide a uniform rate of decay in rooms.

All of the doors of the reverberation chambers can be opened for organ recitals and large choral concerts to extend the reverberation time of the room and create the sense "of

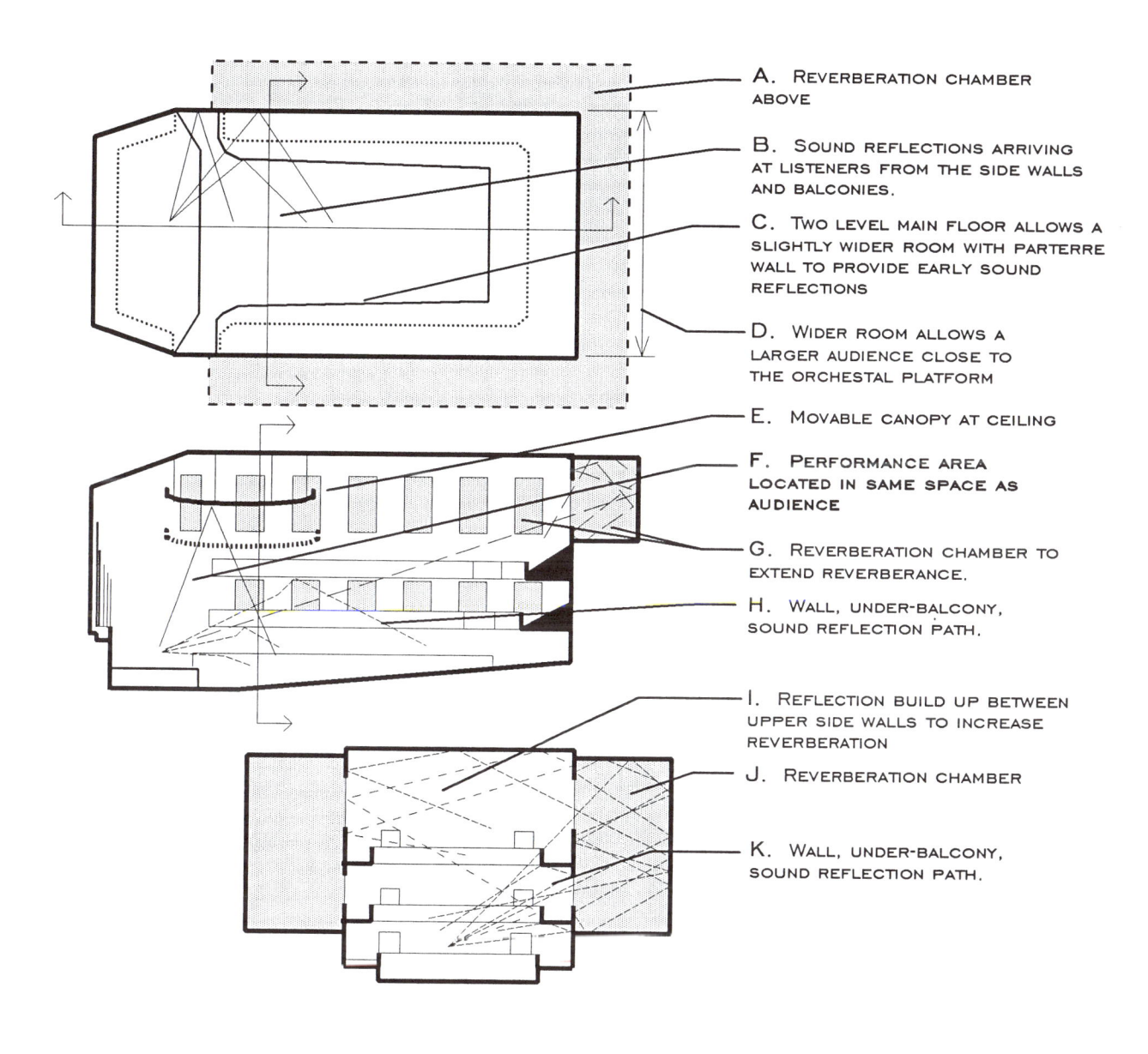

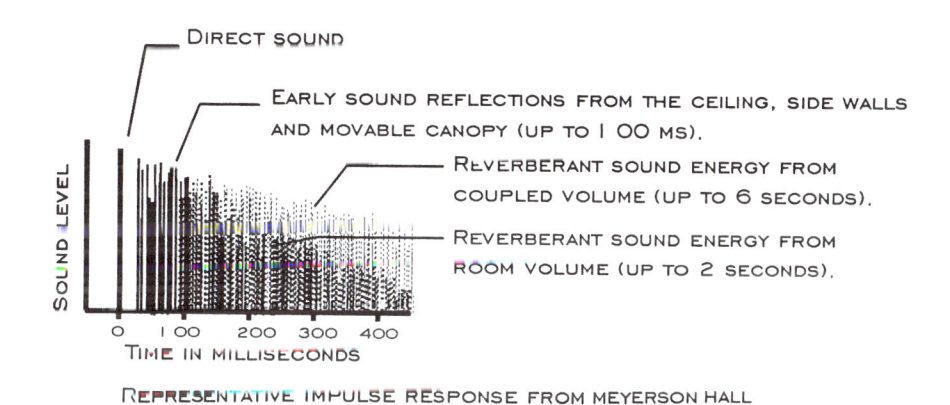

Figure 6.33 *Concept sketches showing acoustical design features of Meyerson Hall. (From Siebein and Gold, 1997.)*

287

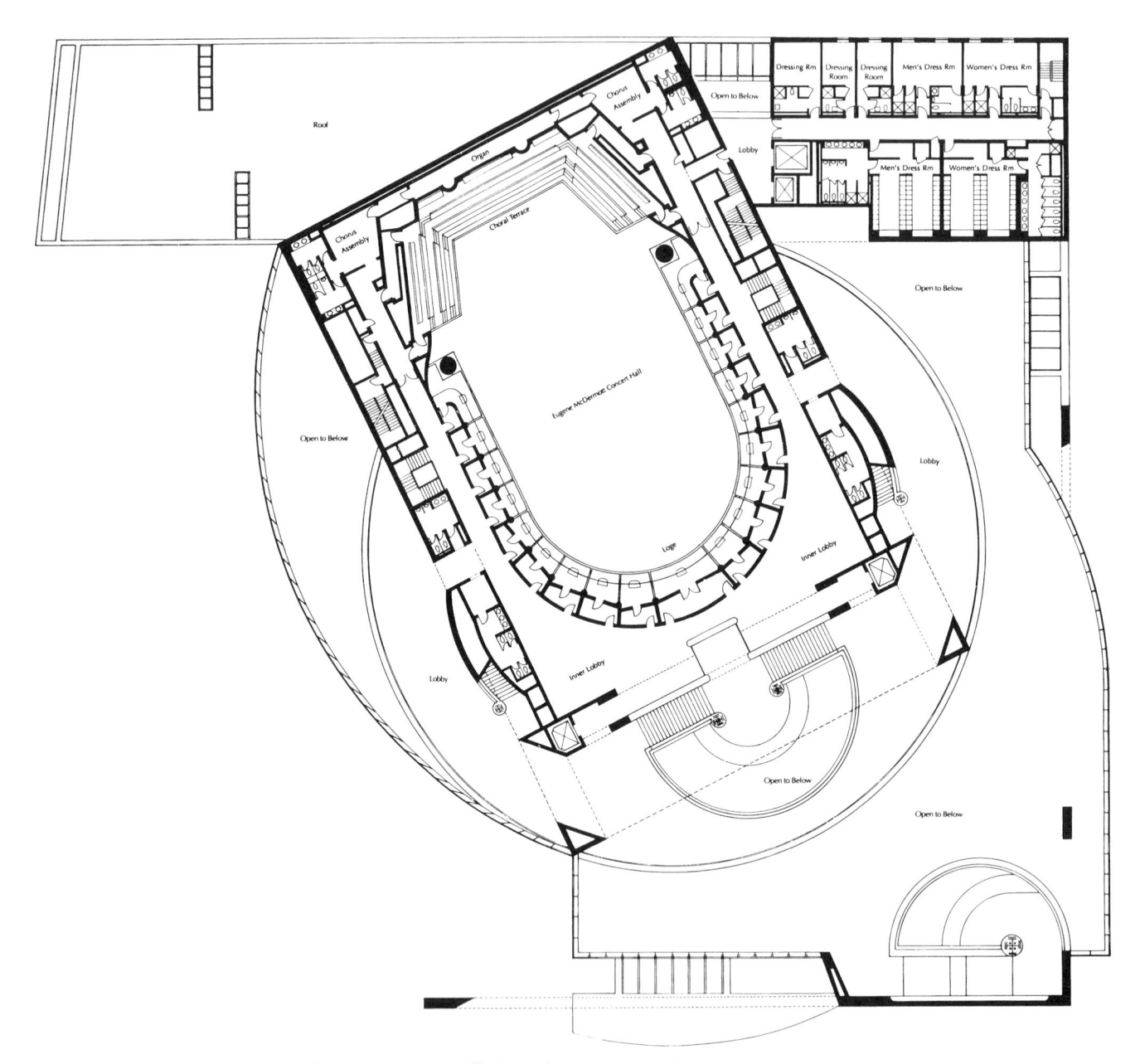

Figure 6.34 Overall plan of Meyerson Hall. (Courtesy Artec, Inc.)

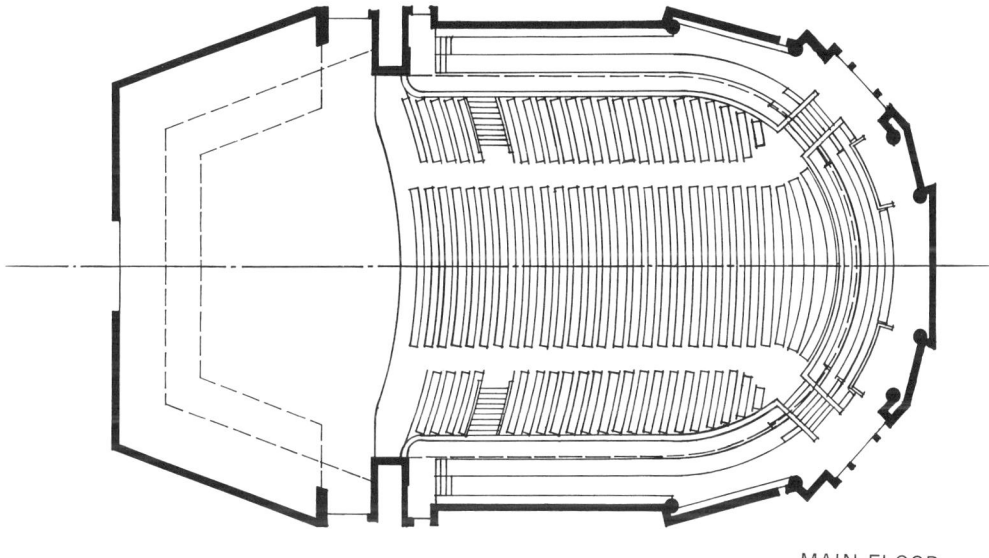

MAIN FLOOR

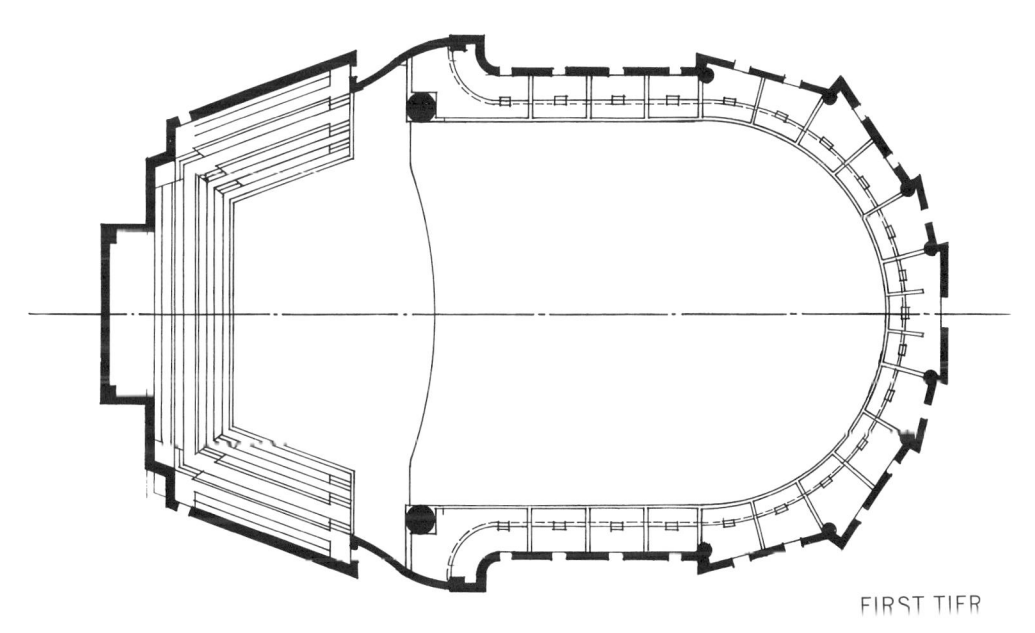

FIRST TIER

Figure 6.34 (*Continued*).

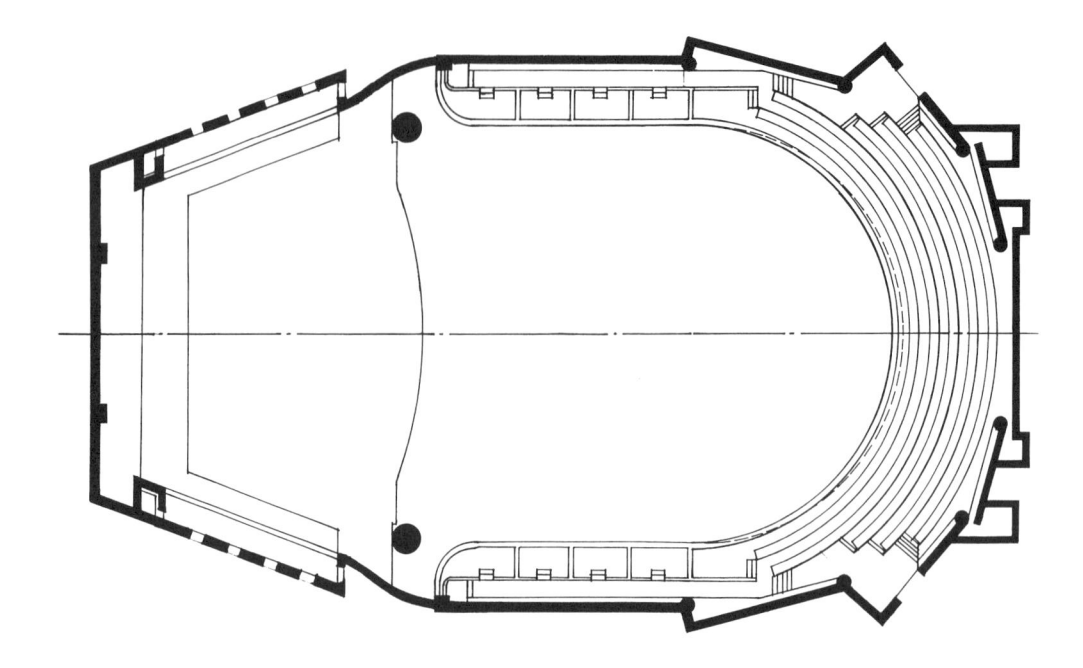

SECOND TIER

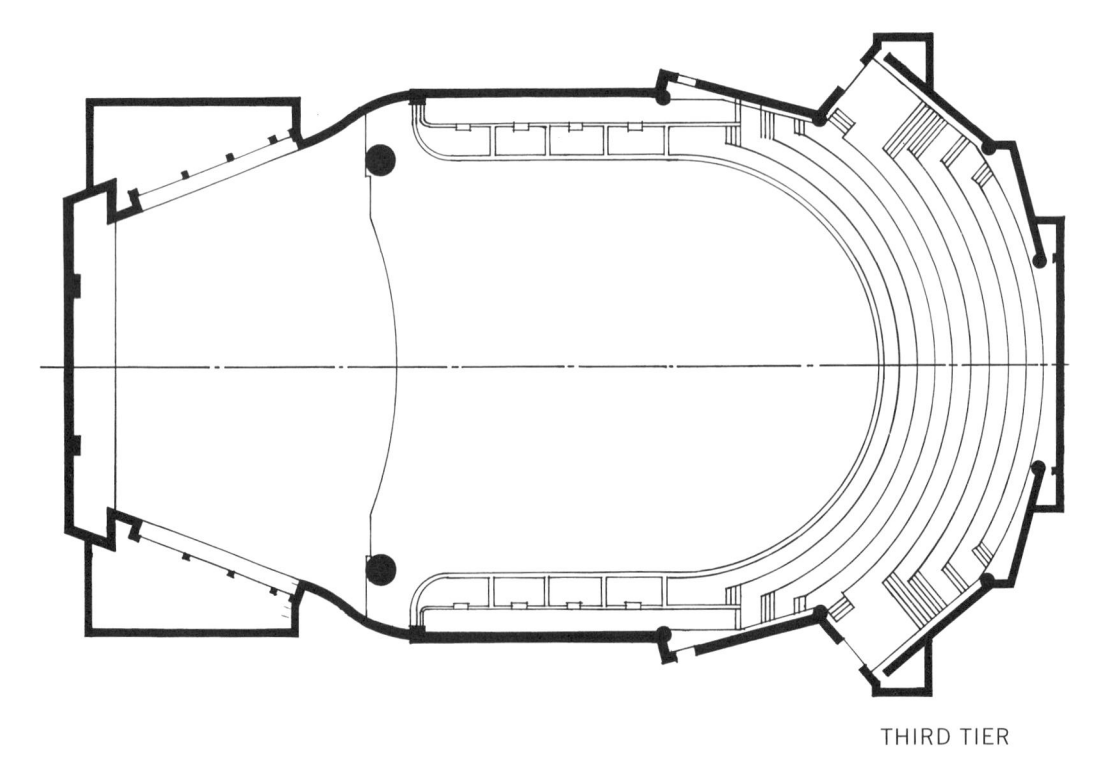

THIRD TIER

Figure 6.34 (*Continued*).

CROSS - SECTION

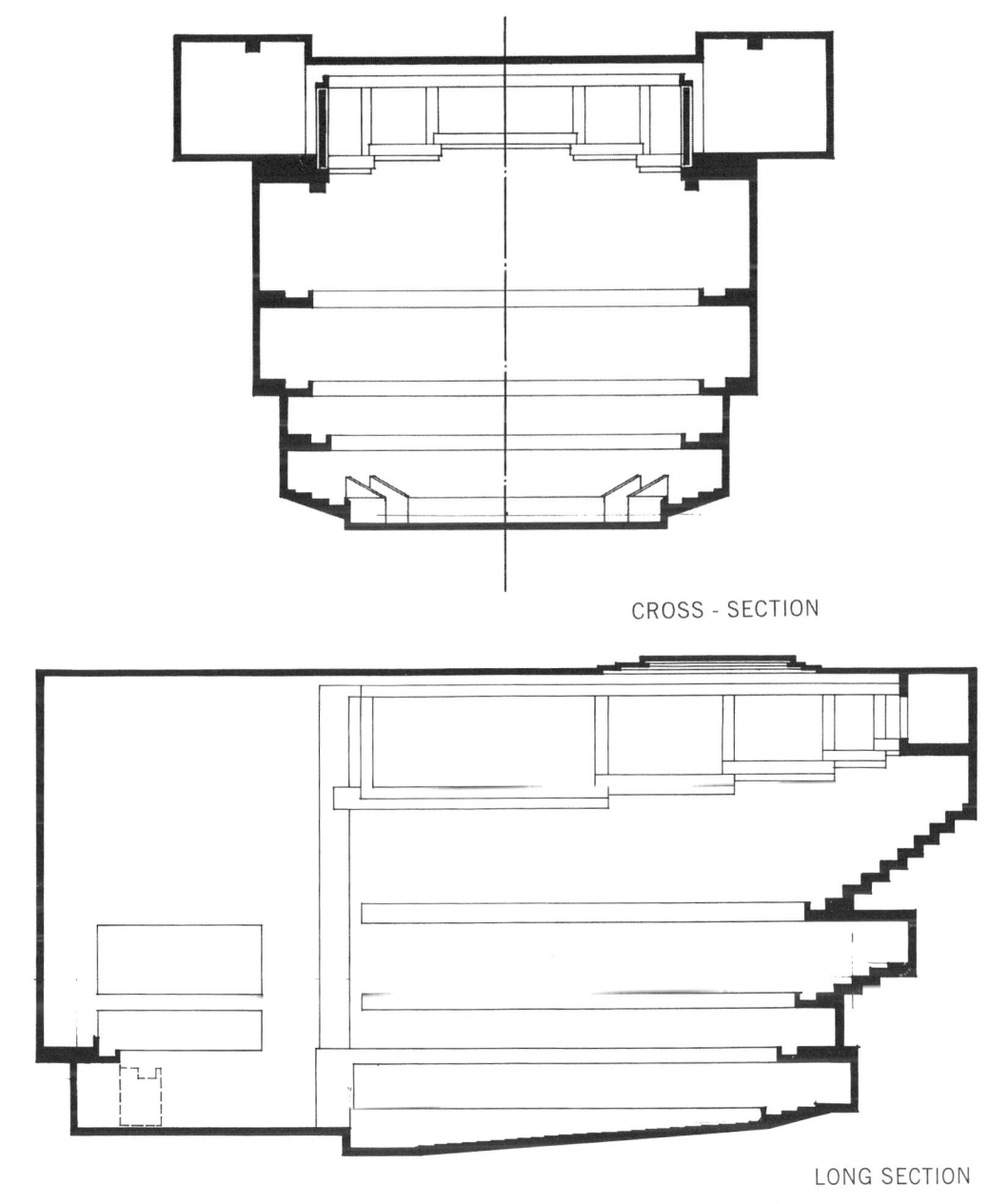

LONG SECTION

Figure 6.35 Sections of Meyerson Hall. (Courtesy Artec, Inc.)

Figure 6.36 *Photographs of Meyerson Hall: with stage canopy raised. (Courtesy Artec, Inc.)*

reverberance that one finds in a large stone cathedral" (Johnson, 1990, p. 4). When the panels are shut for many symphonic pieces or meetings, the reverberance in the room is reduced. The chamber doors are located behind a screen of loudspeaker cloth that allows sound to enter the chambers but prevents people in the audience area from seeing directly into the chambers.

Sound-Absorbing Banners. A system of multilayer cloth banners similar to heavy drapes can be extended from storage pockets along the side walls to cover much of the wall area of the room. The banners are operated by motors. The banners are kept inside their storage pockets during most symphonic works, organ recitals, chamber concerts, and choral concerts. This allows sound to reflect off the walls of the room. The banners are extended to cover the wall surfaces during rehearsals and also during popular music events, cinema, meetings, and other events where speech is an important activity. The room is much less reverberant when the banners cover the walls.

Canopy. A suspended, segmented, sound-reflecting canopy is located above the stage. The 4000 ft^2 canopy is made of five layers of laminated wood to reflect even low-frequency sound. The canopy can be raised and lowered by large motors to change the acoustics of the room. As the canopy is raised, more sound energy can enter the reverberation chambers in the upper portion of the room, which increases the sense of reverberance in the audience chamber. The spatial characteristics of the sound increase when the canopy is in the upper positions because more sound is reflected farther into the room. The noticeable differences in the interaural arrival times of reflections contribute to an increased sense of spaciousness. As the canopy is lowered, it restricts the entry of sound energy into the reverberation chambers, thereby decreasing the sensation of reverberance. The canopy also directs strong early sound reflections into the room, which increase the clarity or articulation of the sound when it is moved to the lower position.

Sound reflections from the canopy are beneficial to performers on stage as well as the audience when it is set at the normal height of approximately 40 ft above the stage. These reflections are loud enough and arrive quickly enough after the direct sound that performers can gauge

Figure 6.36 (Continued). With canopy lowered.

the relative loudness level at which they are playing. This allows musicians to hear each other, contributing to a sense of ensemble, and provides each musician with an accurate sense of his or her sound production so that the musician can control the quality and nuances of his or her music to a fine degree. All of the pipes of the large Fisk organ can speak to the room when the canopy is raised to its highest position.

Control of noise from outside sources, such as sirens, planes, and trucks, and from air-conditioning systems and other building equipment was also a major design concern. Buffer spaces such as lobbies and support spaces line the perimeter of the building so the walls of the concert hall are not exposed directly to the outside. Isolation joints break structural paths of sound transmission between the surrounding building and the hall. Large, sound-attenuating plena and very low air velocities (600 fpm or less) are used in the air-conditioning systems to reduce noise levels in the hall.

Case Study

EVANGELINE ATWOOD CONCERT HALL, ALASKA CENTER FOR THE PERFORMING ARTS, ANCHORAGE, ALASKA

Completed: 1989
Architect: Hardy Holtzman and Pfeiffer Associates, Inc.
Acoustical consultant: Jaffe, Holden, Scarborough, Chris Jaffe, Principal
Room type: Concert Hall
Seat count: 2100
Stage: Open
Maximum room width: 100 ft (30.5 m)
Room volume (total): 668,000 ft^3 (18,916 m^3)
Variable features: Electronic architecture system

Figure 6.36 (*Continued*). View from stage of Meyerson Hall.

While many of the architectural and acoustical design features of the Atwood Hall are similar to those employed in the Meyerson and Segerstrom Halls described above, there are also several significant differences among the rooms. The most notable difference is the use of an electronic architecture system to provide varying acoustical qualities within the room. Similar to the Orange County hall discussed previously, the Atwood Concert Hall was to accommodate a wide variety of performances including drama, opera, symphony, and rock and roll among others. The size and seating capacity of the room was limited to provide a sense of intimacy between performers and audience. The raised parterre that surrounds the main seating area, the two additional balcony tiers that surround the hall from the rear to the stage, and the diffusing ceiling panels to create lateral sound reflections are strategies found in historic halls and in the Dallas and Orange County halls discussed above. The basic architectural acoustics solution for the room was designed to meet the requirements of light opera and amplified Broadway programs.

Architectural Acoustics Design

The horse shoe shape of the plan was derived from the plans of traditional European opera houses. A design reverberation time of 1.4 to 1.6 sec and a volume of 600,000 ft^3 (approximately 285 ft^3 per person) were provided. A "throat" was designed immediately in front of the proscenium to project sound into the room from the stage. The reverse fan-shape plan of the room helps to increase the reflections arriving from the sides (lateral reflections) that contribute to the sense of spaciousness. The highly articulated ceiling design is shaped at the macro scale to project reflections throughout the room. The zig-zag surfaces provide a high degree of sound diffusion that increases the sense of spaciousness and envelopment by directing reflected sound energy toward the audience from many directions. The balcony faces and the parterre fascias are also designed as diffusing surfaces. The walls and the ceiling at the front of the room are one continuous surface. The architectural shaping of the room was also investigated with a scale model in Jaffe's laboratory. Deficiencies in reflected sound coverage of the hall were found in the model tests that were attributed to the shape of the ceiling. The ceiling was flattened somewhat and the tests repeated showing significant improvements in reflected sound coverage throughout the room.

Coupled Spaces

A concert shell made of plywood on a steel frame is also used during orchestral performances. There are large openings between the panels of the orchestral enclosure that allow sound from the stage area to enter the upper volume of the stage house. The sound reverberates within the stage house volume and then reenters the room providing an enhanced sense of warmth on stage and in the orchestra seating in the house. The shell also provides surfaces that reflect sound back to the performers so they can hear each other and achieve a sense of ensemble.

Electronic Architecture System

An electronic architecture system was employed to add early reflections and reverberation for orchestral concerts (see Figure 6.37). Three discrete systems are used in the Atwood Hall. By turning on one or more of the electronic architecture systems, the acoustical qualities of the room can be enhanced subtly or changed dramatically.

1. The early field system provides early sound reflections from a series of loudspeakers located around the proscenium and in the mezzanine and balcony faces. Sound is picked up by a microphone on stage and delayed with digital signal processing units and amplified if necessary. The sound played through the early field loudspeakers is heard by listeners at very specific times after the direct sound. Early reflections arriving within 20 to 30 msec after the direct sound provide a sense of clarity, intimacy, and increase the overall loudness of the sound.

2. The late-field system or warmth system is designed to provide sound reflections below 250 Hz that arrive at listener locations between 75 and 300 msec after the direct sound. These reflections can be provided by a large room volume and hard, sound-reflective materials as in the Dallas hall or electronically as at Atwood. Late-field loudspeakers were located in the ceiling and over the balcony, mezzanine, and parterre. They were also located under the stage house and in return air ducts under the auditorium seating. The sound is picked up by a microphone and duplicated eight times and delayed by a digital signal processing unit before it is played through the loudspeakers.

3. The reverberant field system or reverberation on demand system (RODS) was designed to add almost 0.5 sec to the reverberation time of the room. By designing the room for opera and Broadway plays, the room volume and the corresponding reverberation time was not long enough for most symphonic performances. Therefore it was desirable to prolong the reverberation time in the room during some performances. The RODS processor is an expanded version of the late-field processor. One section adds 48 replicas of the original sound in the middle frequencies. A second processor adds 72 replicas of the original sound in the lower frequencies. The sound is then amplified and played through the reverberant field loudspeakers located in the main ceiling of the room, as well as in the ceilings over the balcony, the mezzanine, and the parterre, the stage house, and the fascias of the balcony and mezzanine at the rear of the room. The multiplicity of loudspeakers provides a reverberant field that surrounds or envelops the audience.

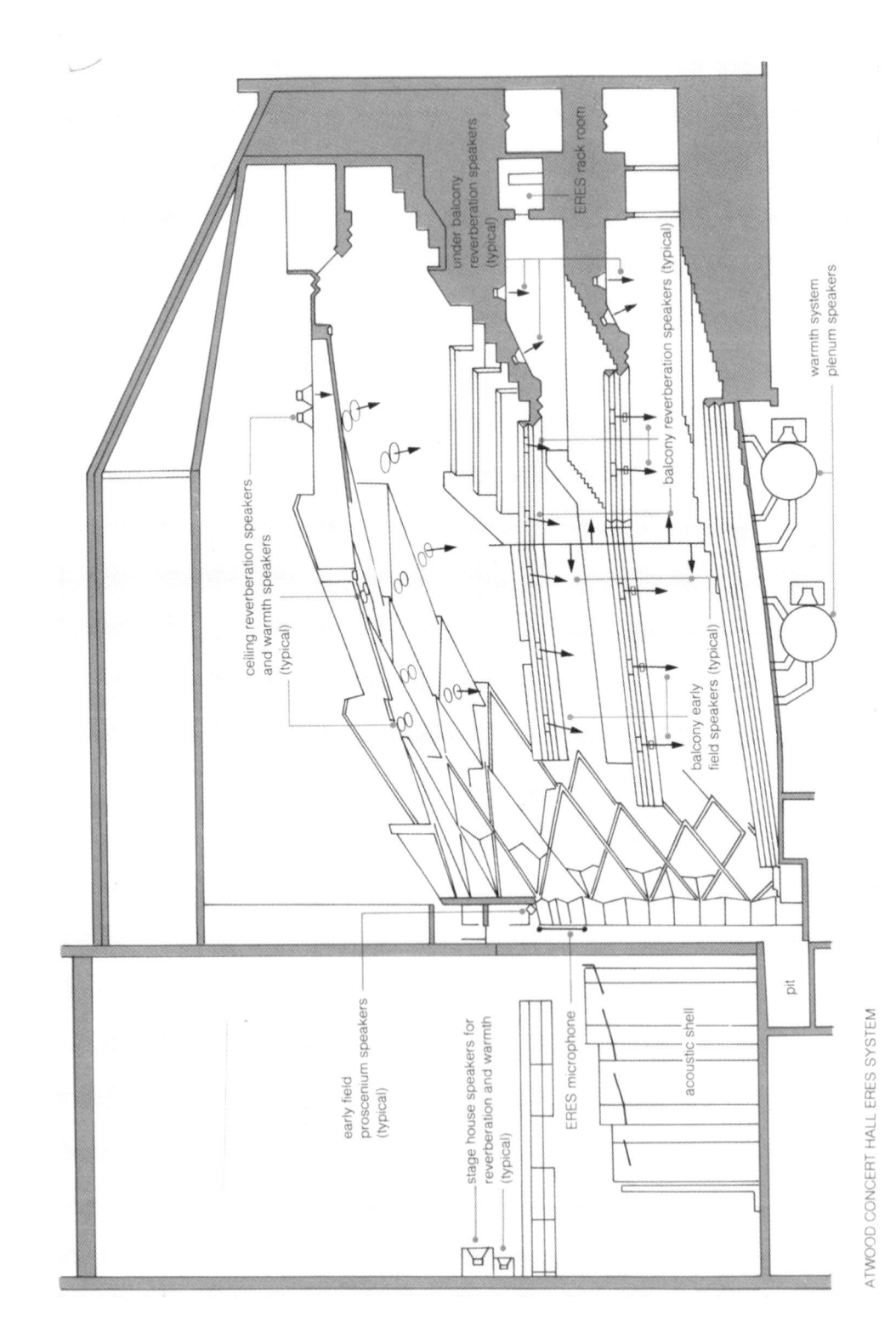

Figure 6.37 Concept sketch of the ERES electronic architecture system. (Architectural Record. Copyright 1989 by the McGraw-Hill Companies. All rights reserved. Reproduced by permission of the publisher.)

Jaffe emphasizes that the electronic architecture systems do not change the character of the sound from the source. But they do provide the opportunity to vary the acoustical qualities of a room by pushing buttons rather than by moving walls and banners. "It is important for people to realize how subtle it is" remarks Mark Holden, one of the acoustical designers on the project. "We are providing additional reflections and additional reverberation, enhancing it by 10 or 15% to change the overall perception of the room's quality. . . . For the system to work well, it must be used in an acoustically sound room in the first place" (Guericio, 1989, p. 44).

Sound isolation and mechanical system noise control were also important concerns in this project. The lobby and supporting spaces surround the main hall providing buffer spaces between the room and traffic noise outside (see Figures 6.38, 6.39, and 6.40). The roof is made of a metal roof over insulation on a concrete deck with two layers of gypsum board suspended below it to isolate the interior of the room from jet aircraft noise. Large ducts and low air velocities were used in the heating, ventilating, and air-conditioning (HVAC) system to meet noise criteria (NC) 20 sound levels in the room. Supply ducts are run through the ceiling where they discharge conditioned air above the finish ceiling. The air is returned through mushroom vents located under the seats that lead to large ducts and tunnels that carry the air long distances back to the air-handling units.

Case Study
ACOUSTICAL MODEL TESTS

Selected studies of design applications of scale model impulse response tests on actual building projects will be presented to illustrate some of the ways that these methods can become an integrated part of the architectural design process. The applications include examples of identifying and solving potential problems attributed to room geometry, exploring the acoustical effects of major design features of rooms, comparing acoustical changes attributed to relatively subtle design features in rooms, exploring the changes in acoustical response due to material selections, and exploring changes in acoustical response at various listener positions due to the location of the sound source. Each of the case studies will illustrate one or more of these points. The instrumentation and model building techniques have been described in Chiang (1992), Siebein et al. (1992), Siebein (1988, 1986), and Lyon and Cann (1975).

Study of Major Design Features of Rooms

The effects of major design features of rooms has been an area of emphasis in many scale model studies. One of the more interesting experiments that has been repeated on many projects involves the addition of a sound-reflecting panel over the front of an auditorium. Often several "new" reflections appear in an impulse response taken after a single panel has been added to a model. A simple experiment has been conducted to determine the source of the new reflections. The ceiling is removed from the model. A small sound-reflecting surface, usually a narrow stick that spans across the model, is placed at the end of the panel nearest the stage. An impulse response is taken. The stick is moved forward by its own width, and another impulse response is taken. This process is repeated until the full length of the panel has been traversed.

Studies were undertaken in the design of a large concert hall concerning the effects of a large wooden screen element designed by the architects to evoke artifacts of the local culture as well as the effects of a large movable ceiling canopy. There were concerns about selective absorption of sounds through what was supposed to be an acoustically transparent screen.

Study of Subtle Design Features of Rooms

There have been many impulse response studies of subtle design features of rooms conducted for actual projects in our laboratory. An example of this type of study occurred during the renovation of a historic hall to accommodate handicapped access by adding a new elevator. The only place an elevator for handicapped access could be located in the building was at the rear of the hall with the doors of the elevator opening directly into the room! There were severe problems with the stability of the soil and foundations that prevented other areas from being considered. First, a computer model of the room was constructed. The subtleties of the differences in the impulse responses at locations in the room with and without the elevator door were not evident in the computer model study. The extremely small size and detail of the door and the slight change in absorption coefficient along with the small number of sound rays that would actually strike it resulted in inconclusive results. A 1:40 scale model of the room was constructed (see Figure 6.41). Tests were conducted with the existing rear wall in place. An elevator door and frame were constructed and added to the model. Additional tests were conducted. The impulse responses with and without the elevator door are compared in Figure 6.42. There are several subtle changes in the impulse response attributed to the changes in framing the door opening that are evident.

Sound Level Variation Due to Source Location

A scale model of a partial enclosure for an outdoor firing range was constructed at a scale of 1:24 to study sound propagation from the range into the surrounding community. The building was to have a concrete floor, slotted concrete block walls, a sawtooth roof covered with metal plates for

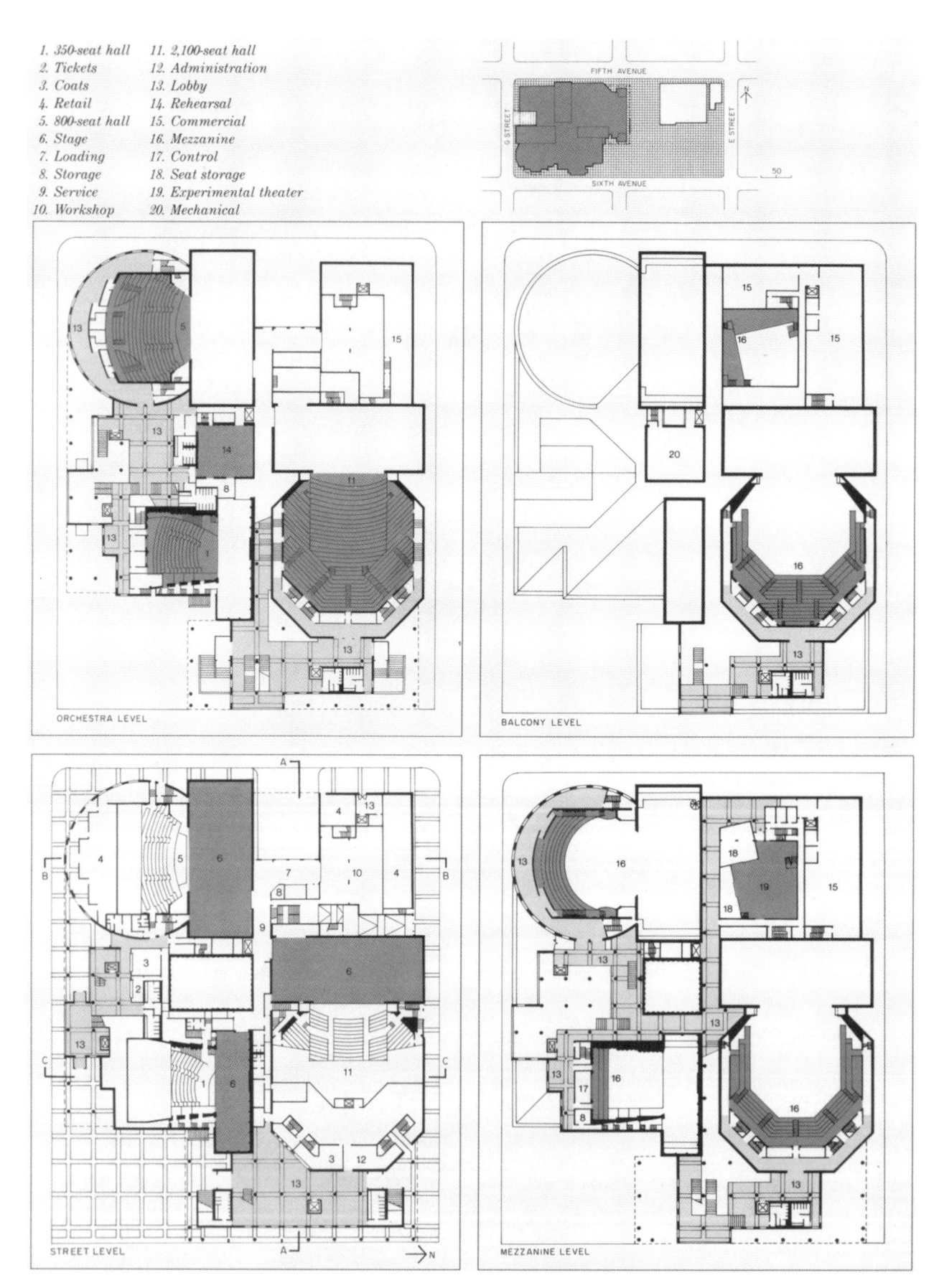

1. 350-seat hall
2. Tickets
3. Coats
4. Retail
5. 800-seat hall
6. Stage
7. Loading
8. Storage
9. Service
10. Workshop
11. 2,100-seat hall
12. Administration
13. Lobby
14. Rehearsal
15. Commercial
16. Mezzanine
17. Control
18. Seat storage
19. Experimental theater
20. Mechanical

Figure 6.38 Plans of Atwood Hall. (Architectural Record. Copyright 1989 by the McGraw-Hill Companies. All rights reserved. Reproduced by permission of the publisher.)

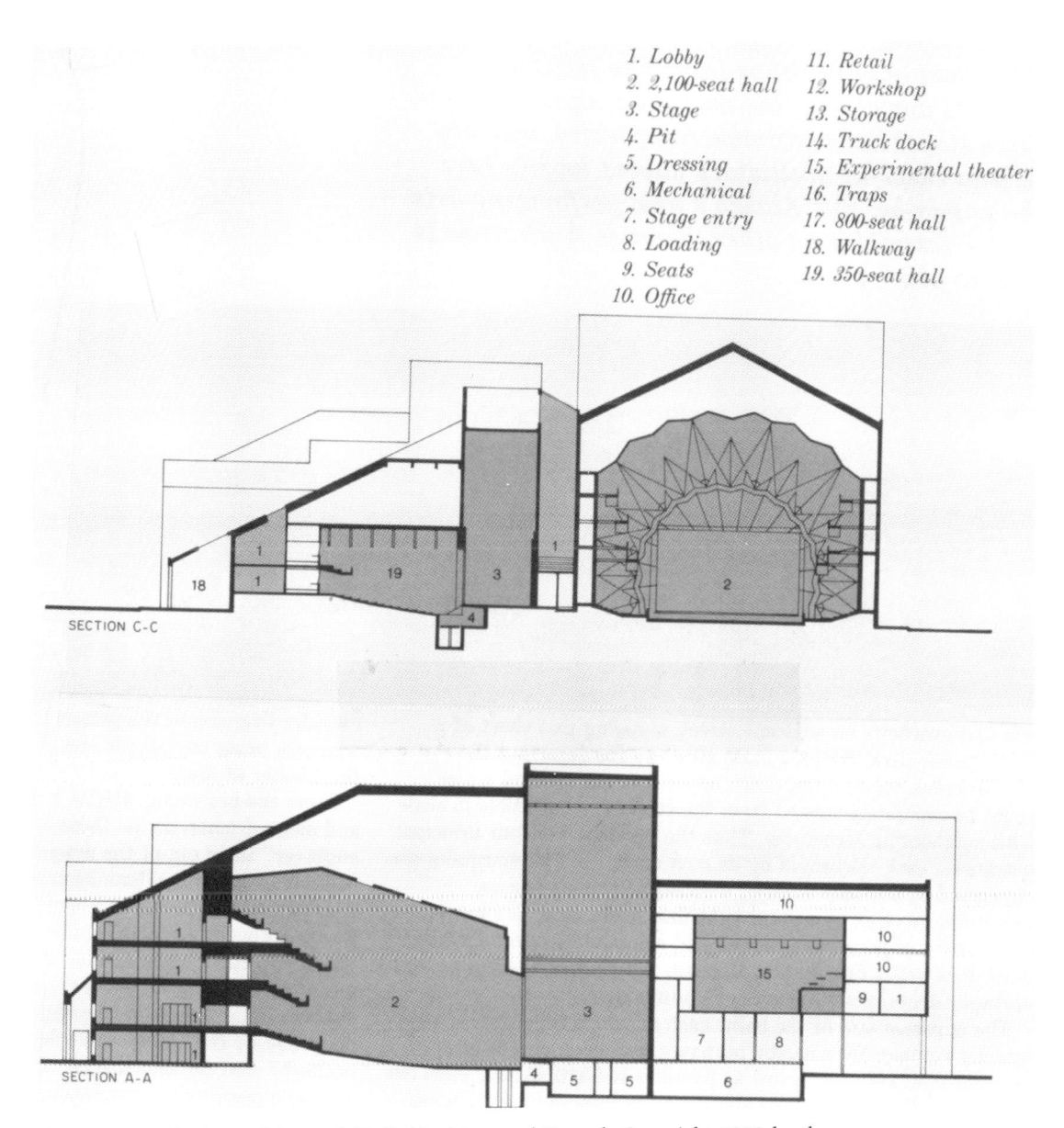

1. Lobby
2. 2,100-seat hall
3. Stage
4. Pit
5. Dressing
6. Mechanical
7. Stage entry
8. Loading
9. Seats
10. Office
11. Retail
12. Workshop
13. Storage
14. Truck dock
15. Experimental theater
16. Traps
17. 800-seat hall
18. Walkway
19. 350-seat hall

Figure 6.39 Sections of Atwood Hall. (Architectural Record. Copyright 1989 by the McGraw-Hill Companies. All rights reserved. Reproduced by permission of the publisher.)

Figure 6.40 Photograph of Atwood Hall. (Architectural Record. Copyright 1989 by the McGraw-Hill Companies. All rights reserved. Reproduced by permission of the publisher. Photo courtesy of Christopher Little.)

Figure 6.11 Photograph of the actual concert hall (top) and the scale model of the concert hall (middle and bottom).

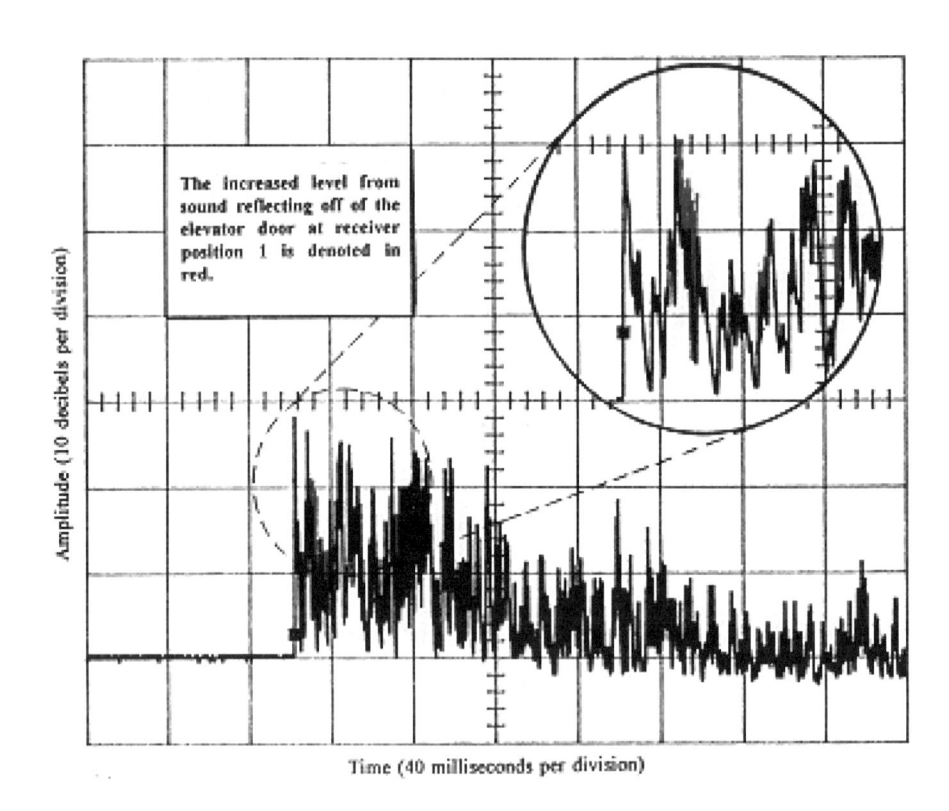

Figure 6.42 Comparison of impulse responses taken with (shaded trace) and without (black trace) the addition of the elevator door. The diffusion of energy in the reflection off the rear wall is similar to that which occurs at seats near existing doors in the house. The change from a specular reflection off the rear wall to a more diffuse reflection from the recessed door and frame is evident in the shaded impulse response.

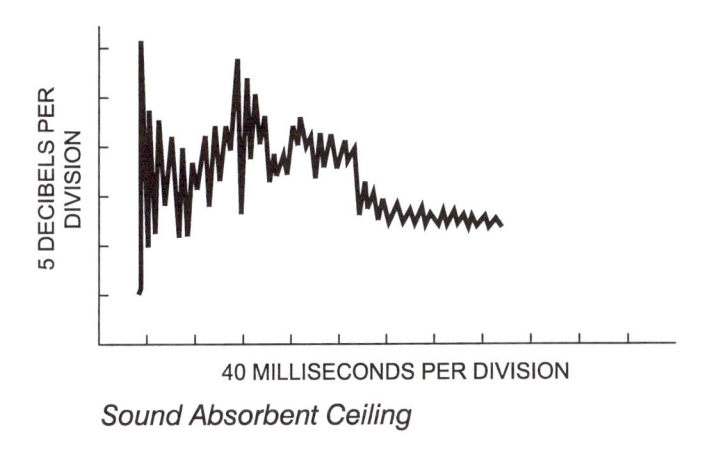

Sound Absorbent Ceiling

Bare Ceiling

Figure 6.43 *Impulse response [without absorbent material on the ceiling panels (bottom) and with absorbent material on the ceiling panels (top)] taken behind the range showing multiple reflections arriving from the single source similar to that which occurs in a room.*

bullet baffles, and possibly some sound-absorbent material located on the ceiling. The vertical sections of the sawtooth were left open to allow air to flow into the range to carry the lead away from the shooters. Field measurements during firearms training in an existing range of similar design by our firm and by several other firms over the years had shown variations of 3 to 5 dB in sound levels at distances of 500 ft when the type of ammunition and number of shooters remained constant. One of the foci of the scale model study was to study the causes of this variation in sound level.

Impulse responses were taken at scale distances of 500 ft away from the firing range on the sides, to the front, and to the rear of the model. The source was located at the 7-yd line and at the 50-yd line for the tests. Sound levels varied by over 4 dB at the same receiver locations as the source was moved from the front position to the rear position, illustrating a direct relationship attributed to the unique source–path–receiver geometry for each case. This is analogous to the situation that occurs in concert halls where very distinct impulse responses are found in different sections of large rooms. It was quite unexpected in this situation where one was not expecting to find a response that looked more like a complete room.

Sound Level Variation Due to Material Selection

Absorbent material was added to the side walls and ceiling of the firing range model incrementally to determine the effects on sound levels away from the range that the absorbent material would have. While it was hypothesized that the addition of the absorbent material would decrease sound levels within the range, it was not understood if it would also decrease sound levels at distances away from the range. The absorption on the side walls contributed less than 1 dB to sound reduction outside the range. The absorbent material on the ceiling contributed from 3 to 9 dB of sound reduction at the locations 500 ft away from the range depending on the source location and the specific receiver location. The differences in the sound reduction were not statistical variation due to inconsistencies in the measurements, as was surmised after several attempts to measure these effects at several existing ranges. They were attributed to the location of the shooters in the range relative to the specific measurement location. At the location behind the range (in the direction that the sawtooth opens), the sound levels were louder when the shooters were closer to the measurement location. At the measurement location outside the front of the range, there was a smaller difference between the sound levels when the shooters were at the front of the range and at the rear of the range due to the diffracted path that the sound must travel to reach this location. A second finding of interest was the impulse responses taken at locations 500 ft from the rear of the range. Sound arrives at this location through several of the clerestory openings, creating an impulse response that looks similar to those found in rooms. There were multiple reflections that arrive from paths through several clerestory openings. The initial arrival from each path was a diffracted sound wave that remains at a similar amplitude with and without the addition of the absorbent material. The subsequent arrivals had to strike the ceiling surfaces and were significantly attenuated after they struck the absorbent material (see Figure 6.43). This is a notably different situation than that which occurs in an outdoor situation where only a direct sound and a ground reflection may be observed.

APPENDIX A

Conversion Factors, Abbreviations, and Unit Symbols

Selected SI Units (Adopted 1960)

QUANTITY	UNIT	SYMBOL	ACCEPTABLE EQUIVALENT
BASE UNITS			
length	meter[†]	m	
mass[‡]	kilogram	kg	
time	second	s	
electric current	ampere	A	
thermodynamic temperature[§]	kelvin	K	
DERIVED UNITS AND OTHER ACCEPTABLE UNITS			
*absorbed dose	gray	Gy	J/kg
acceleration	meter per second squared	m/s^2	
*activity (of ionizing radiation source)	becquerel	Bq	l/s
area	square kilometer	km^2	
	square hectometer	hm^2	ha (hectare)
	square meter	m^2	
density, mass density	kilogram per cubic meter	kg/m^3	g/L; mg/cm^3
*electric potential, potential difference, electromotive force	volt	V	W/A
*electric resistance	ohm	Ω	V/A
energy, work, quantity of heat	megajoule	MJ	
	kilojoule	kJ	
	joule	J	N·m
	electron volt[x]	eV[x]	
	kilowatt hour[x]	kW·h[x]	

305

QUANTITY	UNIT	SYMBOL	ACCEPTABLE EQUIVALENT
*force	kilonewton	kN	
	newton	N	kg · m/s^2
*frequency	megahertz	MHz	
	hertz	Hz	1/s
heat capacity, entropy	joule per kelvin	J/K	
heat capacity (specific), specific entropy	joule per kilogram kelvin	J/(kg · K)	
heat transfer coefficient	watt per square meter kelvin	W/(m^2 · K)	
linear density	kilogram per meter	kg/m	
magnetic field strength	ampere per meter	A/m	
moment of force, torque	newton meter	N · m	
momentum	kilogram meter per second	kg · m/s	
*power, heat flow rate, radiant flux	kilowatt	kW	
	watt	W	J/s
power density, heat flux density, irradiance	watt per square meter	W/m^2	
*pressure, stress	megapascal	MPa	
	kilopascal	kPa	
	pascal	Pa	
sound level	decibel	dB	
specific energy	joule per kilogram	J/kg	
specific volume	cubic meter per kilogram	m^3/kg	
surface tension	newton per meter	N/m	
thermal conductivity	watt per meter kelvin	W/(m · K)	
velocity	meter per second	m/s	
	kilometer per hour	km/h	
viscosity, dynamic	pascal second	Pa · s	
	millipascal second	mPa · s	
volume	cubic meter	m^3	
	cubic decimeter	dm^3	L (liter)
	cubic centimeter	cm^3	mL

* The asterisk denotes those units having special names and symbols.

† The spellings "metre" and "litre" are preferred by ASTM; however "er-" is used here.

‡ "Weight" is the commonly used term for "mass."

§ Wide use is made of "Celsius temperature" (t) defined by

$$t = T - T_0$$

where t is the thermodynamic temperature, expressed in kelvins, and $T_0 = 273.15$ by definition. A temperature interval may be expressed in degrees Celsius as well as in kelvins.

^{x} This non-SI unit is recognized by the CIPM as having to be retained because of practical importance or use in specialized fields.

In addition, there are 16 prefixes used to indicate order of magnitude, as follows:

MULTIPLICATION FACTOR	PREFIX	SYMBOL
10^{18}	exa	E
10^{15}	peta	P
10^{12}	tera	T
10^9	giga	G
10^6	mega	M
10^3	kilo	k
10^2	hecto	h[a]

MULTIPLICATION FACTOR	PREFIX	SYMBOL
10	deka	da[a]
10^{-1}	deci	d[a]
10^{-2}	centi	c[a]
10^{-3}	milli	m
10^{-6}	micro	μ
10^{-9}	nano	n
10^{-12}	pico	p
10^{-15}	femto	f
10^{-18}	atto	a

[a] Although hecto, deka, deci, and centi are SI prefixes, their use should be avoided except for SI unit-multiples for area and volume and nontechnical use of centimeter, as for body and clothing measurement.

Conversion Factors to SI Units

TO CONVERT FROM	TO	MULTIPLY BY
acre	square meter (m^2)	4.047×10^3
angstrom	meter (m)	1.0×10^{-10}†
atmosphere	pascal (Pa)	1.013×10^5
bar	pascal (Pa)	1.0×10^5†
barn	square meter (m^2)	1.0×10^{-28}†
barrel (42 U.S. liquid gallons)	cubic meter (m^3)	0.1590
Btu (thermochemical)	joule (J)	1.054×10^3
bushel	cubic meter (m^3)	3.524×10^{-2}
calorie (thermochemical)	joule (J)	4.184†
centipoise	pascal second (Pa · s)	1.0×10^{-3}†
cfm (cubic foot per minute)	cubic meter per second (m^3/s)	4.72×10^{-4}
cubic inch	cubic meter (m^3)	1.639×10^{-5}
cubic foot	cubic meter (m^3)	2.832×10^{-2}
cubic yard	cubic meter (m^3)	0.7646
dram (apothecaries')	kilogram (kg)	3.888×10^{-3}
dram (avoirdupois)	kilogram (kg)	1.772×10^{-3}
dram (U.S. fluid)	cubic meter (m^3)	3.697×10^{-6}
dyne	newton (N)	1.0×10^{-5}†
dyne/cm	newton per meter (N/m)	1.0×10^{-3}†
fluid ounce (U.S.)	cubic meter (m^3)	2.957×10^{-5}
foot	meter (m)	0.3048†
gallon (U.S. dry)	cubic meter (m^3)	4.405×10^{-3}
gallon (U.S. liquid)	cubic meter (m^3)	3.785×10^{-3}
gallon per minute (gpm)	cubic meter per second (m^3/s)	6.308×10^{-5}
	cubic meter per hour (m^3/h)	0.2271
grain	kilogram (kg)	6.480×10^{-5}
horsepower (550 ft · lbf/s)	watt (W)	7.457×10^2

TO CONVERT FROM	TO	MULTIPLY BY
inch	meter (m)	2.54×10^{-2}†
inch of mercury (32°F)	pascal (Pa)	3.386×10^{-3}
inch of water (39.2°F)	pascal (Pa)	2.491×10^{2}
kilogram-force	newton (N)	9.807
kilowatt hour	megajoule (MJ)	3.6†
liter (for fluids only)	cubic meter (m^3)	1.0×10^{-3}†
micron	meter (m)	1.0×10^{-6}†
mil	meter (m)	2.54×10^{-5}†
mile (statute)	meter (m)	1.609×10^{3}
mile per hour	meter per second (m/s)	0.4470
millimeter of mercury (0°C)	pascal (Pa)	1.333×10^{2}†
ounce (avoirdupois)	kilogram (kg)	2.835×10^{-2}
ounce (troy)	kilogram (kg)	3.110×10^{-2}
ounce (U.S. fluid)	cubic meter (m^3)	2.957×10^{-5}
ounce-force	newton (N)	0.2780
peck (U.S.)	cubic meter (m^3)	8.810×10^{-3}
pennyweight	kilogram (kg)	1.555×10^{-3}
pint (U.S. dry)	cubic meter (m^3)	5.506×10^{-4}
pint (U.S. liquid)	cubic meter (m^3)	4.732×10^{-4}
poise (absolute viscosity)	pascal second (Pa·s)	0.10†
pound (avoirdupois)	kilogram (kg)	0.4536
pound (troy)	kilogram (kg)	0.3732
pound-force	newton (N)	4.448
pound-force per square inch (psi)	pascal (Pa)	6.895×10^{3}
quart (U.S. dry)	cubic meter (m^3)	1.101×10^{-3}
quart (U.S. liquid)	cubic meter (m^3)	9.464×10^{-4}
quintal	kilogram (kg)	1.0×10^{2}†
rad	gray (Gy)	1.0×10^{-2}†
square inch	square meter (m^2)	6.452×10^{-4}
square foot	square meter (m^2)	9.290×10^{-2}
square mile	square meter (m^2)	2.590×10^{6}
square yard	square meter (m^2)	0.8361
ton (long, 2240 pounds)	kilogram (kg)	1.016×10^{3}
ton (metric)	kilogram (kg)	1.0×10^{3}†
ton (short, 2000 pounds)	kilogram (kg)	9.072×10^{2}
torr	pascal (Pa)	1.333×10^{2}
yard	meter (m)	0.9144†

† Exact.

APPENDIX B

Acoustical Societies Throughout the World

The following directory of acoustical organizations throughout the world is presented in two parts: (I) international and multicountry societies and (II) national societies, listed by country in alphabetical order. Capital letters appearing within separate parentheses after some of the society names are the society acronyms.

This directory is reproduced with permission from the 1997 Membership Directory of the Acoustical Society of America (ASA). Further information is available directly from any of the international or national organizations listed. Readers should report any errors or omissions in this list directly to the Acoustical Society of America, Executive Director [ASA, 500 Sunnyside Boulevard, Woodbury, NY 11797-2999, USA; Telephone: (516) 576-2360; Fax: (516) 576-2377; E-mail: asa@aip.org; WWW: http://www.asa.aip.org]

I. INTERNATIONAL AND MULTICOUNTRY SOCIETIES

Acoustical Society of Scandinavia
Sec.: A. Thorbjorn Christensen
Bldg. 352, Dept. of Acoustic Technology
DK 2800 Lyngby, Denmark
Tel.: +45 45 88 16 22
Fax: +45 45 88 05 77
E-mail: atc.das@dat.dtu.dk

European Acoustics Association (EAA)
Managing Dir.: Dr. Manell Zakharia
CPE Lyon, LISA/LASSO, Bat 308
43 Bd du 11 November 1918
BP 2077 69616
Villeurbanne, Cedex, France
Tel.: +33 4 7243 1006
Fax: +33 4 7244 8074
E-mail: zakharia@cpe.fr

Federation of Acoustical Societies of Europe (FASE)
Sec.: Rudolf Martin
Bundesallee 100 (PTB)
Brauschweig 38116, Germany

International Association Against Noise (AICB)
Gen. Sec.: Dr. W. Aecherli
Hirschenplatz 7
CH-6004 Lucerne, Switzerland
Tel.: +41 410 30 13
Fax: +41 410 90 93

International Commission on Acoustics (ICA)
Sec.: Prof. Adriano Alippi
Instituto di Acustica "O. M. Corbino"
via Cassia 1216
I-00189 Rome, Italy
Tel.: +39 6 303 65 766
Fax. +39 6 303 65 341
E-mail: adriano@idac.rm.cnr.it

International Institute of Noise Control Engineering (I/INCE)
Dir.: Prof. Dr. H. Myncke
Catholic Univ. of Leuven
Celestijnenlaan 200 D
B-3001 Heverlee, Belgium
Fax.: +32-16-327984

International Society of Audiology (AUDI)
Sec. Gen.: J. Verschuure
Audiological Centre of the University Hospital Rotterdam
Dr. Molewaterplein 40
NL-3015 GD Rotterdam
The Netherlands
Tel.: +31 10 463 9222
Fax: +31 10 463 3102
E-mail: verschuure@kno.fgg.eur.nl
WWW: www.fgg.eur.nl/fgg/KNO/actif/isa.htm

International Society of Phonetic Sciences (ISPhS)
Sec.: Prof. Dr. Jens-Peter Koester
D-54286 Trier, Germany
Tel.: +49-651-201-2256
Fax: +49-651-201-3940
E-mail: koester@uni-trier.de

II. NATIONAL ACOUSTICAL SOCIETIES

ARGENTINA

Association de Acusticos Argentinos (ADAA)
Sec.: O. Bonello
Cachimayo 544
1424 Buenos Aires, Argentina

AUSTRALIA

Australian Acoustical Society
Gen. Sec.: Mr. D. J. Watkins
Private Bag No. 1
Darlinghurst, N.S.W. 2010, Australia
Tel.: +02 331 6920
Fax: +02 331 7296
E-mail: watkinsd@melbpc.org.au
WWW: www.adfa.oz.au/~mxb

AUSTRIA

Österreichische Gesellschaft fuer Akustik
Institut fuer Allgemeine Physik der
Technischen Universitaet Wien
Wiedner Hauptstrasse 8-10/134
1040 Wien, Austria
Tel.: +43 1 588 01 5711
Fax: +43 1 586 4203
E-mail: benes@iap.tuwien.ac.at

BELGIUM

Assoc. Belge des Acousticiens (ABAV)
Sec. Gen.: ir. D. Soubrier
BBRI
av. Pierre Holoffe 21
1342 Limelette, Belgium
Tel.: +32-2-653 88 01
Fax. +32-2-653 07 29
E-mail: bbri@pophost.eunet.be

BRAZIL

Sociedade Brasileira de Acústica (SOBRAC)
c/o Prof. R. A. Tenenbaum
Acoustics and Vibration Lab.
COPPE-Univ. Federal do Rio de Janeiro
C.P. 68503
21945-900 Rio de Janeiro, Brazil
E-mail: roberto@serv.com.ufrj.br

Brazilian Acoustical Association (ABRAC)
Sec.: A. L. Lisboa de Azevedo
Av. Alvorada 2.150
Bloco B, Salas 225/226
Barra da Tijuca, CEP 22.750
Rio de Janeiro, Brazil

CANADA

Canadian Acoustical Association
P.O. Box 1351
Toronto, Ontario M4Y 2V9, Canada

PEOPLE'S REPUBLIC OF CHINA

Acoustical Society of China (ASC)
Sec.: Prof. Guan Dinghua
17 Zhongguincun Street
Beijing, P. R. China
Fax.: +86 10 255 3898

CROATIA

Croatian Acoustical Society
Sec.: Prof. dr. sc. Branko Somek
Faculty of Electrical Engineering
Dept. of Electroacoustics
Unska 3
HR-41000 Zageb
Tel.: +385-1-62 96 40
Fax.: +385-1-62 98 62
E-mail: branko.somek@etf.hr

CZECH REPUBLIC

Czech Acoustical Society
Sec.: Ondřej Jiřícek
Technicka 2
166 27 Prague 6, Czech Republic
Tel.: +42-2-2435 2310
Fax: +42-2-311 1786
E-mail: csas@feld.cvut.cz

DENMARK

Acoustical Society of Denmark
Sec.: A. Thorbjorn Christensen
Dept. of Acoustic Technology
Bldg. 352-Techn. Univ. of Denmark
DK-2800 Lyngby, Denmark
Tel.: +45 45 88 16 22
Fax: + 45 45 88 05 77
E-mail: atc.das@dat.dtu.dk

EGYPT

Acoustical Group of National Committee of Pure and Applied Physics
Sec.: Dr. M. A. Ibrahim
c/o National Institute for Standards
Tersa St., El Haramel Giza
P.O. Box 136 Giza Code No. 12211
Cairo, A. R. Egypt
Tel.: +386 7452-3867453
Fax: +386 7451-3867454

FINLAND

Acoustical Society of Finland (ASF)
Sec.: Antti Järvinen
Helsinki Univ. of Technology
Acoustics Laboratory
Otakaari 5A
FIN-02150 Espoo, Finland
Tel.: +358 0 451 2499
Fax: +358 0 460 224
E-mail: akustinen.seura@hut.fi

FRANCE

Société Française d'Acoustique (SFA)
Pres.: Bernard Hamonic
23, Avenue Brunetiere
75017 Paris, France
Tel.: +33 1 48 88 90 59
Fax: +33 1 48 88 90 69
E-mail: sfa@cal.enst.fr

GERMANY

Deutsche Gesellschaft für Akustik (DEGA)
Sec.: Dr. Albert Sill
c/o Dept. Physics/Acoustics
Univ. of Oldenburg
D-26111 Oldenberg, Germany
Tel.: +49 441 798 3572 or 3561
Fax: +49 441 798 3698
E-mail: dega@aku.physik.uni-oldenburg.de
WWW: www.itap.physik.uni-oldenburg.de/dega.html

GREECE

Hellenic Acoustical Society
Sec.: K. Psichas

147 Patission Street
11251 Athens, Greece
Tel.: +01-8646509

HUNGARY

Acoustical Commission of the Hungarian Academy of
  Sciences
Sec.: Prof. P. L. Timar
Egry Jozsef u.18
1111 Budapest, Hungary
Fax: +36 1 166 6808

OPAKFI—Scientific Society for Optics, Acoustics, Motion
  Pictures and Theatre Technology
Pres.: Dr. A. Illényi
Fo u 68
1027 Budapest, Hungary
Tel.: +36 1 202 0452
Fax: +36 1 138 7105
E-mail: borsinear@muszak.radio.bu

INDIA

Acoustical Society of India
c/o Dr. M. L. Munjal
Indian Inst. of Science
Centre of Excellence for Technical Acoustics
Dept. of Mechanical Engineering
Banglore 560012, India
Tel.: +91-80-3092303
Fax: +91-80-3341648
E-mail: munjal@mecheng.iisc.enet.in

Ultrasonic Society of India
V.P.: Dr. V. R. Singh
National Physical Laboratory
De K. S. Krishnan Road
New Delhi 110012, India
Tel.: +91 11 578 3303
Fax: +91 11 575 2678

ITALY

Associazione Italiana di Acustica (AIA)
Sec.: Dr. Giovanni Brambilla
CNR/IDAC
via Cassia 1216
00189 Roma, Italy
Tel.: +39 6 30365765

Fax: +39 6 30365341
E-mail: aia@idac.rm.cnr.it
Società Italiana di Audiologia
Sec.: Prof. Salvatore Iurato
Chair of Audiology
Dept. of Ophthalmology & Otolaryngology Policlinico
I-70124 Bari, Italy
Tel.: +39 80 5478354
Fax: +39 80 5478330
E-mail: bioacous@cimedoc.uniba.it

JAPAN

Acoustical Society of Japan (ASJ)
Sec.: Kenji Goto
Ikeda Bldg., 2-7-7 Yoyogi
Shibuya-ku
Tokyo 151, Japan
Tel.: +81 3 3379 1200
Fax: +81 3 3379 1456
E-mail: kym05145@niftyserve.or.jp
WWW: www.soc.nacsis.ac.jp/asj/asj-index.html

Institute of Noise Control Engineering of Japan (INCE/J)
Kobayashi Inst. of Physical Research
Higashimotomachi 3-20-41, Kukubunji
Tokyo 185, Japan
Tel.: +81 423 25 1852
Fax: +81 423 27 3847

KOREA

Acoustical Society of Korea (ASK)
Science Building 302
635-4 Yucksam-dong
Kangnam-Ku
Seoul 135-703, Korea
Tel.: +82-2-556-3513
Fax: +82-2-569-9717

MEXICO

Instituto Mexicano de Acústica
Pres: M. en C. Sergio Beristáin
Eugenia 240-4
Col. Narvarte
03020 Mexico, D. F.
E-mail: sberista@vmredipn.ipn.mx

Sociedad Mexicana de Acústica, A. C.
Pres. L Ing. Rafael tovamala Landa
Apaseo del Alto 21-6
Col. San Bartolo Atepehuacán
Del. Gustavo A. Madero
C. P. 07730, Mexico, D. F.
Tel.: + 752-85-13
Fax: + 752-61-83

NETHERLANDS

Nederlandse Stichting Geluidhinder (NSG)
Gen. Mgr.: Dr. J. Kuiper
Postbus 381
2600 AJ Delft, Netherlands
Tel.: +31 015 256 2723
Fax: +31 015 257 8663

Nederlands Akoestisch Genootschap (NAG)
Sec.: Dr. ir A. C. Geerlings
TNO Institute of Applied Physics
P.O. Box 155
NL-2600 AD Delft, Netherlands
Tel.: +31 15 69 20 00
Fax: +31 15 69 21 11
E-mail: nag@tpd.tno.nl

Netherlands Society for Audiology
Sec.: Dr. Bert G. A. Van Zanten
Dept. Ent/Hearing & Speech Centre
Dr. Molewaterplein 60
3015 GJ Rotterdam, Netherlands

NEW ZEALAND

New Zealand Acoustical Society
Sec.: Dr. George Dodd
CPO Box 1181
Auckland, New Zealand
Fax: +64 9 6233248

NORWAY

Acoustical Society of Norway
Sec.: Truls Gjestland
Acoustics, NTNV
N-7034 Trondheim, Norway
Tel.: +47 73 59 2645
Fax: +47 73 59 1412
E-mail: truls.gjestland@informatics.sintef.no

PERU

Sociedad Peruana de Acustica
Sec.: Richard Moscoso-Bullon
Garcilazo de le Vega 163
Salamanca de Monterrico
Lima 3, Peru
Tel.: +51 1 435 1151
E-mail: rmoscos@pucp.edu.pe

POLAND

Acoustical Committee of the Polish Academy of Sciences
Pres.: Tadeusz Fidecki
F. Chopin Academy of Music
Okolnik 2
00-368 Warsaw, Poland
Tel.: +48 22 827 72 41
Fax: +48 22 827 83 10
E-mail: akustyka@plearn.edu.pl

Acoustical Society of Poland
Sec.: Dr. Tadeusz Pustelny
Institute of Physics
Silesian Technical University
Krzywoustego 2 Str.
44-100 Gliwice, Poland
Tel.: +48 32 37 12 08
Fax: +48 32 37 22 16
E-mail: pustelny@tytan.matfiz.polsl.gliwice.pl

PORTUGAL

Sociedade Portuguesa de Acústica (SPA)
CAPS-IST
Av. Rovisco Paais
1096 Lisboa Codex, Portugal
Tel.: +351 1 841 9393/39
Fax: +351 1 352 3014

ROMANIA

Rumanian Acoustical Society (SRA)
University Politehnica Bucuresti
Splaiul Independentei no 313
77206 Bucuresti, Rumania
Tel.: +401 410 1615
Fax: +401 410 4488
71102 Bucharest, Roumania

RUSSIA

Russian Acoustical Society
Andreyev Acoustics Institute
4 Shvernik Str.
117063 Moscow, Russia
Tel.: +7 095 126 7401
Fax: +7 095 126 8411
E-mail: ras@asu.acoins.msk.su

East-European Acoustical Association
Moskovskoe shosse 44
196158 St. Petersburg, Russia
Tel.: +7(812) 2919981
Fax: +7(812) 126323
E-mail: krylspb@sovam.com

SLOVAKIA

Slovak Acoustical Society (SAS)
c/o Slovak Academy of Sciences
Račianska 75
P.O. Box 95
830 08 Bratislava 38, Slovakia
Tel.: +42 7 254 751
Fax: +42 7 253 301
E-mail: markus@umms.savba.sk

SOUTH AFRICA

Southern African Acoustics Institute (SAAI)
Sec.: Dr. F. Anderson
P.O. Box 912-169
Silverton 0127, South Africa
Tel.: +2712 832857
Fax: +2712 832857

SPAIN

Sociedad Española de Acustica (SEA)
Sec.: A. Calvo-Manzano Ruiz
Serrano 144
28006 Madrid, Spain
Tel.: +34 1 561 8806
Fax: +34 1 411 7651
E-mail: ssantiago@fresno.csic.es

SWEDEN

Swedish Acoustical Society (SAS)
Sec.: Dr. P. A. Hellstroem
Lindholmen Development
Hearing Research Lab.
P.O. Box 8714
402 75 Goteborg, Sweden
Tel.: +46 31 50 7016
Fax: +46 31 23 9796
E-mail: pa.hellstrom@lindholmen.se

SWITZERLAND

Schweizerische Gesellschaft für Akustik
  (SGA)
Sec.: Dr. K. Heutschi
Postfach 251
CH-8600 Duebendorf, Switzerland
Tel.: +41 1 823 4743
Fax: +41 1 823 4793
E-mail: kurt.heutschi@empa.ch

TURKEY

Turkish Acoustical Society (TAS)
Sec.: Prof. S. Kurra
Dept. of Mechanical Engineering
Istanbul Technical Univ.
Gumussuyu Campus
80191 Istanbul, Turkey
Tel.: +90 212 293 13 45
Fax: +90 212 245 07 95
E-mail: kurra@sariyer.cc.itu.edu.tr

UNITED KINGDOM

British Society of Audiology
Admin. Sec.: Mrs. Ann Allen
80 Brighton Road
Reading RG6 1PS, United Kingdom
Tel.: +44 0118 966 0622
Fax: +44 0118 935 1915
E-mail: bsa@cityscape.co.uk
WWW: www.gold.net/users/hh97/index.html

Institute of Acoustics
Sec.: Mrs. C. Mackenzie
P.O. Box 320
St. Albans, Herts. AL1 1PZ, United Kingdom
Tel.: +44 (0)1427 848195
Fax: +44 (0)1727 850553
E-mail: acoustics@clusl.ulcc.ac.uk
WWW: ioa.essex.ac.uk/ioa

UNITED STATES OF AMERICA

Acoustical Society of America (ASA)
Ex. Dir.: Dr. Charles E. Schmid
500 Sunnyside Blvd.
Woodbury, NY 11797-2999, U.S.A.
Tel.: +516-576-2360
Fax: +516-576-2377
E-mail: asa@aip.org
WWW: asa.aip.org

American Speech-Language-Hearing Association (ASHA)
Exec. Dir.: Frederick Spahr
10801 Rockville Pike
Rockville, MD 20852, U.S.A.
Tel.: +301-897-5700
Fax: +301-571-0469
E-mail: fspahr@asha.org
WWW: www.asha.org

Audio Engineering Society (AES)
Ex. Dir.: Roger K. Furnes
60 East 42nd Street
New York, NY 10165-0075, U.S.A.
Tel.: +212-661-8528
Fax: +212-682-0477

Institute of Noise Control Engineering (INCE)
P.O. Box 3206, Arlington Branch
Poughkeepsie, NY 12603, U.S.A.
Tel.: +914-462-4006
Fax: +914-463-0201
E-mail: inceusa@aol.com
WWW: users.aol.com/inceusa/ince.html

YUGOSLAVIA

Acoustical Society of Yugoslavia
Sec.: Prof. Dr. Peter Pravica
Faculty of Electrical Engineering
Bulevar Revolucije 73
11000 Belgrade, Yugoslavia
Tel.: +381-11-322 55 96
Fax: +381-11-324 86 81
E-mail: epravica@ubbg.etf.bg.ac.yu

APPENDIX C

Selection of an Acoustical Consultant

The following statement is reproduced with permission from the 1997 Membership Directory of the National Council of Acoustical Consultants (NCAC). Further information on the NCAC and copies of the current NCAC Membership Directory may be obtained from the Executive Secretary [NCAC, 66 Morris Avenue, Suite 1 A, Springfield, NJ, 07081-1409, USA; Telephone: (973) 564-5859; Fax: (973) 564-7480]. Many other professional organizations such as the acoustical Society of America (ASA), and the Institute of Noise Control Engineering of the United States of America (INCE) also identify members who provide independent acoustical consulting services. (See addresses and contact numbers in Appendix B.)

SELECTION OF AN ACOUSTICAL CONSULTANT

Every project undertaken by a consultant is unique. While many assignments may be similar in nature, no two ever are identical. For this reason, it is essential that a consultant be chosen with deliberate care. In essence, the more experienced and qualified the consultant is to undertake a given project, the more likely the services will be in accord with the goals and objectives of the client. Moreover, provision of consulting services, by definition, implies a close, privileged relationship between the consultant and his client. To give less than full consideration to the selection/retention process, therefore, would be to jeopardize a successful consultant–client relationship before it begins, thereby jeopardizing the successful outcome of the project at hand.

In the event that you have not already established a relationship with an acoustical consultant, the National Council of Acoustical Consultants recommends for your consideration the following method of selection and retention tested through many years of successful application:

1. Determine, to the extent possible, the nature and scope of the problem or assignment involved.
2. Through contact with mutual acquaintances who have previously utilized acoustical consultants, or from directories of qualified independent consulting firms provided by an organization such as NCAC, identify one or more acoustical consultants who, by virtue of previous experience, stated capabilities, availability and proximity of location, as well as other relevant factors, appear to be generally qualified to undertake the project.
3. Provide project details to the consultants so identified and request from each statements of qualification, including a complete description of the firm, previous assignments and clients, names and biographies of persons who would be working on the project, anticipated time schedules involved, and other factors which relate to the quality of working to be performed.
4. After thorough review of applicant firms, credentials and experience data, possibly including direct contact with firm representatives if such can be arranged, identify the firm which appears most qualified to serve your specific requirements.
5. Contact representatives of that firm believed to be most qualified and open negotiations to establish a mutually acceptable consulting fee arrangement and payment methodology. Most consultants are experienced in at least several types of retention agreements, including hourly fixed rate, fixed fee, cost plus fixed fee, percentage of overhead, etc. Usually one of these will be most suited to the type of work involved.
6. If the negotiations prove satisfactory, the client should at this point retain the consultant to ensure availability for the project. If negotiations are not successful, they should be terminated and opened with other qualified firms, one at a time.

It should be noted that NCAC encourages open and frank discussion of financial concerns between the client and consultant. Experience demonstrates that mutually satisfactory client–consultant relationships rest predominantly on the consultant's ability to deliver cost effective services on time and within the scope of the agreement. In fact, most successful consultants pride themselves on their ability to tailor their efforts to the scope of the project and the budget available for services as well as for implementation of the recommendations resulting from their services. It is urged strongly, however, that discussion of fees be divorced completely from the ranking of qualifications to prevent financial considerations from biasing the selection process. True economy results only when services provided are cost-effective in the long-term, helping ensure results which satisfy the client's needs from an overall standpoint. A consultant who is fully competent to undertake the work is the one most likely to provide such results.

Glossary

Absorbing Materials
Materials that dissipate acoustic energy within their structure as heat and/or mechanical energy of vibration. Usually building materials designed specifically for the purpose of absorbing acoustic energy on the boundary surfaces of rooms or in the cavities of structures.

Absorption Coefficient
See Sound Absorption Coefficient.

Acoustical Analysis
The detailed study of all pertinent sound sources, sound transmission paths, and sound receptors in the context of a particular acoustical problem.

Acoustical Environment
The overall environment, interior to exterior, that affects the acoustic conditions of the space or structure under consideration.

Acoustical Treatment
The use of acoustical absorbing or reflecting materials or sound-isolating structures to improve or modify the acoustical environment.

Airborne Sound
Sound transmitted through air as a medium rather than through solids or the structure of a building.

Amplification
Usually the increase in intensity level of an audible signal produced by means of loudspeakers and associated electrical amplification apparatus.

Amplitude
The maximum displacement to either side of the normal or rest position of the molecules or particles of the medium transmitting a sound wave.

Antiphonal Organ
Literally, "echo organ." Typically, the antiphonal organ comprises a division of the main organ in a church but located at the opposite end of the church so it can participate in responsive musical dialog with the main instrument.

Architectural Acoustics
The science and technology of controlling sound in and around buildings.

Articulation
The percentage of speech units correctly perceived by a listener. Similar to "intelligibility" except that articulation applies to meaningless fragments of speech, while intelligibility applies to meaningful material.

Articulation Class (AC)
A classification method that rates building systems and subsystems for speech privacy purposes in accordance with ASTM E1110-86 (reapproved 1994).

Attenuation
Reducing the magnitude of a sound signal by separation of a sound source from a receptor, acoustical absorption, enclosure, active cancellation by electronic means or a combination of these or other means.

Audibility Threshold
The sound pressure at which a typical young, healthy person with normal hearing begins to respond.

Audible
Capable of producing the sensation of hearing.

Background Sound Level
The generally lowest or residual sound level present in a space above which speech, music, or other sounds may be heard.

Ceiling Attenuation Class (CAC)
A laboratory derived rating for this sound attenuation of a suspended ceiling system over two adjacent rooms sharing a common plenum.

Chancel
The part of a church that is the center of the service. Usually includes the altar, lectern, and pulpit and often the choir and organ.

Coloration
The noticeable change in sound quality that occurs when a signal passes through an imperfect transmission system such as enclosed space or a sound reproduction system.

Column Loudspeaker
A linear array consisting of multiple loudspeakers interconnected to perform as a single loudspeaker system. Also called "line source."

Coupling
Any means of joining separated elements in any media so that sound energy is transmitted between them.

Cycle
The entire sequence of movement of a particle (during periodic motion) from rest to one extreme of displacement, back through the rest position to the opposite extreme of displacement, and back to the rest position. The unit for one cycle per second is hertz (Hz).

Damping
Damping refers to energy dissipation in an oscillating system. A damped system cannot oscillate freely.

Decible (dB)
A basic metric for describing the magnitude of sound. A division of a uniform scale based upon 10 times the logarithm to the base 10 of the relative value being compared (sound intensity, pressure squared, or power) to a specified reference value.

Diffraction
The ability of a sound wave to "flow" around an obstruction or through openings with little loss of energy.

Diffusion
Dispersion of sound within an enclosure such that there is uniform energy density throughout the space. A perfectly diffuse sound field is one in which there is an equal flow of sound power in all directions at any point.

Direct Sound
A sound field produced and dominated by the sound waves emanating directly from a source (also referred to as the "near field").

Distortion
Any change in the transmitted sound signal such that the sound received is not a faithful replica of the original source sound.

Distribution
The pattern of sound intensity level within a space; also, the patterns of sound dispersion the sound travels within a space.

Dynamic Insertion Loss (DIL)
The reduction in sound level due to the insertion of a sound attenuating element in the sound transmission path under normal operating conditions of air or fluid flow in the system.

Dynamic Range
The range in decibels between the maximum and minimum signal levels from a device or sound source.

Echo
Any reflected sound of sufficient intensity and delay in arrival to be heard as distinct from the source.

Field Sound Transmission Class (FSTC)
A rating of the field-derived airborne sound transmission loss data for structure determined in accordance with the procedure of ASTM E 336 and E 413.

Flanking Paths
Any paths for sound transmission that bypass or circumvent the primary path through the structure under consideration.

Flutter
A rapid reflection or echo pattern between opposing sound-reflective walls with sufficient time between each reflection to cause a listener to be aware of separate discrete signals.

Focusing
Concentration of reflected acoustic energy within a limited location in a room as the result of reflections from concave surfaces.

Forestage
The portion of stage that exists in front of the proscenium or curtain line.

Frequency
The number of complete cycles per second of a vibration (or other periodic motion). The unit is hertz (Hz).

Frequency Analyzer
An instrument capable of measuring the acoustical energy present in various frequency subdivisions (one, one-third, one-tenth-octave bands, etc.) of a complex sound.

Frequency Band
A subdivision of the frequency range of interest for measurement of analysis purposes (e.g., octave band, one-third-octave band, etc.).

Frequency Spectra
The distribution of sound energy over a frequency range of interest (e.g., the octave band sound pressure levels for a particular sound source plotted over the audible frequency range).

Hearing
The subjective response to sound, including the entire mechanism of the external, middle, and internal ear system and the nervous and cerebral operations that translate the physical stimuli into meaningful signals.

Hertz (Hz)
Cycles per second, used to describe frequency (*see* Frequency).

Horn Loudspeaker
Horns are used to greatly increase the radiation efficiency and directivity of a loudspeaker driver and to radiate that sound energy within a particular solid angle. (Note that horns designed for this purpose have a very different function from horns designed as musical instruments since the latter are built to be effective resonators, while resonances are undesirable in loudspeakers.)

Insulation
A fibrous material that absorbs sound and insulates against heat and cold. Sometimes (esp. in Britain) also synonymous with *Isolation*.

Intensity
The rate of sound energy transmitted in a specified direction through a unit area.

Intensity Level
A measure of the magnitude of sound intensity on a logarithmic scale; a value equal to 10 times the logarithm to the base 10 of the ratio of a sound intensity to a referance intensity. (The reference intensity is usually taken to be 10^{-16} W/cm^2.)

In-the-Round
Refers to an audience–performer relationship in auditoriums in which the audience is seated all around the performing area.

Inverse Square Law
At a distance from a source, under free-field conditions, sound intensity varies inversely with the square of the distance from the source, resulting in a decrease in sound pressure level of 6 dB for each doubling of distance.

Isolation
The ability of materials or constructions to resist that transmission of sound or vibration through them.

Line Level
Level of electric signal encountered in audio transmission lines, generally between -20 and $+30$ dBm.

Loudness
The subjective response of the human hearing mechanism to changing sound pressure.

Masking
The obscuring or covering up of one sound by another.

Microphone Level
Level of electrical signal available on the output terminals of a microphone, generally between -65 and -50 dBm.

Natural Frequency
The frequency at which a spring, a pendulum, a column of air, or other such body or object resonates when set in motion.

Noise
Unwanted sound.

Noise Criteria Curves (NC, RC, NCB, etc.)
A numerical rating system or family of curves used to specify background sound levels over a specified frequency range.

Noise Insulation Class (NIC)
A rating of the measured room-to-room noise reduction determined in accordance with the procedures of ASTME E 336 and E 413.

Noise Reduction
The reduction in level of unwanted sound by any of several means (e.g., by distance in outdoor space, by boundary surface absorption, by isolating barriers of enclosures, etc.).

Noise Reduction Coefficient (NRC)
The arithmetic average of the sound absorption coefficients of a sound-absorbing material at 250, 500, 1000, and 2000 Hz rounded to the nearest 0.05.

Octave Band
A division of the audible frequency range, the center frequency of which is twice that of the preceding band center frequency. The standard acoustical octave bands are centered at 16, 31.5, 63, 125, 250, 500, 1000, 2000, 4000, and 8000 Hz.

Outdoor–Indoor Transmission Class (OITC)
A classification method for rating the sound isolation performance of exterior wall constructions in accordance with ASTM E1332-90.

Pain Threshold
A sound intensity level sufficiently high to produce the sensation of pain in the human ear (usually about 120 dB).

Pink Noise
Noise signal obtained by feeding a white noise signal through a filter that decreases the spectrum level by 3 dB for each octave or doubling of frequency. The spectrum level of pink noise is constant as a function of frequency.

Pitch
The subjective response of the human hearing mechanism to changing frequency.

Reflected Sound
The resultant sound energy returned from a surface(s) that is not absorbed or otherwise dissipated upon contact with the surface(s).

Resilient Mounting
Any mounting or attachment system that reduces the transmission of vibrational energy from a vibrating body or structure to an adjacent structure.

Resonance
The natural, sympathetic vibration of a volume of air or a structure at a particular frequency as the result of excitation by sound energy at that particular frequency.

Resonant-Panel Absorption
The dissipation of sound energy that occurs at the natural frequency of oscillation of a panel having a sealed air cavity behind.

Reverberant Sound
A sound field produced and dominated by reflected sound from room boundary surfaces.

Reverberation
The persistence of sound within a space after the sound source has stopped.

Reverberation Time
The time in seconds required for a sound to decay, roughly speaking, to inaudibility after the source ceases. (Strictly, the time in seconds for the sound level at specific frequency to decrease 60 dB in level after the source stops.)

Ring
A type of repeating echo that occurs as sound reflections persist longer in a particular plane than reflections occurring in all other paths. In a rectangular solid, for example, if reflections are relatively suppressed in one orthogonal direction, sound reflections persisting in the plane of the remaining two directions will have prominence.

Room Mode
A resonant or natural-oscillation frequency in an enclosed space.

Room Shape
The configuration of enclosed space, resulting from the orientation and arrangement of surfaces defining the space.

Room Volume
The cubic volume of space enclosed by the room boundary surfaces.

Sabin
The measure of sound absorption of a surface equivalent to 1 ft^2 of a perfectly absorptive material (named after Wallace Clement Sabine, a pioneer in architectural acoustics).

Sound
Vibrations in a medium, usually in the frequency range capable of producing the sensation of hearing.

Sound Absorption Coefficient
The ratio of sound-absorbing effectiveness (at a specific frequency) of 1 ft^2 of acoustical absorbent to 1 ft^2 of perfectly absorptive material; usually expressed as a decimal value between 1.0 (perfect absorption) and 0 (perfect reflection).

Sound Control
The application of acoustical principles to the design of structures, equipment, and spaces to permit them to function properly and to create the desired environment for the activities intended.

Sound Leak
Any opening in a structure that permits airborne sound transmission with little or no loss compared with the basic structure.

Sound Level
A measure of sound pressure level as determined by instrumentation with standardized frequency-weighting characteristics (e.g., A scale sound level in dBA).

Sound Level Meter
A standard electrical instrument for determining sound pressure level.

Sound Power
The rate at which sound energy is radiated, expressed in watts.

Sound Pressure
The change in pressure at a point due to sound energy relative to the static pressure at that point without the sound wave.

Sound Pressure Level
A measure of the magnitude of sound pressure on a logarithmic scale; a value equal to 20 times the logarithm to the base 10 of the ratio of a sound pressure to a reference pressure (the reference pressure is usually taken to be 2×10^{-5} N/m^2).

Sound Reinforcement
Refers to the beneficial reinforcement of a sound signal provided by sound-reflecting surfaces or by a loudspeaker system.

Sound Transmission
The propagation of sound energy through various media.

Sound Transmission Class (STC)
A rating of the laboratory-derived airborne sound transmission loss data for a structure determined in accordance with the procedures of ASTME E 90 and E 413.

Sound Velocity
The velocity at which a sound wave propagates through a medium (e.g., in air, the speed of sound is approximately 1100 ft/s).

Sound Wave
A disturbance that is propagated in any medium (gas, liquid, or solid) by energy transfer between adjacent molecules.

Structure-Borne Sound
Sound energy transmitted through solid elements of a building structure.

Thrust Stage
A stage that projects into the audience area, thereby increasing intimacy by avoiding the barrier effect of a proscenium frame.

Timbre
The quality of a sound that distinguishes it from other sounds of the same pitch and volume, especially, the distinctive tone of a musical instrument, a voice or a voiced speech sound. Timbre depends on the harmonic or overtone structure of the sound.

Transepts
The part of a cruciform church that crosses at right angles to the greatest length between the nave and the apse or choir.

Transmission Loss (TL)
A logarithmic measure of the decrease in sound power during transmission from one point to another (or through a panel, wall, etc.,); a value in decibels equal to 10 times the logarithm to the base 10 of the ratio of transmitted-to-incident sound power.

Vibration
A cyclical alternation in pressure or direction of movement.

Vibration Isolation
Any of several means of minimizing transmission of structure-borne sound from a vibrating body to the structure in or on which it is mounted.

Volume (Helmholtz) Resonator
A container or cavity having a restricted opening will dissipate sound energy at its natural resonant frequency because of frictional losses occurring as the air vibrates within the constricted opening.

Wavelength
The distance between adjacent regions of a sound wave where identical conditions of particle displacement or pressure occur.

White Noise
Noise signal having a continuous frequency spectrum that has equal power for any frequency band of constant width. The spectrum level of white noise increases 3 dB with each octave or doubling of frequency.

Index

Abbreviations, 305–308
Acoustical and Board Products Association (ABPA) standards, 44
Acoustical concepts, 2–35. *See also* Room acoustic qualities
   cavity absorption in double constructions, 25
   ceiling attenuation class, 31
   combining decibels, 15–16
   composite constructions, 25–26
   decibel scale, 8–9
   direct structure-borne sound, control of, 32–33
   double walls, 24–25
   field sound transmission class, 32
   flanking paths, 27–29
   frequency, 4–6
   frequency bands, 6–7
   frequency-weighted sound levels, 9–12
   impact insulation class, 33–34
   isolation of equipment, 34–35
   magnitude of sound, 7–8
   noise isolation class, 32
   noise reduction coefficient, 20
   outdoor versus indoor sound, 16–18
   overview, 1–2
   reverberation, indoor, 21–22
   single homogeneous walls, 23–24
   sound-absorbing materials, 18–20
   sound and sound control, 3–4
   sound leaks, 27
   sound level change, relative, 16
   sound level reduction, indoor, 20–21
   sound transmission between rooms, 22–23
   sound transmission class, 30–31
   sound transmission loss, 23
      field measurement of airborne, 32
      laboratory measurement of airborne, 29–30
   time-varying sound levels, 12–15
   wavelength, 7

Acoustical consultant, selection of, 316–317
Acoustical design. *See* Design
Acoustical louver, noise control methods, 131
Acoustical materials. *See also* Materials
   case study, Duke University Chapel, 95–99
   design and research innovations, 269
   listing of, 65–71
   performance tables, 76–93
      airborne sound isolation performance data, 80–93
      impact and airborne sound isolation performance data, 89–93
      sound absorption coefficients, 76–77
      sound absorption data for selected, 78–79
Acoustical modeling, 234, 264–268
   aural simulation, 268
   case study, 297, 301–304
   computer modeling techniques, 267–268
   expert systems, 268
   noise control methods, HVAC systems, 136–139
   physical modeling techniques, 266–267
   studies in, 264–266
Acoustical privacy, 101–110
   closed-plan spaces, 101–105
   open-plan spaces, 105–110
Acoustical shielding, sound attenuation, 58
Acoustical societies, listing of global, 309–315
Acoustical thermometer, magnitude of sound, 8, 9
Active noise control systems, design and research innovations, 269–271
Airborne sound attenuation, 56–58

Airborne sound isolation performance data, performance tables, 80–93
Air Diffusion Council (ADC) standards, 44–45
Airflow, noise control methods, HVAC systems, 135–136
Air springs, acoustical properties of, 71
American National Standards Institute (ANSI):
   background sound levels, design criteria, 37
   noise control criteria, 110, 111–114
   standards, 45
American Society for Testing and Materials (ASTM):
   ceiling attenuation class, 31
   field sound transmission class, 32
   impact insulation class, 33
   noise control criteria, 110, 111, 117, 118, 125
   noise isolation class, 32
   sound-absorbing materials, 18
   sound absorption, 61
   sound transmission loss, 29
   standards of, 45–46
American Society of Heating, Refrigerating and Air Conditioning Engineers (ASHRAE):
   noise control criteria, 110, 111, 126, 128, 129, 136–139, 146
   standards of, 45
Amphitheater design, outdoor, 152–153
Announcement system, sound reinforcement system case study, 222–223
Apparent source width (ASW), concert hall design, 160
Architectural features, effects of, room acoustic qualities, 263–264
Assembly hall, sound reinforcement systems, 201–202

327

328 • Index

Athletic facility. *See* Sports facility
Attenuation. *See* Sound attenuation
Auditoriums:
   case studies
      Ben Franklin Hall, 185–186
      Cargill 97 Lecture Auditorium, 206–208
      Concord-Carlisle Regional High School, 205–206
   sound reinforcement systems, 200
Aural simulation, acoustical modeling, 234, 268

Background sound levels, design criteria, 36–39
Balanced noise criteria curve (NCB):
   design criteria, 37
   noise control criteria, 111, 112–114
Barriers. *See* Noise barriers
Bass ratio, based on early decay time, room acoustic qualities measurement, 262
Batts and blankets, fibrous, acoustical properties of, 68
Ben Franklin Hall (Philadelphia, Pennsylvania), 185–186
Beranek, Leo L., 151, 157
Brick, acoustical properties, 63
Broadcast system, sound reinforcement system case study, 222
Building codes, noise control criteria, 114–120, 121
Building exterior envelope design, noise control methods, 125–126
Building floor/ceiling design, noise control methods, 122–125
Building noise control. *See* Noise control
Building Officials and Code Administrators International (BOCA), 110, 115, 116–117
Building partition design, noise control methods, 121–122

Cargill 97 Lecture Auditorium (Northeastern University School of Law, Boston, Mass.), 206–208
Carpet, acoustical properties of, 67
Cathedral design, 175–176
Cavity absorption, in double constructions, acoustical concepts, 25
Ceiling attenuation class, acoustical concepts, 31
Ceiling design, noise control methods, 122–125
Cellulose fiber, acoustical properties of, 67
Channels, resilient, acoustical properties of, 73
Churches. *See* Worship spaces

Clips, resilient, acoustical properties of, 73
Closed-plan spaces, acoustical privacy, 15–16, 101–105
Combined decibels, acoustical concepts, 15–16
Community Church of Vero Beach, Florida, 212–215
Community noise exposure level (CNEL), noise control criteria, 117
Composite constructions, acoustical concepts, 25–26
Computer modeling techniques, 267–268
Concert halls, 155–167
   case studies
      Evangeline Atwood Concert Hall, Alaska Center for the Performing Arts, 293–300
      Hong Kong Cultural Center, 223–226
      Lenna Hall at Chautauqua Institution, 183–184
      McDermott Concert Hall, Morton H. Meyerson Symphony Center, 278–293
      Segerstrom Hall, Orange County Performing Arts Center, 277–278
   electronic concert hall, 166–167
   in-the-round, 161, 165
   multipurpose, 166
   panel-array halls, 161
   partially enclosed, 165–166
   shoebox approach, 160–161
   sound reinforcement systems, 199–200, 223–226
   subjective attributes and objective measures, 157–160
Concord-Carlisle Regional High School auditorium (Concord, Mass.), 205–206
Concrete, acoustical properties, 63
Concrete masonry units, acoustical properties, 63
Conference room:
   design of, 179–180
   sound reinforcement systems, 200, 217, 220
Console:
   control, sound reinforcement systems, 192
   stage manager, sound reinforcement system case study, 230
Convention centers, Queen Sirikit National Convention Center (Bangkok, Thailand), 215–223
Conversion factors, 305–308
Cue light system, sound reinforcement system case study, 230
Cultural events, sports facility design, 178–179
Curtains, acoustical properties of, 67

Day-night average sound level (DNL), noise control criteria, 117
Decibels, combined, acoustical concepts, 15–16
Decibel scale, acoustical concepts, 8–9
Deck, acoustical, properties of, 66
Design, 151–186
   case studies
      Ben Franklin Hall (Philadelphia, Pennsylvania) case study, 185–186
      Hitchcock Presbyterian Church (Scarsdale, N.Y.), 182–183
      Lenna Hall (Chautauqua, N.Y.), 183–184
   concert halls, 155–167
      electronic concert hall, 166–167
      in-the-round, 161, 165
      multipurpose, 166
      partially enclosed, 165–166
      shoebox approach, 160–161
      subjective attributes and objective measures, 157–160
   conference rooms, 179–180
   criteria for, 35–44
      background sound levels, 36–39
      general, 35–36
      high noise level areas, 39–40
      listening and performance rooms, 43–44
      mechanical systems, 41–43
      sound isolation between dwelling units, 40–41
   historical perspective, 151–152
   home listening rooms, 180–181
   lecture rooms, 181
   motion picture theaters, 174–175
   movable stage enclosures, 173–174
   music buildings, 179
   opera houses, 168–170
   outdoor amphitheaters, 152–153
   outdoor/indoor transition, 153–155
   recital halls, 167–168
   reverberation goals, 155
   sports facilities, 178–179
   theaters, 170–171
   worship spaces, 175–177
Design and research innovations, 233–304
   acoustical modeling, 264–268
      aural simulation, 268
      computer modeling techniques, 267–268
      expert systems, 268
      physical modeling techniques, 266–267
      studies in, 264–266
   active noise control systems, 269–271
   case studies
      acoustical model tests, 297, 301–304
      Evangeline Atwood Concert Hall (Anchorage, Alaska), 293–300

McDermott Concert Hall (Dallas, Texas), 278–293
  Segerstrom Hall, Orange County Performing Arts Center, 277–278
  impact noise, 272
  mechanical systems and health and energy issues, 272–273
  overview, 233–235
  plumbing system noise control, 271–272
  room acoustic qualities, 235–264
    architectural feature effects measurement, 263–264
    existing rooms measurement, 263
    general, 235–236
    impulse response, 254–260
    measurement systems, 253–254
    measures derived from impulse response, 261–263
    research summary, 236–246
    sound perception issues, 246–247
    sound quality assessment methods, 247–253
    summary table, 264
  sound-diffusing materials, 269
  sound isolation testing, 271
Direct structure-borne sound, control of, acoustical concepts, 32–33
Discrete reflections control, sound absorption, 62
Distribution amplifiers, signal processing equipment, 194
Double constructions, cavity absorption in, acoustical concepts, 25
Double walls, sound transmission loss, acoustical concepts, 24–25
Duct attenuation, sound attenuation, 59
Duct lining, acoustical properties of, 67–68
Duct silencers:
  acoustical properties of, 71
  noise control methods, 128–134
Duke University Chapel, 95–99
Dwelling units, sound isolation between, design criteria, 40–41
Dynamic insertion loss (DIL), 128, 130, 131. See also Noise control methods

Ear, frequency-weighted sound levels, 9–12
Early decay time:
  bass ratio based on, room acoustic qualities measurement, 262
  room acoustic qualities measurement, 261
  treble ratio based on, room acoustic qualities measurement, 262
Early inter-aural cross correlation coefficient, room acoustic qualities measurement, 262

Early-late ratios:
  concert hall design, 159
  room acoustic qualities measurement, 261
Elastomers, acoustical properties of, 71–72
Electronic concert hall design, 166–167
Electronic delays, signal processing equipment, 193
Emission-type limits, building codes, noise control criteria, 114
Energy issues, mechanical systems noise control, 272–273
Energy-time curve (ETC), concert hall design, 157–158
Environmental Protection Agency (EPA), sound levels, 9, 12
Evangeline Atwood Concert Hall, Alaska Center for the Performing Arts, 293–300
Exhibition hall, sound reinforcement systems, 201–202
Expert systems, acoustical modeling, 268
Exterior envelope design, noise control methods, 125–126

Fans, noise control methods, 126–128
Federal Aviation Administration (FAA), 15, 118
Federal Department of Housing and Urban Development (HUD), 15, 40–41
Federal Highway Administration (FHWA), 15
Federal Housing Administration (FHA), 40–41, 42
Feedback suppressors, signal processing equipment, 193–194
Fiberglass, acoustical properties of, 68
Fibrous batts and blankets, acoustical properties of, 68
Fibrous board, acoustical properties of, 69
Fibrous plank, acoustical properties of, 69
Fibrous spray, acoustical properties of, 69
Field impact insulation class (FIIC), noise control criteria, 118
Field sound transmission class (FSTC):
  acoustical concepts, 32
  noise control criteria, 117
Flanking paths, acoustical concepts, 27–29
Flexible connections, acoustical properties of, 72
Floor design, noise control methods, 122–125
Foam, acoustical, properties of, 66
Fogg Art Museum Lecture Hall (Harvard University), 47–54
Frequency, acoustical concepts, 4–6

Frequency bands, acoustical concepts, 6–7
Frequency-weighted sound levels, acoustical concepts, 9–12
Functional absorbers, acoustical properties of, 72–73

Gaskets, acoustical properties of, 73
General Services Administration (GSA) standards, 46
Glass:
  acoustical properties, 63
  laminated, acoustical properties of, 69
Glass-fiber linings, mechanical systems, design and research innovations, 272–273
Grilles, acoustical properties of, 71
Gypsum board, acoustical properties, 63–64

Hangers, resilient, acoustical properties of, 73
"Hard-of-hearing" systems, sound reinforcement systems, 203
Health risks, mechanical systems noise control, design and research innovations, 272–273
Heating, ventilation, and air conditioning (HVAC) systems:
  background sound levels, design criteria, 37, 38
  duct attenuation, 59
  listening and performance rooms, design criteria, 43
  mechanical systems, design criteria, 41–43
  noise control methods, 126–139
    duct silencers, 128–134
    fans, 126–128
High noise level areas, design criteria, 39–40
Hitchcock Presbyterian Church (Scarsdale, N.Y.), 182–183
Home listening room design, 180–181
Hong Kong Cultural Center, 223–232
Hospital-type duct silencer, noise control methods, 130

Impact attenuation, sound attenuation, 58
Impact insulation class, acoustical concepts, 33–34
Impact noise, design and research innovations, 272
Impact sound isolation performance data, performance tables, 89–93
Impulse response:
  measures derived from, room acoustic qualities, 261–263
  room acoustic qualities, 254–260

Indoor sound:
  outdoor sound versus, acoustical concepts, 16–18
  reduction of levels, acoustical concepts, 20–21
  reverberation, acoustical concepts, 21–22
  transmission between rooms, acoustical concepts, 22–23
Industrial facilities, sound reinforcement systems, 203
Insulation (loose), acoustical properties of, 69
Interaural cross-correlation coefficient (IACC), concert hall design, 160
Intercom system, sound reinforcement system case study, 230, 232
International Conference of Building Officials (ICBO), 110, 115, 116, 117
International Standards Organization (ISO), 46
Interpretation systems, simultaneous, 204, 217, 220
In-the-round concert hall design, 161, 165
Isolation:
  of equipment, acoustical concepts, 34–35
  vibration, noise control methods, 139–147

Laminated glass, acoustical properties of, 69
Lateral energy fraction, room acoustic qualities measurement, 262–263
Lateral fraction (LF) ratio, concert hall design, 160
Lead sheet, acoustical properties of, 69
Lecture halls:
  case studies
    Cargill 97 Lecture Auditorium, 206–208
    Fogg Art Museum Lecture Hall (Harvard University), 47–54
  design of, 181
  sound reinforcement systems, 200
Lenna Hall (Chautauqua, N.Y.), 183–184
Listener envelopment (LEV), concert hall design, 160
Listening environment, outdoor amphitheater, 153
Listening room design, 43–44, 180–181
Loose insulation, acoustical properties of, 69
Loudspeaker systems, sound reinforcement, 189–191, 194–196. See also Sound reinforcement systems
Louisiana Superdome, 208–212

Magnitude, of sound, acoustical concepts, 7–8
Masonry, acoustical properties, 64

Materials, 55–99
  case study, Duke University Chapel, 95–99
  design and research innovations, 269
  listing of acoustical materials, 65–71
  listing of common materials, 63–65
  overview, 55
  performance tables, 76–93
    airborne sound isolation performance data, 80–93
    impact and airborne sound isolation performance data, 89–93
    sound absorption coefficients, 76–77
    sound absorption data for selected, 78–79
  sound-absorbing, acoustical concepts, 18–20
  sound absorption concepts, 59–63
    attenuation improvement, 62–63
    discrete reflections control, 62
    noise control, 62
    principles, 59–61
    reverberation control, 61–62
  sound attenuation concepts, 56–59
    acoustical shielding, 58
    airborne attenuation, 56–58
    duct attenuation, 59
    impact attenuation, 58
    vibration isolation, 58–59
  special devices, 71–76
  worship spaces design, 177
McDermott Concert Hall, Morton H. Meyerson Symphony Center (Dallas, Texas), 278–293
Measurement:
  concert hall design, 157–160
  field measurement of airborne sound transmission loss, 32
  laboratory measurement of airborne sound transmission loss, 29–30
  room acoustic qualities, 233–234, 253–254. See also Room acoustic qualities
    architectural feature effects, 263–264
    existing rooms, 263
    impulse response, 261–263
Mechanical/electrical equipment, isolation of, acoustical concepts, 34–35
Mechanical systems:
  design and research innovations, 272–273
  design criteria, 41–43
Mechanics Hall (Worcester, Mass.), 149–150
Meeting room. See Conference room
Metal pans, acoustical properties of, 69
Metals, acoustical properties, 64
Microphones, sound reinforcement systems, 191
Mineral fiber, acoustical properties of, 69
Modeling. See Acoustical modeling

Motion picture theater:
  design of, 174–175
  sound reinforcement systems, 200
Mounts, resilient, acoustical properties of, 75
Movable stage enclosure design, 173–174
Multipurpose auditoria design, 171–173
Multipurpose concert hall design, 166
Music building design, 179

National School Supply and Equipment Association (NSSEA) standards, 46
Natural frequency, vibration isolation, 141–142
Noise barriers, noise control methods, 120–121
Noise control, 100–150
  acoustical analysis, 100–110
    acoustical privacy, 101–110
      closed-plan spaces, 101–105
      open-plan spaces, 105–110
    systems approach, 100–101
  case study, Mechanics Hall (Worcester, Mass.), 149–150
  criteria, 110–120
    balanced noise criteria curves (NCB curves), 112–114
    building codes, 114–120, 121
    noise criteria curves (NC curves), 111–112
    room criteria curves (RC curves), 114
    standards and organization, 110–111
  overview, 100
  sound absorption, 62
Noise control methods, 120–147
  building exterior envelope design, 125–126
  building floor/ceiling design, 122–125
  building partition design, 121–122
  HVAC systems, 126–139
    acoustical modeling, 136–139
    airflow, 135–136
    duct silencers, 128–134
    fans, 126–128
  noise barriers, 120–121
  vibration isolation, 139–147
Noise criteria (NC) curves:
  background sound levels, design criteria, 36–39
  noise control criteria, 111–112
Noise floor or background sound levels, design criteria, 36
Noise isolation class (NIC):
  acoustical concepts, 32
  noise control criteria, 118
Noise reduction, airborne sound attenuation, 56
Noise reduction coefficient, acoustical concepts, 20
Normalized noise isolation class (NNIC), noise control criteria, 118

Normalized A-weighted sound level difference (DNL), noise control criteria, 118
North American Insulation Manufacturers Association (NAIMA), 129

Objective measures, concert hall design, 157–160
Occupational Safety and Health Administration (OSHA), 39–40
Office spaces, sound reinforcement systems, 200–201
Open-plan spaces:
    acoustical privacy, 15, 105–110
    sound reinforcement systems, 201
Opera houses:
    design of, 168–170
    sound reinforcement systems, 198–199
Orchestra pit, opera house design, 169–170
Organ-choir arrangement, worship spaces design, 177
Outdoor amphitheater design, 152–153
Outdoor-indoor transmission class (OITC), noise control methods, 125
Outdoor sound, indoor sound versus, acoustical concepts, 16–18

Paging systems, sound reinforcement systems, 203, 222–223, 228, 230
Panel-array halls, concert hall design, 161
Partially enclosed concert hall design, 165–166
Partition. *See* Building partition design
Path, source and receiver, noise control, 100–101
Performance rooms, design criteria, 43–44
Performance tables (materials), 76–93
    airborne sound isolation performance data, 80–93
    impact and airborne sound isolation performance data, 89–93
    sound absorption coefficients, 76–77
    sound absorption data for selected, 78–79
Performing environment, outdoor amphitheater, 153
Physical modeling techniques, 266–267
Piping, noise control methods, HVAC systems, 138–139
Plaster, acoustical properties, 64, 66
Plumbing system noise control, design and research innovations, 271–272
Plywood, acoustical properties, 64
Power amplifiers, sound reinforcement systems, 194
Power plants, sound reinforcement systems, 203
Preamplifiers, sound reinforcement systems, 191–192
Privacy. *See* Acoustical privacy

Production communications system, sound reinforcement system case studies, 222, 230
Program monitoring system, sound reinforcement system case studies, 220, 222, 228
Pumps, noise control methods, HVAC systems, 138–139

Quadratic-residue diffusors, acoustical properties of, 73, 269
Queen Sirikit National Convention Center sound system (Bangkok, Thailand), 215–223

Rapid speech transmission index (RASTI), concert hall design, 159
Receiver, path and source, noise control, systems approach, 100–101
Recital hall design, 167–168. *See also* Concert halls
Reinforcement systems. *See* Sound reinforcement systems
Relative loudness or strength, room acoustic qualities measurement, 261–262
Relative sound level change, acoustical concepts, 16
Religious building design. *See* Worship spaces
Research. *See* Design and research innovations
Resilient clips and channels, acoustical properties of, 73
Resilient hangers, acoustical properties of, 73
Resilient mounts, acoustical properties of, 75
Resilient tile, acoustical properties, 64
Reverberation:
    design, 155
    indoor, acoustical concepts, 21–22
Reverberation control, sound absorption, 61–62
Reverberation time, room acoustic qualities measurement, 261
Room acoustic qualities, 233–234, 235–264
    architectural feature effects measurement, 263–264
    case study, 297, 301–304
    existing rooms measurement, 263
    general, 235–236
    impulse response, 254–260
    measurement systems, 253–254
    measures derived from impulse response, 261–263
    research summary, 236–246
    sound perception issues, 246–247
    sound quality assessment methods, 247–253
    actual rooms, 247–249
    laboratory analysis, 249–253
    summary table, 264

Room criteria (RC) curves:
    background sound levels, design criteria, 37
    noise control criteria, 111, 114

Sabine, Wallace C., 151
Sealants, acoustical properties of, 69
Segerstrom Hall, Orange County Performing Arts Center, 277–278
Shoebox approach, concert hall design, 160–161
Signal distribution, sound reinforcement system case study, 222
Signal processing equipment, sound reinforcement systems, 192–194
Silencers. *See* Duct silencer
Simultaneous interpretation systems, 204, 217, 220
Single homogeneous walls, sound transmission loss, 23–24
Slats, acoustical properties of, 71
Sound:
    acoustical concepts, 3–4
    magnitude of, acoustical concepts, 7–8
    outdoor versus indoor, acoustical concepts, 16–18
    wavelength of, 7
Sound-absorbing materials, acoustical concepts, 18–20
Sound absorption, 59–63
    attenuation improvement, 62–63
    discrete reflections control, 62
    noise control, 62
    principles, 59–61
    reverberation control, 61–62
    sports facility design, 178
Sound absorption coefficients, performance tables (materials), 76–77
Sound attenuation, 56–59
    acoustical shielding, 58
    airborne attenuation, 56–58
    duct attenuation, 59
    impact attenuation, 58
    improvement in, sound absorption, 62–63
    materials, 56–59
    vibration isolation, 58–59
Sound isolation, between dwelling units, design criteria, 40–41
Sound isolation testing, design and research innovations, 271
Sound leaks, acoustical concepts, 27
Sound levels:
    reduction of, indoor, acoustical concepts, 20–21
    relative change in, acoustical concepts, 16
    time varying, acoustical concepts, 12–15
Sound quality assessment methods, 247–253
    actual rooms, 247–249
    laboratory analysis, 249–253

332 • *Index*

Sound reinforcement systems, 187–232
  case studies
    Cargill 97 Lecture Auditorium (Boston, Mass), 206–208
    Community Church of Vero Beach, Florida, 212–215
    Concord-Carlisle Regional High School auditorium (Concord, Mass.), 205–206
    Hong Kong Cultural Center, 223–232
    Louisiana Superdome Sound System (New Orleans, Louisiana), 208–212
    Queen Sirikit National Convention Center sound system (Bangkok, Thailand), 215–223
  equipment, 191–197
    control consoles, 192
    loudspeakers, 194–196
    microphones, 191
    power amplifiers, 194
    preamplifiers, 191–192
    signal processing equipment, 192–194
    sound control, 196
  examples, 197–203
    assembly hall, 201–202
    auditoriums and lecture halls, 200
    concert halls, 199–200
    exhibition hall, 201–202
    general, 197–198
    meeting and conference rooms, 200
    motion picture theaters, 200
    office spaces, 200–201
    sports facilities, 202–203
    theaters and opera houses, 198–199
    worship spaces, 201
  loudspeaker systems, 189–191
  overview, 187–189
  special installations, 203–204
    "hard-of-hearing" systems, 203
    paging and voice-alarm systems, 203
    simultaneous interpretation systems, 204
Sound transmission:
  noise control, systems approach, 101
  between rooms, acoustical concepts, 22–23
Sound transmission class (STC):
  acoustical concepts, 30–31
  field (FSTC), acoustical concepts, 32
  materials, 57, 121
  noise control criteria, 117. *See also* Noise control
  sound isolation testing, 271
Sound transmission loss. *See also* Sound absorption; Sound attenuation
  acoustical concepts, 23

  airborne sound attenuation, 56–57
  cavity absorption in double constructions, acoustical concepts, 25
  composite constructions, acoustical concepts, 25–26
  double walls, acoustical concepts, 24–25
  field measurement of airborne, acoustical concepts, 32
  laboratory measurement of airborne, acoustical concepts, 29–30
  single homogeneous walls, 23–24
  sound isolation testing, 271
Source, path and receiver, noise control, systems approach, 100–101
Southern Building Code Congress International (SBCCI), 110, 115, 116, 118, 119, 120
Space units, acoustical properties of, 75
Special devices, listing of, 71–76
Speech intelligibility, room acoustic qualities, 236–247
Speech privacy potential, acoustical privacy, open-plan spaces, 110
Speech transmission index (STI), concert hall design, 159
Sports facility:
  case study, Louisiana Superdome, 208–212
  design of, 178–179
  sound reinforcement systems, 202–203
Stage manager console, sound reinforcement system case study, 230
Standards:
  listing of selected, 44–46
  noise control criteria, 110–111
  sound absorption, 61
  Sound Transmission Class (STC), materials, 57
Static deflection, vibration isolation, noise control methods, 141
Steel decking, acoustical properties, 64
Steel joists and trusses, acoustical properties, 64–65
Steel springs, acoustical properties of, 75–76
Steel studs, acoustical properties, 65
Stone, acoustical properties, 65
Structure-borne sound, direct, control of, acoustical concepts, 32–33
Subjective attributes, concert hall design, 157–160
Support (of reflected sound), room acoustic qualities measurement, 263
Sway braces, acoustical properties of, 76

Tape recording, sound reinforcement system case study, 222
Television system, sound reinforcement system case study, 222
Theaters:
  case study, Hong Kong Cultural Center, 226–228
  design of, 170–171
  sound reinforcement systems, 198–199, 226–228
Tie lines, sound reinforcement system case study, 222
Tile, acoustical, properties of, 66–67
Time-varying sound levels, acoustical concepts, 12–15
Transmission. *See* Sound transmission
Treble ratio, based on early decay time, room acoustic qualities measurement, 262

Uniform Building Code (UBC), noise control criteria, 117, 118
U.S. Environmental Protection Agency (EPA), sound levels, 9, 12
Unit symbols, 305–308

Vibration isolation:
  equipment, acoustical concepts, 34–35
  noise control methods, 139–147
Voice-alarm systems, sound reinforcement systems, 203

Walls:
  composite constructions, acoustical concepts, 25–26
  double, sound transmission loss, acoustical concepts, 24–25
  single homogeneous, sound transmission loss, acoustical concepts, 23–24
  sound transmission between rooms, acoustical concepts, 22–23
  sound transmission loss, acoustical concepts, 23
Wavelength, acoustical concepts, 7
Wood decking, acoustical properties, 65
Wood paneling, acoustical properties, 65
Wood studs and joists, acoustical properties, 65
Worship spaces:
  case studies
    Community Church of Vero Beach, Florida, 212–215
    Hitchcock Presbyterian Church, 201
  design of, 175–177
  sound reinforcement systems, 201